Environmental Impact Analysis
Process and Methods

Environmental Impact Analysis
Process and Methods

James T. Maughan

CRC Press
Taylor & Francis Group
Boca Raton London New York

CRC Press is an imprint of the
Taylor & Francis Group, an **informa** business

CRC Press
Taylor & Francis Group
6000 Broken Sound Parkway NW, Suite 300
Boca Raton, FL 33487-2742

© 2014 by Taylor & Francis Group, LLC
CRC Press is an imprint of Taylor & Francis Group, an Informa business

No claim to original U.S. Government works

Printed on acid-free paper
Version Date: 20130806

International Standard Book Number-13: 978-1-4665-6783-2 (Hardback)

This book contains information obtained from authentic and highly regarded sources. Reasonable efforts have been made to publish reliable data and information, but the author and publisher cannot assume responsibility for the validity of all materials or the consequences of their use. The authors and publishers have attempted to trace the copyright holders of all material reproduced in this publication and apologize to copyright holders if permission to publish in this form has not been obtained. If any copyright material has not been acknowledged please write and let us know so we may rectify in any future reprint.

Except as permitted under U.S. Copyright Law, no part of this book may be reprinted, reproduced, transmitted, or utilized in any form by any electronic, mechanical, or other means, now known or hereafter invented, including photocopying, microfilming, and recording, or in any information storage or retrieval system, without written permission from the publishers.

For permission to photocopy or use material electronically from this work, please access www.copyright.com (http://www.copyright.com/) or contact the Copyright Clearance Center, Inc. (CCC), 222 Rosewood Drive, Danvers, MA 01923, 978-750-8400. CCC is a not-for-profit organization that provides licenses and registration for a variety of users. For organizations that have been granted a photocopy license by the CCC, a separate system of payment has been arranged.

Trademark Notice: Product or corporate names may be trademarks or registered trademarks, and are used only for identification and explanation without intent to infringe.

Library of Congress Cataloging-in-Publication Data

Maughan, James T., 1949-
  Environmental impact analysis : process and methods / James T. Maughan.
      pages cm
  Includes bibliographical references and index.
  ISBN 978-1-4665-6783-2 (hardback)
  1. Environmental impact analysis--United States--Handbooks, manuals, etc. I. Title.

TD194.65.M38 2013
333.7'140973--dc23   2013018217

Visit the Taylor & Francis Web site at
http://www.taylorandfrancis.com

and the CRC Press Web site at
http://www.crcpress.com

Dedication

This book represents a lifetime of environmental consulting, problem solving, research, and teaching. For the experience and knowledge I gained during my career, which enabled me to write this book, I am indebted to a large number of people. The thousands of other environmental professionals I have worked with over the years that have set high standards, and each in some way taught me something important about environmental analysis. The hundreds of students I have taught have similarly challenged me to reflect and explain in a straightforward manner the lessons I have learned about environmental analysis and protection. My family has supported me as I traveled and spent inordinate amounts of time and energy in environmental analysis. My daughter Edwina and son Matthew have been understanding and are tangible reasons to protect the environment for the next generation. But more importantly, my wife Emily has always been a sounding board and put up with my sometimes boring, tedious tales and explanations of environmental issues.

Contents

List of Figures xiii
List of Tables xvii
Preface xix
Author xxi

1. Introduction 1

2. Summary of the National Environmental Policy Act and
   Implementing Regulations 11
   2.1 Development of a National Environmental Policy 11
   2.2 Contents of NEPA 16
       2.2.1 Purpose and Jurisdiction of NEPA 18
       2.2.2 Environmental Policy 18
       2.2.3 Basic NEPA Requirements 19
       2.2.4 The Environmental Impact Statement 22
       2.2.5 NEPA Consultation and Public Disclosure 26
       2.2.6 International Implementation of NEPA 27
       2.2.7 Council on Environmental Quality 31
       2.2.8 Bureaucratic Culture under NEPA 32
   2.3 Summary of CEQ Regulations Implementing NEPA 32
       2.3.1 Public Involvement 34
       2.3.2 CEQ Regulations' Relationship to Other Statutes 35
       2.3.3 Making NEPA User Friendly 38
       2.3.4 Commenting and Coordinating on NEPA Products 39
       2.3.5 Record of Decision 40
       2.3.6 Frequently Asked NEPA Questions 41
   2.4 Clean Air Act as It Relates to NEPA 44
   2.5 NEPA Implementation 52
   References 53

3. NEPA Process and Specific Requirements 55
   3.1 NEPA Process 55
       3.1.1 Defining the Action 56
       3.1.2 Categorical Exclusion (CATEX) as Part of
             the NEPA Process 58
       3.1.3 Environmental Impact Statement (EIS)
             and Record of Decision 58
             3.1.3.1 Planning and Structuring the EIS 58
             3.1.3.2 Draft EIS Environmental Analyses 61
             3.1.3.3 Draft EIS Contents 62

vii

		3.1.3.4	Review and Comment on Draft EIS 63
		3.1.3.5	Final EIS 65
		3.1.3.6	Record of Decision 66
	3.1.4	Environmental Assessment and Finding of No Significant Impact 68
3.2	Purpose and Need 71
	3.2.1	Purpose and Need Overview 71
	3.2.2	NEPA Compliance Related to Purpose and Need 73
		3.2.2.1	*City of New York v. U.S. Department of Transportation.* 715 F.2d 732 (2d Cir. 1982), cert. denied, 465 U.S. 1055 (1984) 74
		3.2.2.2	*Natural Resources Defense Council v. Morton* 458 F. 2d 827 (D.C. Cir. 1972) 74
		3.2.2.3	*Isaak Walton League of America v. Marsh*, 655 F.2d 346 (D.C. Cir 1981) 75
		3.2.2.4	*Methow Valley Citizens Council v. Regional Forester*, 833 F.2d 810 (9th Cir. 1987); Reversed *Robertson v. Methow Valley Citizens Council* 490 U.S. 109 S.Ct. 1835, 104 L.Ed.2d 351 (1989) 75
		3.2.2.5	*City of Angoon v. Hodel*, 803 F.2d 1016 (9th Cir. 1986) 76
		3.2.2.6	*National Parks & Conservation Association v. U.S. Bureau of Land Management*, 586 F.3d 735 (9th Cir. 2009) 77
3.3	Purpose and Need Case Study: Washington Aqueduct Water Treatment Residuals 77
	3.3.1	Background 77
	3.3.2	Washington Aqueduct Water Treatment Residuals Purpose and Need 78
3.4	Purpose and Need Case Study: U.S. Coast Guard Rulemaking for Dry Cargo Residue Discharge in the Great Lakes 81
	3.4.1	Background 81
	3.4.2	USCG Rulemaking for DCR Discharge in the Great Lakes: Purpose and Need 82
3.5	Purpose and Need Conclusions 85
3.6	Categorical Exclusion as an Efficiency Approach to Environmental Analysis 86
3.7	NEPA Enforcement 96
3.8	NEPA Conclusion 99
References 103

4. Overview and Initiating the Environmental Impact Analysis and Assessment 105
4.1 Environmental Impact Analysis Approach 106
4.2 Need for Environmental Impact Analysis 108
    4.2.1 Magnitude and Type of the Proposed Action 108
    4.2.2 Environmental Setting 110
    4.2.3 Preliminary Assessment 110
4.3 Scoping 113
    4.3.1 Scoping Topics 115
        4.3.1.1 Potential Areas of Impacts and Concern 116
        4.3.1.2 Environmental Setting 121
        4.3.1.3 Alternatives for Consideration 122
        4.3.1.4 Methods for Environmental Impact Analyses 123
        4.3.1.5 Environmental Approvals 125
    4.3.2 Scoping Targets 126
        4.3.2.1 Social Scoping 126
        4.3.2.2 Technical Scoping 128
    4.3.3 Scoping Logistics and Statement 131
        4.3.3.1 Basic Scoping Approach 131
        4.3.3.2 Enhanced Scoping Approach 132
        4.3.3.3 Scoping Statement 136
4.4 Public Outreach 138
    4.4.1 Public Outreach Commitment and Extent of Involvement 139
    4.4.2 Benefits of a Public Outreach Process 143
    4.4.3 Public Outreach Tools 145
4.5 Development and Preliminary Evaluation of Alternatives 148
    4.5.1 Proposed Action and Alternatives Treatment 150
    4.5.2 Development of Alternatives 154
    4.5.3 Screening of Alternatives 157
    4.5.4 Comparison and Selection of Proposed Action 158
References 161

5. Conducting the Environmental Impact Analysis and Assessment 163
5.1 Environmental Impact Analysis Components 163
5.2 Affected Environment 164
    5.2.1 Environmental Resources Comprising the Affected Environment 164
    5.2.2 Mandatory First Step in Describing the Affected Environment 165
    5.2.3 Original Investigations to Describe the Affected Environment 169

		5.2.4	Estuarine Wastewater Discharge-Affected Environment Investigation Case Study 171
		5.2.5	Over-the-Horizon Radar-Affected Environment Investigation Case Study 178
		5.2.6	Dry Cargo Discharge to the Great Lakes Affected Environment Investigation Case Study 182
		5.2.7	Affected Environment Investigation Summary 184
	5.3	Impact Prediction 185	
		5.3.1	Impact Prediction Conceptual Model 186
			5.3.1.1 Model Development 186
			5.3.1.2 Impact Prediction Conceptual Model Example: USCG DCR Environmental Impact Analysis 188
		5.3.2	Impact Prediction Process 189
		5.3.3	Impact Significance Criteria 192
			5.3.3.1 Environmental Standards as Significance Criteria 193
			5.3.3.2 Significance Criteria for Specific Environmental Impact Analyses 194
			5.3.3.3 Programmatic Significance Criteria: Clean Water Act Section 201 Wastewater Management 195
		5.3.4	Impact Prediction of DCR Management in the Great Lakes 202
			5.3.4.1 Impact Prediction Approach 203
			5.3.4.2 Prediction of Invasive Mussel Impacts 206
			5.3.4.3 Comparison of DCR Impact Prediction to North Fork of the Hughes River EIS 210
		5.3.5	Impact Prediction of Boston Harbor Cleanup Alternatives 213
	5.4	Impact Mitigation 216	
		5.4.1	Classic Approach to Mitigation 216
		5.4.2	Integrated and Proactive Approach to Mitigation 219
		5.4.3	DCR Management Mitigation Case Study 221
		5.4.4	RWA Lake Whitney Mitigation Case Study 221
		5.4.5	Programmatic Mitigation 223
	5.5	Long-Term Productivity, Irreversible Commitment of Resources and Cumulative Impacts 226	
	References 227		
6.	Multilevel Environmental Impact Analysis 229		
	6.1	Overview 229	
	6.2	Multilevel Environmental Impact Analysis Approaches 231	
		6.2.1	Inclusion by Reference 231
		6.2.2	Supplemental Environmental Impact Analysis Documents 232

        6.2.3   Programmatic and Tiered Environmental Impact Analyses 233
        6.2.4   Piggyback Environmental Analysis 236
  6.3   Multilevel Environmental Impact Case Studies 238
        6.3.1   Boston Harbor Cleanup 238
        6.3.2   Fort Campbell Programmatic Environmental Assessment 245
        6.3.3   U.S. Coast Guard Dry Cargo Residue Management Tiered EIS 249
  6.4   Strategic Environmental Assessment 252
  6.5   Conclusions 259
  References 260

7. Environmental Analysis Tools 261
  7.1   Overview 261
  7.2   Ecological Risk Assessment 261
        7.2.1   History and Development 262
        7.2.2   Process and Approach 264
        7.2.3   Initial Problem Formulation 267
        7.2.4   Screening Level Assessment 271
        7.2.5   Final Problem Formulation and Detailed Ecological Investigation 276
        7.2.6   Exposure Characterization 277
        7.2.7   Effects Characterization 280
        7.2.8   Risk Characterization and Input to Remediation Decisions 281
        7.2.9   AJ Mine Ecological Risk Assessment as Part of Environmental Analysis Case Study 283
                7.2.9.1   Ecological Endpoints 284
                7.2.9.2   Exposure Characterization 289
                7.2.9.3   Effects Characterization 289
                7.2.9.4   Risk Characterization 291
        7.2.10 Ecological Risk Assessment and Environmental Impact Analysis 294
  7.3   Net Environmental Benefit Analysis (NEBA) 295
        7.3.1   The NEBA Process 296
        7.3.2   Quantification of Environmental Value 298
        7.3.3   Environmental Currency 299
        7.3.4   NEBA Examples 302
                7.3.4.1   Contaminated Sediment Remediation 302
                7.3.4.2   Pollution Abatement in the Green River 304
  7.4   Adapting Tools to Environmental Impact Analysis 309
  References 310

8. International and Individual State Environmental Impact Analysis Programs 313
    8.1   Introduction 313
    8.2   Canadian Environmental Program 313
    8.3   Australian Environmental Program 316
    8.4   European Union Environmental Program 317
    8.5   Japanese Environmental Program 320
    8.6   U.S. State Environmental Analysis Programs 324
        8.6.1   Massachusetts Environmental Policy Act 325
        8.6.2   California Environmental Quality Act 332
            8.6.2.1   CEQA Process 333
            8.6.2.2   Unique Features of CEQA 335
    8.7   Summary 337
    References 338

9. Coordinating and Managing the Environmental Impact Analysis Processes 341
    9.1   Introduction 341
    9.2   Integrating Environmental Approvals and Analysis 342
        9.2.1   Environmental Permits and Approvals 342
        9.2.2   The Role of Environmental Permits in Environmental Analysis 345
        9.2.3   Environmental Analysis in Support of Environmental Permitting 346
    9.3   Managing the Environmental Impact Analysis 350
        9.3.1   Focused Scope of Work and Truncated Analyses 351
        9.3.2   Project Plan and Delegation of Work by Discipline 353
        9.3.3   Monitoring Progress and Bad News 355
    9.4   Environmental Impact Analysis Critical Success Factors 357

10. Background on Case Studies 359
    10.1   Boston Harbor Cleanup Environmental Impact Statement 359
    10.2   U.S. Coast Guard Dry Cargo Residue Management EIS 363
    10.3   AJ Mine Tailings Disposal EIS 365
    10.4   Washington Aqueduct Water Treatment Residuals Management EIS 366
    References 369

Index 371

List of Figures

Figure 2.1	Environmental impact statements (EISs) filed by federal agencies: June 2007 to June 2012	48
Figure 2.2	U.S. EPA ratings for a subset of federal agency submittals: June 2007 to June 2012	50
Figure 2.3	Environmental impact statements filed by federal agencies with a rating of Lack of Objection	51
Figure 3.1	Summary of the NEPA compliance process	56
Figure 4.1	Simplicity of an environmental impact prediction	108
Figure 4.2	Area potentially impacted by USCG live fire training	112
Figure 4.3	Comparison of safe levels to calculated sediment concentration resulting from USCG live fire training	112
Figure 4.4	Comparison of upper Lake Whitney management alternatives: all survey respondents and by subgroup	160
Figure 5.1	North River and nearshore estuarine system	172
Figure 5.2	Video survey, sediment camera imaging, and sediment sampling locations	176
Figure 5.3	Sediment imaging camera	176
Figure 5.4	Scituate offshore affected environment	177
Figure 5.5	Over-the-horizon radar study area antelope/deer migration routes	180
Figure 5.6	Over-the-horizon radar study area cultural resource (obsidian flake) sites	181
Figure 5.7	Over-the-horizon configuration to avoid cultural resource and mammal migration impacts	181
Figure 5.8	Sonar L-3Klein System 3000 towfish used to map Great Lakes affected environment	183
Figure 5.9	Great Lakes sediment sample with discharged iron ore (taconite) dry cargo residue	184
Figure 5.10	A DCR discharge impact prediction conceptual model for natural resources	189

Figure 5.11	Impact significance criteria for defining relative level of impact	192
Figure 5.12	Comparison of impacts using significance criteria among alternatives for hypothetical highway bypass project	198
Figure 5.13	Marine Ecosystem Research Laboratory (MERL) experimental tank	200
Figure 5.14	MERL nutrient and solid enrichment experiment and pollutant loading	201
Figure 5.15	Attachment of invasive mussels (zebra and quagga) to DCR and Native Sediment Phase II Investigation: U.S. Coast Guard DCR EIS	208
Figure 5.16	Percent attachment by immature (veliger) mussels to DCR with 1 mm and 3 mm sediment cover compared with native sediments	209
Figure 5.17	Areas of Boston Harbor exceeding impact significance criteria for benthic habitat	215
Figure 5.18	Areas of Boston Harbor exceeding impact significance criteria for nutrient enrichment	216
Figure 7.1	An ecological risk assessment process summary	266
Figure 7.2	A simplified ecological risk assessment process	266
Figure 7.3	An example conceptual site model	267
Figure 7.4	Initial food web as part of AJ Mine EIS conceptual site model	286
Figure 7.5	Herring food web adapted for ecological risk assessment conceptual model	287
Figure 7.6	Marine mammal reproduction ecological endpoints	287
Figure 7.7	AJ mine ecological risk assessment physical stress exposure characterization	290
Figure 7.8	Food loss prediction process	293
Figure 7.9	Calculation of food quantity loss	294
Figure 7.10	Hypothetical cost versus ecological value resulting from waste cleanup	296
Figure 7.11	Application of net environmental benefit analysis to natural resource damage	297

List of Figures

Figure 7.12 Environmental deficit from contaminated sediment with no active remediation 303

Figure 7.13 Environmental deficits from contaminated sediment dredging remediation 304

Figure 7.14 NEBA-derived hot spot removal with restoration contaminated sediment remediation 305

Figure 7.15 Nitrogen loads to Green River under existing conditions and the best practical treatment under the national pollutant discharge elimination systems (BPT/NPDES) 306

Figure 7.16 Costs for nitrogen removal for each source for the best practical treatment under the national pollutant discharge elimination systems 306

Figure 7.17 Cost of nitrogen removal per kg for sources to the Green River 307

Figure 7.18 Green River nitrogen removal (kg per day) from each source under the NEBA alternative 308

Figure 7.19 Nitrogen removal total costs for the NEBA alternative 308

Figure 8.1 A schematic diagram summarizing the process to determine whether the Canadian Environmental Assessment Act applies 315

Figure 8.2 A threshold screening process under the Massachusetts Environmental Policy Act (MEPA) 326

Figure 9.1 A simplified example of a project plan tool 354

Figure 10.1 Boston Harbor cleanup program overview 361

List of Tables

Table 2.1	Example Methods Developed by Federal Agencies to Facilitate Environmental Consideration as Required by NEPA	23
Table 2.2	Select Federal Regulations to Be Considered during the NEPA Process	36
Table 2.3	Summary of 40 Most Asked NEPA Questions	42
Table 2.4	Monthly Filings of Environmental Impact Statements, June 2007 to June 2012	47
Table 2.5	Federal Agencies Filing Environmental Impact Statements from June 2007 to June 2012	49
Table 3.1	NEPA Environmental Assessment (EA) Compared with Environmental Impact Statement (EIS)	70
Table 3.2	USCG Rulemaking Dry Cargo Residue Discharge Alternative Screening Criteria to Meet Purpose and Need Statement	84
Table 3.3	Department of the Navy's List of Categorical Exclusions	89
Table 3.4	Summary of USCG's List of CATEXs	93
Table 4.1	Issues, Concerns, and Their Relative Importance (with 100 being the most important) for Lake Whitney Management	119
Table 5.1	Environmental Resources Eliminated from Detailed Evaluation in the U.S. Coast Guard DCR EIS	166
Table 5.2	Example Impact Significance Criteria Used in the U.S. Coast Guard DCR EIS	196
Table 5.3	Impact Significance Criteria for Estuarine Benthic Communities Derived from Environmental Impact Research	202
Table 5.4	Investigations to Predict Impacts from DCR Management Alternatives	204
Table 5.5	Haiti USAID Watershed Initiative for National Natural Environmental Resources (WINNER) Programmatic Example of Mitigation Measures	225

Table 7.1	Examples of Ecological Endpoints and Related Attributes	272
Table 7.2	Example Contaminant Screening for Ecological Risk	275
Table 7.3	Common Investigations Conducted as Part of Detailed Ecological Risk Assessments at Sites with Contamination of Aquatic Systems	278
Table 7.4	AJ Mine Assessment Endpoints and Their Relation to Selection Criteria	285
Table 7.5	AJ Mine Ecological Endpoints and Risk Thresholds	288
Table 7.6	Benthic Fauna Risk Characterization of AJ Mine Tailings Discharge	292
Table 7.7	Representative Species Food Loss from AJ Mine Tailings Discharge	293
Table 8.1	Levels of Environmental Analysis for Canada and the United States	315
Table 8.2	Example Projects Subject to Environmental Assessment under European Union Directive 2011/92/EU (Annex I)	319
Table 8.3	Categories of Projects Subject to Japan's Environmental Assessment Law	322
Table 8.4	Examples of Massachusetts Environmental Analysis Thresholds	328
Table 9.1	Environmental Analysis and Permits: Comparisons and Contrasts	350

Preface

In the late 1960s and early 1970s, a proactive segment of the American population became increasingly aware and concerned that we were ignoring and damaging our critical environmental resources. Many of these proactive "environmentalists" believed that the damage resulted from a combination of development with little or no control, profit motive, and general lack of understanding and concern for the environment. Many in the movement, particularly those with technical expertise in ecology, environmental engineering, and related fields, felt much of the damage resulted from a scarcity of techniques and regulatory tools available to identify, analyze and control environmental consequences. The lack of adequate tools frequently resulted in uninformed decisions that produced unintended consequences and produced severe and irreversible environmental damage.

In the early stages of the environmental movement, there was an uninformed electorate and limited awareness of the long-term implications of environmental damage. Consequently, as environmental damage issues were initially addressed through regulations and legislation, there was little appetite for command and control measures to constrain development based on environmental concerns. However, the collective consciousness had been raised and there was strong support for "doing something." That "something" was the development of a national policy, which in essence acknowledged that awareness and maintenance of essential environmental resources were in the national interest. More precisely, the National Environmental Policy Act (NEPA) of 1969 sought to "encourage productive and enjoyable harmony between man and his environment." NEPA also initiated a national research effort to understand ecological systems and how our societal activities interacted with these systems.

Recognizing the lack of broad and deep support for punitive or constraining environmental legislation, the U.S. Congress structured NEPA to focus on the understanding and analytical aspects of our harmony with the environment. For similar reasons, NEPA was also limited to control actions by U.S. federal government agencies. In other words, NEPA forced all federal agencies implementing or sanctioning an action (e.g., permitting a private action, constructing facilities or infrastructure, funding, or issuing regulations) that could affect the environment, to understand the environmental consequences of their proposed actions prior to implementation. With an understanding of the consequences and acknowledgement, documentation and publication of the findings, the agency was free to take whatever action they deemed necessary to fulfill their mission. However, they could not make a decision and act without investigating the environmental consequences or impacts and making these findings known to the public. Over time, this increased understanding and exposure to public scrutiny resulted in substantial environmental protection, well beyond that expected by many early NEPA critics.

When NEPA was enacted in 1969, investigating and analyzing environmental consequences was a challenge, since environmental science as a discipline did not exist and there were no established tools or methods to analyze or assess environmental impacts. In fact the tools and methods we now have, as well as environmental science in general, developed largely in response to the mandates created by NEPA and the associated legislation and regulations. Thus, over the last 40 years a wide variety of methods and tools have been developed: some have been discarded; some have been refined and perfected; and new approaches are continually put forward. This book describes the processes, methods, and tools that have proven successful in assessing environmental impacts. The focus of the book is to illustrate how these tools can be integrated to produce effective environmental evaluation documents, such as a federal Environmental Impact Statements, Environmental Assessments or similar documentation to meet state or international requirements.

In my over 35 years of environmental consulting, education, and research, I have learned that reading a textbook will not transform a biologist into an engineer or turn an anthropologist into a water quality modeler. Thus, this book will not attempt to train or instruct the reader in the multitude of disciplines required to conduct a successful environmental analysis. This book will describe alternative methods to apply and integrate the information from the engineer, biologist, anthropologist, and water quality modeler to produce a successful interdisciplinary environmental impact analysis. It demonstrates multiple examples of qualitative and quantitative techniques to address the impacts on various environmental resources. It also presents approaches to compare the impacts to each environmental resource resulting from various alternative actions on a common basis to support an informed and accepted decision. The book focuses on the assessment process and how to make the process and description of impacts clear to stakeholders with various technical and policy backgrounds.

A common criticism of environmental studies is that they take too long and cost too much. This book addresses these issues by providing approaches to organize and plan environmental studies and prepare environmental documents that are efficient with respect to both schedule and level of effort. The book also discussed approaches to maintaining schedule and budget during the execution of the studies and preparation of environmental documents.

The book is heavily populated with case study examples that I have been directly associated with over the years. From these examples, the reader will gain an appreciation for the level of technical detail required to fully assess the impacts, make informed decisions, and gain public/regulatory/legal acceptance. From the description of assessment methods supported by numerous case studies, the engineer mastering the book will not become a biologist but the engineer will know the essential questions to ask the biologist. Similarly, the biologist will know how to supply the engineer or planner with information they can understand and use directly in the impact assessment. The reader will learn not only how to produce an environmental document that meets regulations but also one that results in a more useful product with strong stakeholder support.

Author

I have a long history in environmental education, research, consulting, regulation, and advocacy and I apply this wealth of experience in authoring this book. My career parallels the environmental movement, beginning with participating in the original Earth Day demonstrations as an undergraduate. My entire academic, professional, and personal lives have been and continue to be dedicated to understanding the implications of human's activities on the environment.

I have been an environmental consultant since 1974 and risen to the position of vice president at CH2M HILL, one of the world's largest and most respected environmental consulting firms. During this career, I have worked on well over 200 environmental projects for a balanced mixture of federal government, local government, industrial, commercial, and environmental advocacy clients. I have prepared environmental analyses for major infrastructure projects, including the Boston "Big Dig" and the controversial Boston Harbor cleanup. I have also managed the environmental aspects of major national environmental programs for the U.S. military, including the U.S. Coast Guard's Aid to Navigation Program, and the U.S. Air Force's coast-to-coast Over-the-Horizon Radar Program. I led CH2M HILL's ecological risk assessment efforts, conducted assessments at dozens of Superfund sites, and authored one of the first reference/textbooks on the subject (*Ecological Assessment of Hazardous Waste Sites*: 1993, Van Nostrand Reinhold). Most of the work during my consulting career has been in planning and execution of environmental impact statements and related environmental studies. This has involved scoping the required investigations, analyzing the results and presenting the information in a form and venue suitable to all stakeholders, including project proponents, regulators, affected citizens and environmental advocacy groups. It has also including completing the projects on time and within budget.

Concurrently with my consulting, I have continued environmental education and research. I have published several articles on the topic and an early book on ecological risk assessment. I taught a graduate course in Methods in Environmental Impact Assessment at Tufts University for nine years and I am currently teaching a similar course at the University of Massachusetts, Lowell. I have also given lectures on the topic at over half a dozen other colleges and universities, including Harvard and Brown universities.

I have been heavily involved in environmental advocacy and regulation over the last 35 years. I served for 10 years on the Conservation Commission in my community, which was charged with regulating wetlands and other environmental management issues. I also was a volunteer, interim executive director, and board member for Global Village Engineers, a nonprofit organization dedicated to promoting and assisting environmental and culturally sustainable infrastructure projects in the developing world.

James T. Maughan, PhD

1

Introduction

Today in the 21st century, the United States and most of the rest of the world are environmentally aware societies. The popular press and social media are ripe with discussions about thinking and acting green, environmental sustainability, climate change, threatened marine mammals, toxic chemicals, and hazardous waste. Even advertisers in capitalistic societies play to our environmental awareness by touting the low environmental impact of their products and the zero-landfill factories that produce them.

However, environmental awareness alone cannot achieve environmental protection, much less improvement in environmental conditions or correct the cumulative environmental insults from over a century of human development, "progress," and inattention to the environment. Environmental legislation, regulation, analysis tools, dedicated professionals, activism, and behavioral changes are some of the components necessary to actually move from simple environmental awareness to the actual protection of our environment and remediation of past environmental insults and inattention. This book focuses on some of these necessities (environmental legislation, regulation, analysis tools, and instruction for dedicated professionals), describing how they relate to environmental impact analysis, achievement of environmental protection, and society's self-view of environmental awareness and concern.

Our current environmental awareness has not always been the case. Following World War II, the United States was a country of vast natural resources with seemingly limitless human resource potential and energy, and a strong sense of mission and confidence that they could advance their civilization and society beyond anything imagined in the 19th and early 20th centuries. This attitude and commitment to progress led to a complexity of civilization and infrastructure never before approached in human history with little or no consideration of environmental resources. Take as evidence the nationwide Interstate Highway System, the drastically expanded automobile industry, the annual production of household appliances at an annual rate exceeding total production of all commodities prior to the 1930s, vastly increased domestic military presence, and the extraction and processing of materials (e.g., metals, forest products, and petroleum) that support these advances in society and lifestyle.

In reaction to the exponential growth in American civilization, consumerism, and infrastructure during the late 1950s and 1960s, a proactive segment of the American population became increasingly aware and concerned that these activities were ignoring and damaging the nation's critical environmental resources. Also, the initiation of space exploration in the 1960s resulted in observations of the entire planet from afar for the first time in human existence. These observations brought home the fact to many that the planet is finite, fragile, and the only place suitable for human habitation and that they had better be nice to Mother Nature. Many of these proactive "environmentalists" believed that the damage resulted from a combination of development with little or no control, profit motive, and general lack of understanding and concern for the environment. As a result of the increased environmental awareness and pressure to do something about it, the U.S. Congress passed in late 1969 and the President signed on January 1, 1970 the National Environmental Policy Act (NEPA). This Act established a policy of environmental awareness and commitment to not ignore the environment. It was directed at actions by the U.S. federal government and mandated openness, and active solicitation of input from the public whenever the government took actions that could affect the environment.

Following the enactment of NEPA, there was a flurry of new or enhanced environmental protection activities including passage of legislation, promulgation of regulations, and creation of agencies. Almost immediately, several other countries followed suit with their own environmental policy statements and environmental protection regulations. The United States and other countries also followed with resource-specific environmental protection acts and regulations, such as the Clean Water Act, enhancement of the Endangered Species Act, and the Clean Air Act. States within the United States and in some cases, municipalities, also started to pass legislation and promulgate regulations to expand environmental protection beyond the scope of NEPA and tailor environmental analysis and protection to conditions and activities important and specific to their jurisdiction. The passage of legislation and promulgation of regulations continues to be an ongoing process, as evidenced by the current movement to require strategic environmental assessment or SEA (see Chapter 6). SEA encourages a strong focus on broad issues and proactive environmental protection rather than the general focus on individual actions and specific projects typical of NEPA and related environmental regulations and policies.

The environmental protection approach and regulatory infrastructure that evolved in the early 1970s from the passage of legislation and enactment of regulations was a three-pronged strategy. The first was an official statement of overall environmental policy at the local, state/provisional, national, and in some cases international level. In general, the policy statement was unenforceable and lacked any forcing function for compliance, but in most cases it did set the stage and provide the mission statement for subsequent development of the other, more enforceable prongs of the environmental protection and regulatory infrastructure.

The second prong of the strategy was the requirement for an environmental analysis process for specific actions. The environmental analysis process has evolved and matured, as described throughout this book, but the original intent was multifaceted. The requirement for environmental analysis was designed to open the process to the public, predict changes in environmental resources as consequences of a proposed actions, consider alternatives and other measures to mitigate any adverse consequences, and include the results of environmental analysis both early in the project formulation and in the final decision processes. Carried out well, environmental analysis also takes a holistic view of the environment, integrating effects on multiple environmental resources across varying lengths of time. The analysis was critical to implementing the policy statement because it forced a technical evaluation of the environmental repercussions of proposed actions and made adherence to the stated environmental policy enforcible. More often than not, with NEPA being the prime example, the mandate for environmental analysis did not set standards or require environmental protection. However, in almost every case the environmental analysis mandate included a public disclosure and an active public input component. These aspects of environmental analysis opened the process to the public and promoted environmental accountability by decision makers. Given the transparency of the process, it was difficult for a responsible official to announce in the public eye that she or he had decided to implement a project that caused irreparable environmental damage because the alternative courses of action made life more difficult for the decision makers and their agencies, or that the environmentally preferable alternative was marginally more expensive. The requirement for environmental impact analysis and assessment also brought to the forefront the need for environmental standards to measure and express the significance and relative intensity of impacts. This set the stage for the enactment of multiple environmental resource protection measures such as the Clean Air Act and Clean Water Act, which are the basis for the third prong in the environmental protection strategy.

The third and final prong in the environmental infrastructure was a permitting or approval system based on protection of a single environmental resource (e.g., air, water, historic properties, and endangered species). These statutes, acts, and regulations typically establish standards or criteria that set the allowable level of resource alteration that will not compromise the sustainable values of the resource. Water quality limits (e.g., the allowable copper concentration, temperature, or pH of a water body or effluent) are classic examples of such standards. Thus, environmental legislation such as the U.S. Clean Water Act not only establishes such protections (by requiring the states to set enforceable water-quality standards) but also requires a permit to allow discharges into the receiving waters, monitoring of the discharge quality, and a demonstration that the discharge quality will not result in a violation of the standards.

The integration of these three prongs is perhaps more critical to environmental protection than any one individually. The policy alone does not force specific environmental protection or enhancement action and the analysis requirement generally does not mandate specific environmental protection. Environmental permits can force environmental protection but they are narrowly focused so that protection of one resource can often be accompanied by even greater degradation of another or a combination of resources. A full and proactive integration of these three prongs can achieve holistic environmental protection, including attention to individual resources, and help move forward in a positive direction of meeting the national environmental policy.

Early in the environmental movement, many who were concerned about the degradation of the environment, particularly those with technical expertise in ecology, environmental engineering, and related fields, felt that although lack of awareness and concern for the environment were at the heart of the problem, much of the damage resulted from a scarcity of techniques and regulatory tools available to identify, analyze, and control environmental consequences. Even with good intentions and adherence to newly enacted laws and regulations, the lack of adequate tools frequently produced uninformed decisions that resulted in unintended consequences and created severe and irreversible environmental damage. The lack of such tools and environmental analysis methods became more apparent as analyses were conducted and reviewed by the public, academicians, technical experts, environmental advocacy groups, and natural resource agencies. Thus began the process, which continues today, of developing environmental analysis techniques to meet the objectives of the national environmental policy and similar environmental acts and regulations.

This book looks back, over 40 plus years, at the techniques and methods developed and used in environmental analysis. The ones that have been successful are highlighted but the ones less so are not ignored and the lessons taken away are reported. Although the methods and techniques are the focus of the book, they are not presented in isolation. They are discussed in terms of the comprehensive environmental analysis process from the inception of project, program, and policy to implementation of the selected alternative to accomplish the stated purpose and need. The environmental analysis methods and techniques are also discussed in relation to environmental laws and other requirements and perhaps most importantly in relation to making a better decision, not only from an environmental perspective, but also from the angle of accomplishing the purpose and the need for action and generating public support and comprehension.

The book is designed to assist in the preparation of environmental analyses that accomplish the goals discussed, including:

- Achieving a better decision
- Implementing a better project, program, and policy
- Meeting procedural and technical expectations of environmental regulations
- Informing the public and facilitating their support regarding environmental problems and solutions

Introduction

- Providing a platform to achieve actual environmental protection and enhancement
- Moving forward stated environmental policies including sustainability and protection

Attainment of these and similar goals is not a simple process that can be specified in detail and then followed to the letter to achieve or, hopefully exceed the desired goals. This book describes the general processes, techniques, and methods, and provides case studies as to how they have been adapted in specific situations. Through these case studies, the reader will experience the types of successful adaptations of general procedures to specific cases and learn how they were developed. This will facilitate similar development of approaches to environmental impact analysis as specific situations demand.

An attempt to instruct on every environmental resource and technical discipline required in all types of environmental analysis would be an endless and hopeless exercise. Seasoned environmental scientists, planners, and policy specialists have years developing their technical skills. One book cannot accomplish that level of knowledge or relay that level of experience. But this book does provide information that can assist in identifying particular areas of technical expertise that are needed for a project, plan, or program and, primarily through case studies, illustrate the type of studies and level of detail that have been used successfully in environmental impact analysis.

Another emphasis of this book is that environmental analysis is just this—analysis. It is not quite as technical, prescribed, or accurate as the analysis of an environmental sample that needs a gas chromatograph and mass spectrophotometer to determine the concentration of chlorinated organic compounds down to the parts per trillion levels, but it is an analysis and not solely a "discussion." If conducted well, an environmental impact analysis should bear some similarities to the chemical analysis and be just as defensible to stakeholders.

Scientifically defensible impact analysis guarantees a hard look at the consequences of a proposed action, which as discussed in Chapters 2 and 3, forms the basis for NEPA compliance. The methods used do not have to strictly follow a formal scientific method, nor do the methods always have to meet the strict standards of peer-reviewed scientific articles. However, it is essential that the methods used and scientific basis for arriving at decisions are clearly described in the environmental analysis document. It is also advantageous to explain the limitations of the methods and where appropriate, the rationale for not using more rigorous scientific methods and tools.

A first step in a sound and defensible environmental analysis, similar to the procedures used to conduct a chemical test, should be the establishment of a method to analyze the environmental impacts. This method should be either a standard method or more commonly a combination of standard methods and ones developed to meet the specific needs of each action and environmental setting. In the latter case the methods and approach should be reviewed with the decision makers and key stakeholders before they are implemented to ensure acceptance of the results reached from the environmental analysis. The environmental impact analysis methods should be strictly followed with clear and full documentation of any variance. Also, as with the chemical analysis, some level of quality assurance and quality control is necessary.

This book presents environmental impact analysis as an approach embodying the concepts of science and technical procedure. Each analysis should be designed to address a specific question and most often should test a series of hypotheses. For example:

- Which pollutants could be discharged from a chemical plant?
- Which are the environmental receptors found in the vicinity of the chemical plant?
- Will the pollutants reach the environmental receptors?
- If they do reach the receptors, at what concentration?
- What is the sensitivity of the receptors to the predicted concentrations?
- If the receptors are to be affected, what are the implications on an ecosystem level?
- Which alternatives and mitigation measures best meet the purpose and need, while balancing all environmental and economic considerations?

The approach to environmental impact analysis presented in this book is not strictly science, because the analysis is always constrained by funding, schedule, flexibility of the proposed action, and stakeholder's interests. Even if environmental impact analysis cannot reach the height of pure science it should never resort to "voodoo science" relying on undocumented assumptions, subjective and qualitative opinions, hand waving, or adherence to untested preconceived notions.

Once a reader has mastered this book she or he should be ready to jump into the fray of a multidisciplinary environmental analysis and make a meaningful contribution. Readers should also learn how their particular technical discipline and other areas of expertise can be integrated into the environmental analysis. Similarly, they should be able to determine whether the necessary expertise is represented or available to the team and then ask the right questions to conduct a successful environmental analysis that meets the goals expressed in this chapter.

This and similar books documenting continued development and improvement in environmental analysis techniques and policies are sorely needed. Although great progress has been made in environmental awareness, policy, regulation, and environmental impact analysis tools since 1970, there are still several large gaps. One well-documented gap is the rating of current environmental impact analysis documents (discussed in Chapter 2, Section 2.4). Under the Clean Air Act, the U.S. Environmental Protection Agency (EPA) is obligated to evaluate environmental impact statements (EISs) submitted under NEPA. The EPA has rated more than half the submitted EISs as containing "insufficient information." It is not just the quality of the analysis that is deficient; the environmental impacts resulting from a similar number of proposed actions are rated as having an "Environmental Concern," meaning the proposed action may result in substantial impacts that could be avoided. Reviews done by others (see Chapter 2, Section 2.4) have reflected the same trends, both under NEPA and other international environmental programs. It is especially disturbing is that these reviews have not detected any substantial improvement in the quality of environmental analysis or reduction in environmental impacts over the last 30 plus years.

The other gap in environmental impact analysis progress is less quantifiable than the rating of EISs and similar environmental analyses, but more severe in hindering the achievement of environmental policy goals. From its inception, environmental analysis was intended by supporters and advocates to be an integral part of the project, plan, and policy development process. The vision was to initiate the environmental analysis and consideration of environmental resources at the very first stages of the planning process and incorporate environmental considerations at each and every stage of project development and implementation. Embodied in this vision is a truly integrated and interdisciplinary team, planning the proposed action, developing the implementation details, and considering environmental implications.

Experience has shown that some U.S. federal agencies take this vision to heart and implement projects, plans, and policies that reflect comprehensive environmental consideration, stakeholder support, and ultimately better results. Unfortunately, experience has also demonstrated that this is the exception rather than the rule. A goal of this book is to clearly demonstrate that, both in theory and in practice, early and often, incorporation of environmental consideration, clearly stated as an objective in the national environmental policy, is a path to a successful, efficient, appreciated, and environmentally sustainable project, plan, or policy.

This book is organized to further the goal of better environmental analysis producing better projects, plans, and policies. Chapters 2 and 3 focus on the history of the environmental movement and its creation of U.S. NEPA. NEPA is introduced early in the book because it was the first comprehensive effort at incorporating environmental consideration into the national discussion and has served as the model for many parallel U.S. state and international policies and regulations. To fully understand the current practice of environmental impact analysis, an appreciation of NEPA and its underpinnings is essential. Chapter 2 briefly addresses the history and summarizes the Act and its implementing regulations. The chapter also assesses the quality of analyses performed under NEPA and similar environmental programs. The NEPA process and a description of each step are presented in Chapter 3 with a detailed presentation of NEPA-specific requirements, such as the purpose and need for action and enforcement of NEPA and its regulations. Chapter 3 focuses on the analysis process as it applies specifically to NEPA with universal aspects of environmental analysis processes, such as environmental impact prediction covered in detail in subsequent chapters. NEPA is given special attention, not because it is a perfect, or even a near-perfect environmental analysis process, but because it was the first. An understanding of the steps in the NEPA process provides a better appreciation of other regulations and analysis processes that were developed based on the NEPA experience.

Chapters 4 and 5 describe the heart of environmental impact analysis: the steps leading up to and including the prediction of impacts and how the analysis is integrated into the planning, development, and implementation of projects, plans, and policies. Multiple approaches to each of the steps are described by presenting multiple case studies to illustrate how the process can be adapted to individual cases. The steps required to plan and support an environmental impact analysis are presented in Chapter 4 and implementation of the analysis is described in Chapter 5. These chapters are not the cookbook equivalent of how to prepare an environmental analysis, but they do describe the process, discuss each step, and present important considerations and potential pitfalls at each step.

One of the historic and continuing criticisms of environmental analyses is that they are long, boring, and repetitive. The boredom criticism is hard to address but large strides have been made in controlling the length and repetitiveness. The progress has included numerous approaches to multilevel environmental impact analyses using information developed, decisions made, and a better understanding of issues developed in early stages to streamline subsequent levels of analysis. Chapter 6 presents several of these multilevel environmental analysis concepts and illustrative case studies. The chapter emphasizes the use of multilevel environmental analysis to achieve better decisions, projects/plans/policies, and maximum efficiency.

Early in the history of environmental analysis, the goal was well known, but the path to achievement was uncharted. Since that time numerous methods and tools have been developed and are continually being perfected to predict impacts, weigh environmental trade-offs, and provide environmental input effectively and efficiently to projects, plans, and policies. None of these methods or tools are universally appropriate to all actions, environmental resources, or environmental or geographic settings; however,

Chapter 7 presents two of these methods and case studies where they have been applied. In general, the methods are not described in the level of detail to elevate a novice to an expert, but the concepts are presented with references provided to support a deeper mastering of the methods and tools. The discussion of methods is also intended to stimulate the creativity of environmental professionals to adapt and apply the techniques used in their areas of expertise to environmental analysis.

Chapter 8 addresses environmental analysis policies, legislations, and regulations in numerous jurisdictions. The analysis process is described, for example, in countries and two landmark U.S. state programs (California and Massachusetts). All of these programs have been developed, at least to some extent from the NEPA experience so the discussion is presented in a more or less comparative basis to NEPA. The discussion also emphasizes the evolution of environmental analysis in response to site-specific needs and conditions as well as lessons learned over time.

As discussed above, current environmental strategy is a three-pronged approach based on policy, analysis, and prescribed protection. Chapter 9 addresses the coordination of the analysis and protection legs of the strategy. Incorporating the mandated environmental protection measures, usually in the form of environmental permits and approvals, into the analysis process is a critical component of successful environmental protection, enhancement, and compliance. This takes the form of integrating permit requirements into various components of environmental impact analysis, particularly the alternative comparison and decision aspects of the analysis. The chapter also explores the importance and approaches for laying the groundwork for an environmentally permittable project early in the planning stages.

Equal to, or perhaps greater than, the criticism of wordy and repetitious environmental analyses is the condemnation of the cost and schedule implications of the analyses. Chapter 9 presents techniques to minimize cost, maximize efficiency, and stay on schedule when preparing EISs, EAs, and similar documents. These techniques reflect lessons learned over 35 years of managing environmental analyses to a strict budget and schedule. All too often these techniques were learned the hard way from working 80 hours a week to meet schedule, and all too frequently much of it not charged in order to maintain budget.

Numerous examples are used throughout this book to illustrate environmental analysis concepts and approaches, and to relate lessons learned. Most of these examples are drawn from personal experience, and the thought process, perspective, hindsight, and consideration, often not explicit in public record, are integrated into the case studies. Chapter 10 presents some of the background of these case studies to enrich the appreciation of the limited discussion presented in the preceding chapters where the case studies are used to illustrate only specific environmental analysis concepts or approaches.

2

Summary of the National Environmental Policy Act and Implementing Regulations

2.1 Development of a National Environmental Policy

In the early 1960s, the U.S. Congress responded to the increased environmental awareness and pressure for action discussed in Chapter 1, with the exploration of options to address environmental degradation and lack of sustainability. The effort to respond to the environmental threat and public pressure for environmental action was led by academics and members of Congress from districts reliant on the economic and recreational opportunities afforded by natural resources that were being threatened by continued environmental neglect and insult. An option considered in response to the threat and public demand was a comprehensive mandate requiring every entity, both public and private, to achieve a specified level of environmental protection, or at least implement the alternative that was the least damaging to the environment. Such an approach with specific requirements and penalties for lack of compliance would have had some similarities to the landmark Civil Rights Act of 1964 and the Voting Rights Act of 1965, which had been recently passed by Congress.

These critical human rights acts were specifically directed and enforced, and resulted in some of the most important cultural and social shifts of the century by initiating a change in the U.S. lifestyle and culture as experienced by many citizens. The Acts specifically outlawed certain practices including an unequal application of voter registration and racial segregation. A similar approach to environmental policy, which would outlaw development that resulted in a net loss of environmental value or require mitigation of impacts below a specified level, was considered by legislators and environmental advocates.

An environmental protection approach similar to civil and voters' rights protection was rejected for a number of reasons. First, the ongoing environmental damage was not as immediate, obvious, blatant, or damaging to individuals on a daily basis, as compared with the violation of civil and voting rights in certain sectors of the country in the 1960s. Thus, environmental issues were not a standard feature on the television nightly news, which instead showed compelling images of outrageous violation of citizens' civil rights. Congress recognized that environmental awareness was not as broad or deep as the concern and disgust for civil rights violations and there was no political support for a universal command and control environmental law parallel to those created by the Civil and Voting Rights Acts.

In addition, the constitutionality of an environmental law limiting or controlling actions by private entities was an unresolved question. The Civil Rights legislation was supported by the constitutional guarantees of equal protection of the laws for all citizens under the 14th Amendment and protection of voting rights under the 15th Amendment. Because of these constitutional amendments, the Acts were applied to private as well as government actions without the threat of a constitutional challenge. There were no such constitutional amendments addressing the environment and for reasons outlined below, there was no real possibility of enacting an environmental amendment to the constitution.

As discussed above, there was no broad support for stringent environmental control, and the projected time table for passage of an amendment was viewed as unacceptable by those working on developing a solution to the immediate national environmental problems. The environmental neglect was growing at an accelerated pace and passage of an amendment would take many years after its proposal and then could be rejected. Just a few years later, the Equal Rights Amendment was passed by Congress in 1972 but after the allotted 10 years it was not ratified by a sufficient number of states and it was not adopted. Also Congress, as well as environmental advocates, saw the proposal of a constitutional amendment as a diversion of energy and resources away from formulating and adopting immediately implementable environmental policies and regulations. Thus, the passage of a new amendment was highly unlikely and Congress rejected the option.

Rejection of command and control legislation and pursuit of a constitutional amendment left the Congress with limited and constrained options for environmental protection. An option was needed that was not challengeable for constitutionality and could be quickly enacted. Up to the point of proposing legislation in the late 1960s, the environmental push had been led by academics and well-informed and environmentally committed members of Congress and their staff. This group had successfully avoided creating an active and strong opposition movement and in order to gain support and ultimate passage of legislation, the selected approach had to maintain this trend. Stimulation of a strong antienvironmental movement, likely led by those concerned that private commerce and property rights would be compromised, was seen as a significant constraint to be avoided if meaningful legislation was to be passed in a timely manner.

The results of these considerations and constraints were compromises that created both the concept and content of the National Environmental Policy Act (NEPA). The compromise included creation of a "warm feeling" element for environmental advocates; avoided strong opposition by private concerns; did not usurp authority and mission of federal agencies, and laid the foundation for the evolution of a very strong national environmental protection strategy. The warm feeling element was a national policy statement in both the Purpose (Sec. 2) and Title I (Sec. 101) of the NEPA: (42 U.S.C. §4321 et seq.): *"The Congress, recognizing the profound impact of man's activity on the interrelations of all components of the natural environment... and recognizing further the critical importance of restoring and maintain environmental quality to the overall welfare and development of man, declares that it is the continuing policy of the Federal Government, in cooperation with State and local governments, and other concerned public and private organization, to use all practicable means and measures, including financial and technical assistance, in a manner calculated to foster and promote the general welfare, to **create and maintain conditions under which man and nature can exist in productive harmony**, and fulfill the social, economic, and other requirements of present and future generations of Americans* (emphasis added)."

Who could argue with this statement? The policy subtly acknowledges that we have not been good environmental stewards without laying the blame. It also implies it is not too late to change our ways and even reverse the impact of past sins. It specifically points to cooperation with other jurisdictions, thus avoiding a state rights issue and includes "private organizations," like environmental advocacy groups, as active partners in the policy. It fosters support for these groups with a commitment to help them financially as they tackle environmental problems. The policy addresses fears of an environmental watchdog by emphasizing harmony (i.e., balance) between environmental, social, and economic concerns.

The Act avoided a constitutional challenge and energized private concern opposition by limiting the actual requirements to actions by the federal government. Specifically (42 U.S.C. 4332 Sec. 102) *"...all agencies of the Federal Government shall: (A) utilize a systematic, interdisciplinary approach... in planning and in decision making which may have an impact on man's environment; ... (C) include in every recommendation or report on proposals for legislation and other major Federal actions significantly affecting the quality of the human environment, a detailed statement by the responsible official on (i) the environmental impact of the proposed action, ... (iii) alternatives to the proposed action...."* There is no similar clause, requirement, or indication that actions taken by private concerns or nonfederal government agencies fall under the jurisdiction of NEPA. The federal agencies work for and under the direction of the president and Congress, thus placing requirements on the agencies is not a question of legislative authority granted to Congress under the constitution. It is simply a matter of agencies doing their job, as instructed by their "chain of command."

The restriction of NEPA to federal agency actions was not as limiting to environmental protection as it might appear, and the force of NEPA has increased with time. The number, scope, influence, and size of federal agencies grew immensely beginning with President Franklin D. Roosevelt's

New Deal in the 1930s and accelerated through World War II and into the 1960s. With this expansion, federal agencies are involved in more and more activities that could adversely affect the environment, and thus NEPA applies.

As these agencies grew in number and responsibility, each had its mission, constituency, funding, and authority. Each agency had become accustomed to operating semi-independently and dedicated to carrying out only their mission without regard to broader national issues and concerns, including environmental sustainability and protection. An overarching environmental law or an environmental agency with veto power over their actions would have been extremely threatening to the agencies, their standard operating procedure, special interest constituents, and ensconced bureaucracies. This could result in confusion and resistance that would paralyze or at least delay actions and decisions by the agencies and also constrain achievement of the environmental goals.

Thus, in addition to limiting NEPA to federal agencies as a concession to private industry and commerce, a compromise on federal agency mandate in the environmental legislation was necessary for broad support and to reach the intended goal. The language of NEPA is somewhat vague on what is actually required of federal agencies, using such language as, *"In order to carry out the policy set forth in the Act, it is the continuing responsibility of the Federal Government to use all practicable means, consistent with other essential considerations… to improve and coordinate Federal plans, functions, programs and resources to the end that the Nation may (1) fulfill the responsibilities… as trustee of the environment… (3) attain the widest range of beneficial uses of the environment without degradation… (5) achieve a balance between population and resource use"* (42 U.S.C. 4332 Sec. 101). This language was not overly threatening to the federal agencies and they did not mount a strong resistance. Thus there was no strong opposition for the bureaucratic establishment, private parties did not feel threatened by the sole focus on federal agencies contained in the proposed legislation, and the National Environmental Policy Act was passed overwhelmingly in both houses of Congress and signed by President Nixon on January 1, 1970.

In reality, the scope of NEPA was broader than a superficial interpretation would imply. As the U.S. society, commerce, and government became more complex, the role of federal agencies broadened and became more multifaceted. By the mid-to-late 1970s, the role and activities of federal agencies drastically increased with such actions as the passage of the Clean Air and Clean Water Acts, federal funding of infrastructure projects (e.g., roads, rail, bridges, airports, and wastewater treatment), federal funding for economic development, licensing of electrical power-generating facilities, federal housing, leasing of federal lands, fossil fuel exploration on federal lands, and aid to foreign nations. Thus, more and more development projects involved federal agency approval or funding. Each of these activities was considered a federal action and was subject to NEPA. Thus, there were many more activities with the potential to change the landscape and generate significant environmental impacts than were envisioned when NEPA was enacted and thus fell under the environmental policy goal and were subject to environmental analysis requirements specified in the act.

NEPA also became more of a force as the ambiguous language of environmental requirements by federal agencies was clarified. Specifically, the Council on Environmental Quality (CEQ) Regulations (see Section 2.3 of this chapter for a discussion of the regulations) promulgated to implement NEPA and NEPA litigation history defined federal agency responsibility under NEPA. The intent of the regulations and multiple decisions by the courts have concluded that the agencies must take a "hard look" at the environmental implications of their actions, include the consideration of the implications in their decisions, and for every action provide to the public full and transparent documentation of the process used. Although not a mandate overriding an agency's mission, this interpretation did place a specific and challengeable requirement on each agency to consider environmental implications before they took action.

Also the requirements for federal agencies to *"include in every recommendation ... and other major Federal action ..., a detailed statement by the responsible official on (i) the environmental impact of the proposed action"* (42 U.S.C. 4332 Sec. 102), evolved into a forcing function and specific (and initially considered onerous) requirements for all federal agencies. NEPA regulations promulgated by CEQ made it clear that preparation of the "detailed statement," which we now know as the Environmental Impact Statement (EIS) was to be a transparent process. Not only was a full disclosure of the detailed statement and all information supporting the statement required, but also agencies were to actively solicit input and review from the public and concerned organizations. These requirements changed the decades-old culture in most agencies where decisions were made in isolation by senior bureaucrats with only the goals and interests of the specific agency considered in the decisions. Many such decisions and subsequent federal agency actions, such as the development of the interstate highway system or "harnessing" of the Tennessee River by the Tennessee Valley Authority (TVA), had permanent and irreversible impacts on the environment. In the case of the interstate highway systems, the impacts were not only limited to construction, but also created massive changes in land use patterns, including facilitating an exodus to the suburbs. Yet these changes were not analyzed, documented, considered publicly, or presented to the people most directly affected by the action for their input before the decision to construct the highways was finalized. In all likelihood, if NEPA had been in place prior to the planning and implementation of the TVA or the interstate highway system, these activities would have still taken place. However, they would most likely have been substantially different from their current form. But perhaps more importantly, the adverse and unanticipated consequences of these and similar large-scale federal programs could have been identified early in the process.

Minor modifications, not affecting the underlying purpose and need for the programs, could have been made during planning if NEPA had been integrated into the process. Similarly, the adverse impacts could have been at least partially mitigated with plans and programs put in place to address the unintended and unanticipated consequences such that they were not issues for the following decades.

NEPA, and specifically the EIS hard look requirement and the public input mandate, made drastic changes to the longstanding practice of decisions based solely on an agency's mission and eventually in the culture of many federal agencies. The constraints of limiting NEPA (and thus environmental protection) to only federal actions and considering environmental implications when making a decision were eventually accepted and even embraced by many agencies. It has evolved to represent little or no constraint on the agency missions as the culture changed. As federal agencies that embraced NEPA became more and more involved in actions that could have far reaching environmental impacts, there has been a major reduction of adverse environmental impacts and unintended environmental consequences from federal actions. Unfortunately, NEPA and environmental analysis in general have been viewed in some agencies as only an after-the-fact effort to comply with a regulation. In these cases the environmental protection has been constrained but as discussed later (Section 3.7 of the book), it has been left to the federal courts to painfully and slowly force realistic environmental consideration by federal agencies.

2.2 Contents of NEPA

Procedures developed for NEPA and the evolution of these procedures, both through practice and court decisions, form the original basis for the environmental analysis process as it currently exists and as it is addressed in this book. Thus the NEPA scope and procedures are summarized in this section of the book. They are presented to convey several messages: a historic perspective; the foundation of an objective approach to environmental analysis; a starting point for most international and individual state environmental analysis requirements; and the legal and procedural requirements necessary for a NEPA document and procedure to withstand litigation.

NEPA is truly landmark legislation. Its passage was driven by knowledgeable and influential people with a strong and legitimate concern over ongoing environmental degradation. In turn, NEPA instigated the passage of new regulations and legislation in the United States which promoted protection and enhancement of natural resources and quality of life. NEPA accomplished these protections and enhancements partially by providing a platform and infrastructure for development and administration of new environmental protection regulations and procedures. It also provided an open and transparent venue for public debate and implementation of environmental protection measures at the level of individual projects, agency-wide programs, and national policy. Also, NEPA is considered by many as the "grandfather" of environmental legislation and regulations internationally and in individual states. With this pedigree and breadth of scope, understanding NEPA, how it was developed, judged, implemented, and successful is important background for anyone wishing to practice or review environmental analysis in any arena. Thus this book devotes two chapters to NEPA, not so much for readers to become NEPA experts but so that they fully appreciate the starting point of environmental analysis, generally understand NEPA requirements, and take home the lessons learned from 40 years of NEPA implementation.

This chapter and the next summarize NEPA scope and procedures so that a practitioner planning and implementing an environmental analysis can conduct the process in compliance with requirements. Also as discussed above, NEPA began the evolution of environmental analysis and these chapters track that process and present the analysis guidance provided by the Act. This summary covers the basic scope and procedures that apply to all proposed federal actions. However, every project, plan, and/or policy is unique and there are details in NEPA policy and procedures that come into play only in certain cases. This book does not cover each and every NEPA detail, and at some point during the planning and implementation of an action subject to NEPA, the practitioner is strongly advised to consult and review the Act and its regulations, which are summarized below. These documents can be found at the CEQ website (www.whitehouse.gov/administration/eop/ceq) and are also presented in several NEPA-specific reference books:

- *NEPA Effectiveness, Mastering the Process* (March 1998)
- *Mastering NEPA: A Step-by-Step Approach* (Bass and Herson 1993)
- *The NEPA Reference Guide* (Reinke and Swartz 1999)

There are also several reference books that not only address all the specifics of NEPA but examine the subtleties, present case studies, and discuss implications. *NEPA and Environmental Planning: Tools, Techniques, and Approaches for Practitioners* (Eccleston 2008) is one such text. The same author and publisher published another useful NEPA reference book on environmental assessments (EAs) (Eccleston and Doub 2012), which covers aspects of NEPA-mandated analyses for actions with only minor environmental consequences.

In addition to the CEQ NEPA Regulations (40 CFR 1500 et seq.; see Section 2.3 of the book for a full discussion of the CEQ Regulations), each federal agency has its own regulations and guidance for implementing the CEQ Regulations. As the practitioner initiates an environmental analysis under NEPA, the specific agency regulations should be consulted and rechecked as a quality assurance/quality control measure in the final stages of the process. These regulations can be found on each agency's web page.

2.2.1 Purpose and Jurisdiction of NEPA

The broad purposes of NEPA, as stated in the preamble of the Act (Public Law 91-190, 42 U.S.C. 4321-4347, as amended by Public Laws 94-52, 94-83, and 97-2580) are to:

- Establish a national policy encouraging a "productive and enjoyable harmony between [humans] and their environment"
- Promote efforts to protect the environment
- Advance the understanding of ecological systems and their importance to the nation
- Establish a CEQ

The substantive aspects of these purposes, relative to environmental impact analysis are discussed under three titles:

- Title I Sec. 101: Statement and explanation of the policy
- Title I Sec. 202: Requirements of federal agencies to address the policy
- Title II: Establishment and description of the CEQ

NEPA jurisdiction, or what is covered by and subject to the act, is established within the Act. As discussed earlier, the provisions of the Act only apply to federal agency actions that can potentially affect the environment. Actions by Congress, the president (including staff functions in the executive office), and the judiciary are not subject to NEPA because the Act specifies actions by federal "agencies" alone and not other federal entities. "But for" is a good test to determine NEPA jurisdiction. If the proposal could not occur "but for" the action of a federal agency, it is almost certainly subject to NEPA under the federal action criterion. Thus, if a proposal could not go forward "but for" the direct action, funding, approval, or permitting of a federal agency, it is subject to NEPA. If none of these actions by a federal agency are necessary to implement the project, then it is not subject to NEPA.

2.2.2 Environmental Policy

The national environmental policy is set forth in Title I, Section 101 of NEPA. As presented in Title I, essence of the policy is to "...*create and maintain conditions under which [humans] and nature can exist in productive harmony...."* Section 101 provides the background and specifics on productive harmony which are to:

- "...fulfill the responsibilities of each generation as trustees of the environment for succeeding generations."
- "...assure for all Americans safe, healthful, productive, and esthetically and culturally pleasing surroundings."
- "...preserve important [resources]."

Section 101 describing the policy also makes it clear that (1) NEPA is a compromise and (2) although it gives weight to environmental concerns where none had previously existed, environmental protection does not take precedence over other issues. This is evident in statements such as:

- "...maintain where ever possible an environment which supports .... a variety of individual choice[s]."
- "...attain the widest range of beneficial uses of the environment without degradation, risk to health or safety or other undesirable and unintended consequences."
- "...fulfill the social, economic and other requirements of present and future generations of Americans."

Thus the policy statement of the Act sets forth goals and desires, but appropriately for a policy statement, it is not specific and does not indicate how the goals will be achieved. That is left to other sections of the Act, implementing regulations, trial and error by agencies, and litigation.

2.2.3 Basic NEPA Requirements

A Congressional national environmental policy statement coupled with a mandate for all federal agencies to implement the policy alone would have produced little, if any, change in how agencies carried out their activities. The result would have produced at best only limited environmental protection and enhancement. In practice, each government agency would have remained narrowly focused on its primary mission and there would have been little or no coordination among agencies with regard to environmental issues and preservation. For example, NASA would have been preoccupied with space exploration and would have no incentive to consult with the U.S. Fish and Wildlife Service or National Marine Fisheries when they selected and constructed launch sites and other facilities adjacent to sensitive marine habitats or in proximity to endangered species breeding areas.

In order to go beyond a "feel good" policy statement to actually "create and maintain conditions under which [humans] and nature can exist in productive harmony" and achieve actual environmental protection, forcing functions and specific requirements are necessary. NEPA makes provisions for requirements, such as public disclosure, mandatory consultation with resource agencies and states, and perhaps most importantly a detailed statement of environmental analysis and impacts. These requirements are established in Title I; Section 102 of NEPA, which sets forth the most basic specifications and requirements to implement the Act. Most of these requirements come into play when implementing NEPA and are discussed in detail in Chapter 3; however, several warrant separate discussion (see Sections 2.2.3–2.2.4). Also some of the requirements are conceptual and nonspecific with implications to the intent of NEPA and environmental analysis in general. These conceptual requirements are outlined and the implications discussed subsequently in this section.

The first conceptual requirement specified under NEPA is for federal agencies to *"utilize a systematic, interdisciplinary approach which will insure the integrated use of the natural and social sciences and the environmental design arts in planning and in decision making which may have an impact on man's environment"* (Sec. 201 A). There is a very good reason that the CEQ Regulations (§1501.2[a]) use the term "interdisciplinary" rather than multidisciplinary. CEQ recognized the importance of looking at all disciplines, such as air, water, ecology, hydrology, traffic, community services, noise, etc., in a multidisciplinary approach. But they also recognized the necessity of bringing the potential effects and benefits in each of these areas together in an integrated fashion and interdisciplinary approach. This is the only rational and efficient approach to even begin making comparisons and trade-offs among critical environmental resources and making complex decisions that can affect multiple environmental resources.

The integration of disciplines, a novel idea in the early 1970s, was in its infancy, but NEPA recognized the critical necessity to incorporate it into environmental analysis. The primary goals and ultimate benefits of NEPA and any environmental analysis are to make better decisions, taking into account environmental resources. The goal is not to consider only single or selected environmental resources in making the decision but rather the environment as a whole. Another advantage of an interdisciplinary approach results from a primary law of ecology that "everything affects everything else." Thus considering water quality as an example, recreational use can affect water quality, which in turn can affect aquatic biota habitats. The integration of these topics and coordination of the technical professionals working in each area are necessary to understanding these relationships and taking them into account while predicting impacts and comparing alternatives. Without an integration of all the environmental resources in an "interdisciplinary" approach the goal of adequately considering the environment cannot be achieved.

The comparison of alternatives and selection of a proposed action is perhaps where the benefit of an interdisciplinary approach is most apparent. For example, in even a moderately complex project and associated environmental analysis, comparisons such as air quality benefits versus adverse effects on community noise levels, or highway traffic versus brook trout habitats are inevitable. Using a degree of common methodology, such as significance criteria (see Chapter 5), is an important analysis approach that facilitates an integration of disciplines to compare alternatives and make decisions that balance effects on critical environmental resources.

Using a similar level of detail, accuracy, and precision to investigate resources with similar levels of impact is another aspect of an interdisciplinary approach that facilitates comparison of alternatives and decision making, and adds to the objectivity of an environmental analysis. If two parallel environmental attributes (e.g., endangered species and historic properties) have the potential to be similarly impacted by a proposed action, the level of investigation of each should be similar. If one attribute is analyzed using extensive data collected from existing conditions, laboratory tests, and mathematical modeling of predicted impacts while the other is evaluated only qualitatively, the formulation of alternatives, impact comparisons, trade-offs, and mitigation affecting these resources will be biased.

The next NEPA requirement (Sec. 102 B) is unique in that it is not project-, policy-, or program-specific but addresses agency-wide obligations. NEPA includes a requirement for all agencies of the federal government to develop methods and procedures to ensure that "presently unquantified environmental amenities and values" (42 U.S.C. 4332 Sec. 102 B) are considered appropriately when the agencies make decisions that could affect the environment. Since environmental analysis was in its infancy when NEPA was enacted, there was no accepted overall approach for a comprehensive, integrated, and interdisciplinary environmental impact analysis and for incorporating environmental concerns into agency decisions. The requirement to develop methods and procedures was intended to address the deficiency. This has been interpreted as requiring some level of research into the environmental effects of actions within an agency's jurisdiction: for example, the health effects of radiation by the Nuclear Regulatory Commission, wildlife effects of prescribed burns by the Forest Service, or noise effects on communities by Air Force flight training missions.

The research requirement specified consultation with CEQ in developing these methods, but in practice each agency has proceeded independently with responsible agencies taking the lead in certain areas. Environmental resource agencies, such as the Fish and Wildlife Service, National Marine Fisheries Service, and state historic preservation officers have developed methods for quantifying and evaluating resources under their protection and jurisdiction. The U.S. EPA has also taken the lead in developing investigation and evaluation methods in the aquatic environment, primarily due to their responsibility in issuing National Pollution Discharge Elimination System (NPDES) permits and other responsibilities under the Clean Water Act. The U.S. EPA did extensive toxicity testing and developed protocols to assist states in meeting their Clean Water Act obligation of developing water quality standards. Also as presented in the discussion of programmatic impact significance criteria (Section 5.3.3.3) the U.S. EPA funded research that was critical in evaluating marine biota and water quality impacts and was a determining factor in decisions involving the level of municipal wastewater treatment and discharge location. Since the passage of NEPA, numerous method documents have been produced by EPA and other federal agencies to advance and standardize technical impact analysis methodology. These documents reflect all aspects of analysis including describing existing conditions, predicting impacts, determining level of significance, and developing mitigation. A number of methods were developed by federal environmental and resource agencies initially after the passage of NEPA (Table 2.1) and the development of analysis procedures has continued. The list presented is not intended to be exhaustive but rather to give examples of the early attempts at developing environmental analysis methods and the last entry reflects the current practice of including the documents on agency web pages. As part of the environmental analysis process, the agencies with jurisdiction and/or technical expertise should be contacted to obtain guidance on the latest analysis methods for resources potentially affected.

2.2.4 The Environmental Impact Statement

The most visible and best known aspect of NEPA is the EIS. The foundation for the EIS is a "detailed statement" as required in Title I, Section 102 C: "*...include in every recommendation or report on proposals for legislation and other major Federal actions significantly affecting the quality of the human environment, a **detailed statement** by the responsible official...* [emphasis added]." Most people familiar with environmental analysis, including many NEPA practitioners, consider EIS as the only meaningful aspect of NEPA. Although this may be an exaggeration and represent incomplete knowledge of environmental science and policy, the EIS certainly is the aspect of NEPA that has had the greatest exposure, the most influence on environmental analysis, and been a very large factor driving environmental analysis procedures, protection, and enhancement.

NEPA contains very little guidance or specific requirements for an EIS. It is only addressed in Section 102 C and in that section the only requirement is that the detailed statement address:

- "the environmental impact of the proposed action...."
- "any adverse environmental effect which cannot be avoided..." (this is redundant with the first requirement).
- "alternatives to the proposed action."

TABLE 2.1
Example Methods Developed by Federal Agencies to Facilitate Environmental Consideration as Required by NEPA

Title	Reference	Applicability to NEPA Environmental Analysis
Techniques for Sampling and Analyzing the Marine Macrobenthos	EPA/600/3-78-030. EPA March 1978	Standardized description of existing aquatic invertebrate resources and techniques to determine impacts on the resource from past activities.
Guidance on Use of Habitat Evaluation Procedures and Suitability Index Models	Fish and Wildlife Service; PB 86-100151 U.S. Fish and Wildlife Service.1987	Determination of relative value and characteristics of ecological resources before and after proposed action.
Rapid Bioassessment Protocols for Use in Streams and Rivers	EPA/444/4-89/001. EPA 1989	Presents methods for EIS/EA level analysis of aquatic biological existing conditions and impacts from past actions.
Plant Toxicity Testing with Sediment and Marsh Soils	NPS/NRWRD/NRTR-91/03. Department of Interior 1991	Provides methods to determine what level of contamination would impact vegetation.
Methods for Measuring the Acute Toxicity of Effluents and Receiving Waters to Freshwater and Marine Organisms	EPA/600/4-60/027. E[A. 1991	Establishes standardized methods to determine impacts from discharges on surface waters and can be used not only in impact prediction but development of alternatives.
Framework for Ecological Risk Assessment	EPA/822/R-92/001. EPA 1992	Establishes a conceptual approach to ecological risk assessment, which can be a basic tool in evaluating impacts to biological resources.

(*Continued*)

TABLE 2.1 (Continued)

Example Methods Developed by Federal Agencies to Facilitate Environmental Consideration as Required by NEPA

Title	Reference	Applicability to NEPA Environmental Analysis
Fish Field and Laboratory Methods of Evaluating the Biological Integrity of Surface Waters	EPA/600/R-92/111. EPA March 1993	Standardized description of existing fisheries resources and techniques to determine impacts on the resource from past activities.
Ecological Risk: A Primer for Risk Managers	EPA/734/R-95/001. EPA 1995	Nontechnical presentation on use of ecological risk assessment to evaluate impacts.
Evaluation Guidelines for Ecological Indicators	EPA/620/R-99/005. EPA May 2000	Establishes protocol to select appropriate indicators of ecological conditions and predicted impacts. This focuses and streamlines investigation phase of environmental analysis.
Methods for Collection, Storage, and Manipulation of Sediments for Chemical and Toxicological Analyses; Technical Manual	EPA-823-B-002. EPA October 2001	Establishes standardized methods for evaluation of aquatic sediment conditions and impacts.
A National Park Guide: Protecting and Enhancing Soundscapes	www.nature.nps.gov/naturalsounds. National Park Service Oct. 2012	Methods and criteria for noise impact evaluation.

Abbreviations: EA, Environmental Assessment; EIS, Environmental Impact Statement; EPA, Environmental Policy Act; NEPA, National Environmental Policy Act.

- "the relationship between local short-term uses of man's environment and the maintenance and enhancement of long-term productivity."
- "any irreversible and irretrievable commitments of resources..." (this is also redundant with the first requirement).

It is not surprising that the definition, requirement, and guidance for EISs are so brief, nonspecific, and ambiguous. Prior to NEPA, few if any environmental analyses had ever been prepared so there was little experience in the field, very few examples to use, and expectations varied widely based on the level of environmental concern and the agenda of the environmental analyst. So EIS preparation and content evolved by trial and error until a watershed court ruling in 1972 over the Calvert Cliffs Nuclear Power Station on the western shore of the Chesapeake Bay.

The Atomic Energy Commission (AEC) conducted an environmental review as part of their licensing procedure for the Calvert Cliffs station and it was challenged by environmental groups under NEPA. The court ruled that AEC's "detailed statement" and other aspects of their process did not meet the minimal requirements set forth in NEPA. Following the ruling, AEC embarked on an interdisciplinary effort to develop and implement environmental impact analysis procedures that fully met NEPA. This endeavor set the basic structure and expectation for what we now know as the EIS and environmental analysis in general (Gustafson 1993; Eccleston 2008).

Environmental impact statements are not documentations of scientific studies; they are legal documents. They are prepared to meet the regulatory requirements of NEPA and thus must satisfy the procedural mandates within NEPA and the implementing regulations promulgated by CEQ and individual federal agencies. The requirement resulting from numerous court rulings to take a "hard look" at environmental implications of a proposed action is the primary NEPA mandate that links an EIS to scientific defensibility. The federal agencies and courts constantly work to develop a threshold for the level of scientific investigation and rigor that constitutes the "hard look" as prescribed in early litigation. However, since every action and environmental setting combination is unique, the criterion for the hard look is continually evolving.

In contrast to classical science, the degree of impact prediction detail, peer review, accuracy, and precision considered acceptable in an EIS is not absolute. It is dependent on numerous factors including the practicability of obtaining information, the magnitude of the potential impact, the likelihood the impact will occur, the technical sophistication and experience of the agency proposing the action, the level and type of stakeholder involvement, and the implications to societal values if the impact does occur. In other words, the degree of impact investigation and analysis necessary is dependent on the type and intensity of the potential impact, real or perceived.

As important as the environmental statement requirement is to the goals of NEPA, there are exceptions. During the legislative debate over NEPA, Senator Muskie from Maine insisted on an exception to the mandate for environmental statements on certain regulatory actions designed to provide environmental enhancement and protection (Andrews 1997). The exceptions were proposed because they would slow down the implementation of the intended environmental improvement and create no added value. Drafters of the legislation agreed, recognizing the intent of the regulations was environmental protection and thus exclusion of environmental regulation was consistent with the evolving national environmental policy. They also recognized the need for a significant and immediate action to protect the rapidly diminishing value of threatened environmental resources such as air quality, endangered species, water quality, and wetlands and did not want to delay or compromise environmental protection legislation through the NEPA process.

Thus such regulations and other actions designed to protect environmental resources, which are now promulgated primarily by the U.S. EPA, are not required to adhere to the procedural tenets of NEPA. However EPA has retained much of the intent of NEPA as they promulgate environmental regulations. Their internal procedures mandate extensive public exposure and provisions for comments on both draft and final regulations. Also as a practical matter, virtually all EPA-proposed regulations are supported by the best available science and reviewed by an independent congressionally established EPA Science Advisory Board. Thus it meets the NEPA mandates of a "hard look," public involvement, and incorporation of environmental consideration in decision making.

2.2.5 NEPA Consultation and Public Disclosure

Although not as visible to the public as the detailed statement or EIS document, another NEPA requirement has had as much or greater effect on federal agency actions and environmental protection. NEPA requires in Section 102 C, "*Prior to making any detailed statement the responsible Federal official shall **consult** with and obtain the comments of any federal agency which has jurisdiction by law or special expertise with respect to any environmental impact involved*" [emphasis added]. Although the term "consult" did not convey veto power or even an active role in the decision, there was an additional requirement as part of the section: "*Copies of such statement and the comments and views of the appropriate Federal State and Local agencies ... shall be made available to the president, the CEQ, and to the public.*" This public disclosure requirement has forced the agency proposing an action to take concerns and interests of other agencies seriously and provide a public explanation if they chose to ignore the concerns. This has imposed a high dose of common sense and objectivity to federal agency decisions which potentially affect the environment.

During the mid-to-late twentieth century, the United States moved to a government system dominated by agencies and bureaus. Each entity had its own mission, and worked typically in close coordination and cooperation with the constituents of the services or resources under the jurisdiction of the agency. For example, the Federal Power Commission worked closely with corporations generating electrical power and the Bureau of Land Management had strong working relationships with ranchers who grazed cattle on federal lands. Under this system of strong agencies and bureaus, there was a tendency to focus very narrowly on the mission of the agency, and even sometimes a relatively small subsection of the agency that listened only to the users of the agencies' services and resources. This process tended to limit adherence to the overall national agenda and the participation and influence of other stakeholders, including other federal agencies.

NEPA changed this practice of governance with respect to environmental issues and by extension to other actions and procedures by many agencies and bureaus (Andrews 1997). The most immediate change caused by NEPA in this regard was to mandate an internal review of an agency's actions. This review forced examination by the highest levels of the agency and allowed those in other subsections of the agency to express their opinion and an opportunity to alter decisions made based solely on the narrow interests of those directly involved with the action. NEPA also required coordination and comments by other federal and state agencies with jurisdiction over environmental resources that could potentially be affected by the agency's proposed action. This provision further broadened the review of the action and consideration of issues outside the direct mission of the proposing agency.

Perhaps the greatest change produced by NEPA in a governance system heavily relying on agencies and bureaus was to require public disclosure. This requirement subjected the officials responsible for decisions potentially affecting the environment to public scrutiny. Thus, although NEPA did not legislate against stupid or solely self-serving decisions, it did require officials making such decisions to explain themselves, which resulted in better decisions. This NEPA requirement was expanded in practice to include extensive public input during scoping and review of draft and final EIA documents which has also resulted in generally better projects and decisions. The public exposure requirement also made the legal challenge to procedural requirements of NEPA possible. If all of the information related to actions, investigations, and considerations of a federal agency decision potentially affecting the environment were not available to the public it would be very difficult, if not impossible, to challenge in federal court the adherence to NEPA procedure for any federal proposal or to critically examine adherence to the "hard look" criterion established by the courts.

2.2.6 International Implementation of NEPA

There is another NEPA Section 102 requirement that has implications for environmental impact analysis. Section 102 F states that the federal government shall *"recognize the worldwide and long-range character of environmental problems and where consistent with the foreign policy of the United States lend appropriate support* *to the initiatives.*" There was much confusion created by this ambiguous NEPA clause and CEQ implementing regulations (see Section 2.3 for a discussion of NEPA implementing regulations) did little to address the confusion in the applicability of NEPA for international federal actions that affect the environment of sovereign nations. In order to add clarity to the international aspects of NEPA, President Carter issued Executive Order 12114.* The intent of the order was to give federal agencies, involved in international activity, guidance for compliance with NEPA for overseas actions. The order does not address impacts on foreign nations (primarily Mexico and Canada) from actions within the borders of the United States (transboundary effects). CEQ has issued separate guidance on transboundary effects (CEQ Guidance on NEPA Analyses for Transboundary Impacts, July 1, 1997).

Executive Order 12114 did give some clarity, but it is limited, flexible, and gives deference to the state department. Any federal agency proposing international action is required to coordinate with the state department and work with them to determine the NEPA applicability and specific procedures to follow for each proposed action. The limitations of the executive order and deference to the state department are acknowledgment that international relations and the sovereignty of a foreign nation take precedence over NEPA.

There are three general situations identified in the executive order requiring a federal agency to consider NEPA in international situations. The first is whether the action could affect the global commons outside the jurisdiction of any nation (e.g., Mediterranean Sea, oceans, or Antarctica) or environmental resources protected by U.S. treaties or international agreements signed by the United States. The second condition of creating NEPA jurisdiction outside of U.S. borders is action taken by a federal agency that could affect the environment in a foreign nation that is not participating or involved in the action. This second type of action would rarely, if ever, apply because except for armed conflicts, which are exempted from NEPA (see below), U.S. federal agencies rarely, if ever, take overt actions in other countries without their involvement or participation.

The final condition specified in the executive order requiring application of NEPA procedures for international activity is the most complex. If a federal agency is providing a product, facility, infrastructure, or service to a foreign government with their participation, then NEPA applies under some circumstances. If a similar action was carried out within the United States and it was prohibited, strictly regulated by federal law, or it could create environmental conditions posing a risk to human health, NEPA would apply to the action carried out in a sovereign foreign nation. For example, if the U.S. State Department, Agency for International Development (USAID) were proposing to plan, design, and construct a comprehensive sewage treatment system in a developing nation and if the effluent from the wastewater treatment process could pose a risk to human health, USAID would be required to implement NEPA procedures.

---

* Executive Order No. 12114: 44 Fed. Reg. 1957, 1979.

Summary of the National Environmental Policy Act

Even with the limited applicability of NEPA for international situations, there are important exemptions in Executive Order 12114 (applicable sections of the executive order are noted for each exception):

- Only the natural and physical environments are to be considered. Social, economic, and other environments cannot trigger the need for NEPA and if NEPA otherwise applies, these environmental resources do not have to be addressed (Sec. 3.4).
- International actions taken by the U.S. president are not subject to NEPA (Sec. 2.5 ii).
- International actions taken for national security, intelligence activities, arms transfers, emergency relief, or in the course of an armed conflict (Sec. 2.5 iii through vii) are not subject to NEPA.

In addition to the above-mentioned limitations, NEPA procedures for international action can be modified for a number of reasons. If the agency determines that a strict application of NEPA procedures will prohibit prompt action in a time-sensitive critical situation, adversely affect foreign relations, or appear to infringe on another nation's sovereign responsibilities, the agency can modify its established NEPA procedures on a case-by-case basis. The executive order also includes a modified set of NEPA procedures (some practitioners consider them "NEPA light") to be used in international situations.

Of all federal agencies, the USAID has had to deal with this issue the most because (with the possible exception of the Department of Defense, which is excluded from NEPA in many foreign actions) it is the agency with the greatest international presence and promulgation of actions potentially affecting the environment. They fund and conduct extensive infrastructure programs for roads, water, wastewater, and agriculture, all of which can have environmental consequences. This overseas presence and settlement of litigation prompted USAID to develop a policy and guidance (USAID 2005) to address NEPA when proposing a federal action. Consistent with Executive Order 12114, the guidance allows for numerous exceptions including international disaster assistance, other emergency circumstances, and (not surprisingly) circumstances involving exceptional foreign policy sensitivities. USAID also has a list of categorical exclusions (CATEX) in their guidance (see Section 3.6 of the book for a full discussion on CATEX), one of which acknowledges working in developing nations and states that NEPA does not apply if the local authorities are heavily involved and USAID does not *"have knowledge of or control over the details of the specific activities."* The guidance also exempts many activities, most of which are related to training, education, scientific investigation, and nutrition. Several types of projects commonly supported by USAID are categorically included, such as irrigation or water management projects, including dams and impoundments, penetration road building or road improvement projects, and large-scale potable water projects and sewerage projects.

USAID employs a phased approach to addressing NEPA requirements by first preparing an "Initial Environmental Examination." This examination results in a "Threshold Decision" concluding: (1) no significant environmental effect and no further NEPA action or (2) there is significant effect and either an EA or EIS must be prepared. Generally for other federal agencies and for actions within the United States, the potential for a "significant environmental effect" is the trigger for a full EIS, and an EA will suffice if there is no indication of significant impact (see Section 3.1 of the book). However, the unique circumstances governing NEPA compliance in international situations and USAID's mission prompted the agency to identify a different criterion defining when to prepare an EA and when to prepare an EIS. For USAID actions, a formal NEPA document (an EA or EIS) is prepared only if categorically included or if there is a "significant environmental impact" on the physical or natural environment (i.e., not the built environmental or social resources). An EIS is required only in one of the three specific situations:

- If there is an impact on the global environment (e.g., international ocean areas, Mediterranean Sea) or areas outside the jurisdiction of any nation (e.g., Antarctica).
- If the agency action would have a significant impact on the environment of the United States.
- At the discretion of the USAID administrator.

In all other cases of significant impact, an EA is the designated environmental impact analysis procedure and document under the USAID NEPA program.

The intent of this system is to force consideration of the natural environment within a host country when USAID takes action in that country but is not to be subject to the full public participation and disclosure requirements of a standard EIS prepared for a project in the U.S. This is because in many countries where USAID provides support, the democratic principles and practices have not matured to the point of making public participation and disclosure a common practice. Implementing such procedures for environmental concerns when they are not part of the host country's treatment of other issues could cause difficulty for the government of the host country as well as the U.S. State Department.

Does NEPA apply in international situations? Yes, but there are several limitations, restrictions, exemptions, and constraints. There is a legitimate effort to protect the global commons and also a strong commitment to prevent NEPA from interfering in critical and sensitive relationships with foreign countries. Experience has shown that if NEPA procedures would apply if the federal action was implemented within the boundaries of the United States and if the country where the action would occur sees the benefits of applying NEPA, then the federal agency proposing the action will conduct a thorough environmental analysis. If there are implications to foreign affairs or relations with the host country, particularly as determined by the U.S.

State Department, then the proposing federal agency will conduct a limited environmental analysis or forego NEPA altogether.

2.2.7 Council on Environmental Quality

Title II of NEPA addresses the administration of the Act, among other mandates, by establishing the CEQ. As described below, CEQ was not intended to be a NEPA police or enforcement entity but more a resource for environmental information and coordination. When NEPA was enacted, one of the primary roles assigned to the CEQ was the responsibility of generating an annual report on the environment, namely the environmental quality report. Initially the annual report was primarily a "to do" list and subsequently evolved into a progress report and "report card." CEQ published these reports from 1970 to 1997. However as individual agencies took responsibility for NEPA implementation and compliance, the comprehensive annual report on the environment diminished in importance and visibility. In 1995 Congress passed the Federal Reports Elimination and Sunset Act (Public Law 104-66) which eliminated specific reporting by federal entities, including the environmental quality report. In spite of eliminating the annual report, CEQ continues to provide similar information regarding the environment annually on their web page.

NEPA was originally very clear on the limited authority of CEQ. It was chartered as a source of environmental information made available to federal agencies and not an enforcing or even an aggressive coordinating agency. Sections 202 through 209 of NEPA make it very clear that CEQ is not a "command and control" entity and it does not even have primary oversight responsibility for EISs that are generally considered the center piece of NEPA (see Section 2.4 of this chapter for oversight of EISs). These sections of NEPA even set a very small maximum budget for CEQ ($300,000 for the first year) to emphasize and enforce the limited role of the council.

The role of CEQ has evolved as individual agencies have gained environmental expertise and additional environmental laws have been enacted. In the early years of NEPA, CEQ did provide some guidance to agencies as they struggled with the meaning, intent, and requirements of NEPA and specifically with the preparation of EISs (see Section 2.3 of this chapter, Summary of CEQ Guidelines Implementing NEPA). Through the years, CEQ's most important role has been to resolve conflicts among agencies. If a federal agency feels environmental resources or requirements related to their jurisdiction or within their area of expertise are threatened by the action of another federal agency, their NEPA recourse is to refer the conflict to CEQ. CEQ has the authority to resolve the conflict either by ruling in favor of one or the other agencies or by forging a compromise. In practice, this "elevation to CEQ" is rarely used in comparison with the hundreds of EISs filed annually, because the agency proposing the action realizes this could result in delays at best or worse, cancellation of the proposed action and bad publicity. Also the necessity for an environmental permit for subsequent implementation of most proposed actions is an incentive to reach a compromise with sister agencies that issue or must concur with the permits (see Chapter 9 of the book for an explanation of the relationship between NEPA and environmental permits). Thus the elevation provision serves as an incentive for agencies to work together to reach mutually acceptable compromises consistent with NEPA, and CEQ has even acted in advance to defer elevation. In the case of the U.S. Coast Guard dry cargo residue (DCR) management program (see Chapters 5 and 10) where multiple resources, jurisdictions, and agencies were involved, CEQ acted in advance. CEQ requested a presentation by the U.S. Coast Guard on the status of their rule promulgation addressing discharge of DCR into the Great Lakes. CEQ also requested their comments on how this rule promulgation related to similar and simultaneous action by the U.S. EPA. The system of CEQ using a carrot and stick approach to force federal agencies to cooperate on environmental issues has worked well and the U.S. Congress could take a lesson from this approach to resolve conflict and to advance other national goals similar to the national environmental policy.

2.2.8 Bureaucratic Culture under NEPA

NEPA had another, perhaps unanticipated, consequence that has resulted in significantly better environmental decisions and protection associated with federal actions. NEPA requirements and subsequent court decisions forced agencies to prepare defensible analysis that could withstand challenges and thus avoid lengthy and costly delays of redoing inadequate studies and documents. The agencies quickly realized that hiring environmental professionals to conduct or oversee NEPA compliance, ecological studies, and environmental documentation was a major step in producing defensible products and thus withstanding NEPA challenges. As these environmental professionals became integrated into the organizations and assumed the responsibilities previously held by long-term, agency-mission-focused bureaucrats, they gained respect and influence within the agency. They brought with them an understanding and some level of commitment to the national policy on the environment expressed in NEPA and thus brought agency policy, procedures, projects, and decision more in line with the goals of NEPA and environmental protection in general.

2.3 Summary of CEQ Regulations Implementing NEPA

When NEPA was passed there was little or no experience in environmental analysis. Also as discussed earlier, in order to reach an acceptable compromise among government agencies, industrial stakeholders, development interests, and environmental advocates, the NEPA-specific requirements (Sec. 102) were intentionally designed to be general and nondirective. Consequently, the Act gave no specific or detailed direction on how federal agencies should comply with NEPA, so President Nixon issued Executive Order 11514

(March 5, 1970) giving CEQ authority to issue guidance to federal agencies in implementing NEPA. CEQ issued final guidelines in August 1973 (38 Fed. Reg. 10856 [1973]) and the guidelines were used for a number of years until their limitations became apparent.

The major limitations of the 1973 CEQ NEPA guidelines were in the extensive level of discretion left to officials within the agencies, who had little understanding of NEPA and no commitment to the new national environmental policy or environmental awareness. In the early years following passage of NEPA, senior management in most federal agencies had these limitations and saw the Act and the CEQ guidelines as a threat to their agencies' mission. As a result of minimal experience and appreciation, at best there was inconsistency in the application of the CEQ guidelines among agencies and at worst the guidelines were ignored by many agencies. The courts did little to rectify the situation because they viewed CEQ's efforts as guidance and not mandates. Also the court decision on the Calvert Cliffs Nuclear Power Station (see Section 2.2.3 of this chapter) gave agencies some insight into the NEPA primary product (EIS), but little guidance on how to conduct the environmental impact analysis. This situation prevented attainment of NEPA objectives and severely inhibited productive public participation because the public could not determine how to become involved due to the inconsistencies and variability among agencies and even different factions within the same agency.

This situation was seen as undermining the intent of NEPA, and President Carter issued Executive Order 11991 (May 24, 1977) giving CEQ authority to issue binding requirements on all federal agencies. The resulting regulations (CEQ 43 Fed. Reg. 559990,1978 amended as 51 Fed. Reg. 15625,1986 and Public Law 91-190, 42 U.S.C. 4321-4347, as amended by Public Laws 94-52, 94-83, and 97-2580) became the mandatory "gold standard" for implementing NEPA. In accordance with the CEQ Regulations (§1505 and 1507), each federal agency has since issued their own NEPA guidance that adheres to CEQ Regulations and meets the specific mission and organizational structure of each agency.

The CEQ and agency procedures implementing NEPA have withstood the test of time and numerous court challenges. In more than 25 years, the regulations have only been amended one time, and that was to delete the clause, requiring an "analysis of worst case." This change was not because it was seen as inconsistent with the intent of NEPA but rather because in the complex environmental analysis field with almost infinite variables, the worst case cannot be reasonably defined. The CEQ Regulations have also established a strong framework for technically sound environmental analysis and active public input and participation. Environmental analysis in the United States now operates under the umbrella of these regulations, and state and international programs have been developed in light of the experience gained and lessons learned from the CEQ Regulations.

The CEQ Regulations added meat to the skeletal requirements in Section 120 of NEPA. They cover all aspects of NEPA compliance from applying NEPA in the early stage of an agency's strategy development and planning process (§1501.2) to procedures and content for their final Record of Decision (ROD) (§1501.2). The regulations are over 30 pages long and it is not the intent of this book to repeat every detail or guide the reader through a step-by-step cookbook approach to CEQ regulatory compliance. But the major requirements are summarized below with discussion of the concepts that relate to the practice of environmental impact analysis and comments on how the CEQ Regulations have advanced the practice of environmental analysis. Anyone embarking on a project, program, or policy subject to NEPA (i.e., a federal action with potential environmental impacts) is more than encouraged to not only read, but study the CEQ Regulations and specific NEPA agency guidelines applicable to the action under consideration.

2.3.1 Public Involvement

Section 102 of NEPA specifies public inclusion in the process but the degree, procedure, and scope of public participation are not addressed. The CEQ Regulations (Purpose section, §1500.1) remove any ambiguity regarding the necessity for the public to participate in the process: *NEPA procedures must insure that environmental information is available to public officials and citizens before decision are made...and public scrutiny are essential to implementing NEPA*. The Policy section (§1500.2) also emphasizes the importance of active public involvement: *Encourage and facilitate public involvement in decisions which affect the quality of the human environment*. This requirement has had a great effect on achieving the goals set forth in NEPA as any other aspect of the Act or implementing procedures. At the onset of a federal action, the public is made aware of any action an agency is anticipating, which has had several beneficial results in environmental protection and enhancement. The intended result of the public involvement requirement is that the public can have input early in the process (see Scoping, Section 4.3 of the book) and identify their concerns. Public involvement also provides an avenue for the directly affected stakeholders, who frequently have lived side by side with a problem for many years, to share their ideas regarding potential solutions with federal agencies and project proponents. Even if the stakeholders do not have extensive technical expertise, they frequently have an understanding of the problems and the existing conditions not apparent to an outside expert. This early input also allows stakeholders to organize, contact their elected representatives, make their voice heard, and potentially influence federal agency decisions regarding the environment.

The mandatory public involvement aspect of the CEQ Regulations also has implications beyond influencing the scope of environmental analysis and influencing agency decisions. The requirement to make all environmental analyses and the backup information readily available to the public exposes the analysis to the cleansing effects of daylight. This exposure and accountability improves the quality, accuracy, and completeness of the information used to evaluate environmental impacts. Equally important, the exposure to daylight and availability of all information used in the decision facilitate any litigation efforts following an agency's decision. As discussed later (in Section 3.7, NEPA Enforcement, of the book) the courts have played a major role in defining NEPA procedures and the publicly accessible information required by the CEQ Regulations and NEPA are frequently the critical factors in facilitating challenges.

The lack of public input and readily available information can create a very different environmental protection picture. Many authoritarian governments have environmental regulations in place, and frequently they are stricter and more in command and control than NEPA. However, they frequently lack the "daylighting" requirement.* In some cases the environmental analysis document is held confidentially by the environmental agency and the public is not even allowed to see the backup for the analyses or even the final document. In such situations public input is often not part of the process initially or sometimes it is never part of the process. This lack of public exposure and limited input creates a culture where agencies and private entities can further their objectives without meaningful consideration of the natural or social environment if they can gain the support of a select few influential people or groups.

2.3.2 CEQ Regulations' Relationship to Other Statutes

CEQ Regulations recognize the integrated nature of environmental protection and as described in §1502.25 the regulations specify procedures for compliance with other environmental legislation in addition to NEPA including:

- Environmental Quality Improvement Act of 1970 (42 U.S.C. 4371 et seq.)
- Section 309 of the Clean Air Act (42 U.S.C. 7609)
- Executive Order 11514 Protection and Enhancement of Environmental Quality

The CEQ Regulations (§1502.25[a]) also address the integration of environmental protection by citing other federal environmental requirements that must be addressed during the NEPA process, including:

- Endangered Species Act of 1966 (16 U.S.C. 1534 et seq.)
- National Historic Preservation Act of 1966 (16 U.S.C. 470 et seq.)
- Fish and Wildlife Coordination Act (16 U.S.C. 661 et seq.)

In addition, there are numerous other regulations and environmental permit requirements that should be considered and addressed as part of the NEPA process (Table 2.2).

---

* Author's personal experience in the Middle East, Asia, and South America.

TABLE 2.2
Select Federal Regulations to Be Considered during the NEPA Process

Regulation	Reference	Applicability to NEPA Environmental Analysis
Floodplain Management	Executive Order 11988	Requires determination of relationship of any federal actions and Federal Emergency Management-delineated floodplain. If floodplain potentially involved, must demonstrate level of impact and requires mitigation.
Environmental Justice	Executive Order 12898	Must identify minority and disadvantaged population potentially affected by federal action and demonstrate degree of impact and requires mitigation.
Protection of Wetlands	Executive Order 11990	Requires determination of relationship to any federal actions and wetlands (meeting U.S. Army Corps of Engineers and EPA definition). If wetlands potentially involved must demonstrate level of impact and requires mitigation.
Invasive Species	Executive Order 13112	Prohibits federal agency authorization or funding actions that may contribute to the introduction or spread of invasive species.
Clean Water Act	33 U.S.C. §1251 et seq. (1972)	Compliance with numerous provisions including Section 404 (protection of wetlands), Section 402 (National Pollution Discharge Elimination System), and state water quality standards must be demonstrated.
Clean Air Act	42 U.S.C. §7401 et seq. (1970)	Compliance with numerous provisions for ambient air quality, point-source emissions, and non-point-source emissions. Also Section 309 EPA review of EISs (see Section 2.4 of this chapter).
Rivers and Harbors Act	33 U.S.C. §403; Chapter 425, (1899)	Must demonstrate compliance with act, originally passed to protect navigation but since interpreted to prevent interference with surface water hydraulics and water quality.

Summary of the National Environmental Policy Act

Pollution Prevention Act of 1990	42 U.S.C. §13101 et seq. (1990)	Must consider reducing the amount of pollution through cost-effective changes in production, operation, and raw materials to achieve source reduction.
Farmland Protection Policy Act of 1981	7 U.S.C. §4201 et. seq. (1981)	Must demonstrate that federal actions take measures to minimize the unnecessary conversion of farmland to nonagricultural uses.
Wild and Scenic Rivers Act of 1968	16 U.S.C. §1271-1287 (1968)	Requires the determination if federally designated Wild and Scenic Rivers are potentially affected by proposed action and if so demonstrate there will be no action that would substantially change their wild or scenic nature.
American Indian Religious Freedom Act of 1978	42 U.S.C. 1996	Federal actions cannot restrict access to or adversely affect native American sacred sites and objects considered sacred. As part of the NEPA process the agencies must consult with potentially affected native Americans, identify any sacred resources and take measures to avoid impacting the resources.
Coastal Zone Management Act of 1972	16 U.S.C. §1451 et. seq. (1972)	The act requires each affected state to develop a Coastal Zone Manage plan and federal agencies to demonstrate, within the NEPA process compliance with the plan for each proposed action.
National Historic Preservation Act	16 U.S.C. §4703 et. seq. (1966)	The act requires identification of historical properties and other features and federal agencies to demonstrate, within the NEPA process, protection of the properties for each proposed action.

Abbreviations: EPA, Environmental Policy Act; NEPA, National Environmental Policy Act; U.S.C, United States Code.

2.3.3 Making NEPA User Friendly

Many early EISs and other NEPA documents were encyclopedic, lengthy, nonfocused, and technical beyond the comprehension of all but highly trained stakeholders. On more than one occasion, courts have ruled EISs as inadequate simply because the analyses and project descriptions were so technical in both nature and reporting that they were incomprehensible to the general public and the majority of stakeholders. CEQ recognized these deficiencies and their impediment to advancing the goals of meaningful public input and integration of environmental considerations into agency decisions. Numerous measures (CEQ Regulations, Sec. §1500.25 through §1501.8), as summarized below, were incorporated into the regulations to address these concerns. Most of these requirements not only advance the purposes of NEPA, but they also are important as environmental analysis methods and are discussed in detail elsewhere in this book. NEPA documents and processes should:

- Be analytical rather than encyclopedic, deemphasizing background, including the affected environment (see Chapter 5) information, and focusing on issues and analyses useful to decision makers and the public.
- Use plain language understandable to the nontechnical stakeholder. Many agencies informally express this requirement as NEPA documents should read like *Newsweek* or *Time* magazine. If information is only understandable to the technical audience it should be relegated to an appendix. The courts have overturned EISs and ordered them redone because the use of technical terms and concepts are beyond the comprehension of the intended audience, including the judge.
- Maximize the scoping process (i.e., procedure to determine the issues, their significance, and evaluation methods to be used in the NEPA Process) to focus the environmental analysis on critical issues (see Chapter 4).
- Assist decision makers in the incorporation of environmental considerations by ensuring issues are addressed in adequate detail and quantitatively to the extent possible so that they can be compared with engineering/construction and economic analyses.
- Address common issues once to avoid duplication and maintain focus on the issues ripe for decisions using such techniques as:
    - Environmental analyses of entire agency policies, programs, and plans before embarking on analyses of each individual action addressing the policies, programs, and plans. This approach is often considered tiered or supplemental (see Chapter 6).
    - Jointly conduct environmental analyses to expedite the process and avoid repetitive discussion. This would include multiple

federal actions (e.g., one agency is funding and another agency is issuing an environmental permit) and actions subject to both federal and state/local regulations.
- Develop and maintain a list of agency actions that have been determined not to have an impact. For example, CATEX (CEQ Regulations §1508.4; see Section 3.6 of this book).
* Begin environmental analysis and the NEPA process early in agency action planning and integrate it throughout development. This includes interagency cooperation and dialogue.

The CEQ Regulations (§1501) establish a process for NEPA compliance that is discussed in detail in Chapter 3. Similarly the NEPA centerpiece, the EIS, is described in CEQ Regulations §1501 and discussed in detail elsewhere in this book (Chapter 3 for NEPA-specific processes and Chapters 4 and 5 for processes generally typical to environmental analysis under any program).

2.3.4 Commenting and Coordinating on NEPA Products

As discussed above, daylighting and public access to information are critical components of NEPA and essential to advancing the nation's environmental policy. CEQ Regulations §1503 on "commenting" implements the openness and solicitation of public comment on the NEPA process implied in the Act. This section specifically requires solicitation of comments on a draft EIS and provides the option for comments on the final EIS. In addition, §1506.6 deals directly with public involvement and includes requirements for each agency to *"Make diligent efforts to involve the public in preparing and implementing their NEPA procedures."* The section also requires each agency to provide adequate public notice of meetings and information availability and hold public meetings and hearings whenever appropriate during the execution of the NEPA process. The requirement for scoping (§1501.7) also emphasizes the importance of involving the public, other agencies, and even *"including those who might not be in accord with the action on environmental grounds."*

The CEQ Regulations also seek maximum public involvement by requiring a broad notification of actions potentially affecting the environment. A notice of intent (NOI) must be prepared and issued if an agency intends to prepare an EIS. The notice must include a clear and concise statement addressing the purpose and need for the action (see Section 3.2 for a full discussion of purpose and need). In addition, the NOI should include a brief description of the proposed action and any alternatives the agency has considered or plans to consider. The NOI is not expected to contain an exhaustive or complete list because alternatives will be developed, refined, and eliminated as an integral part of the scoping and environmental evaluation process. The NOI must also summarize the agency's proposed scoping process, including identification of opportunities for public input, and contact information of an agency representative who is available to respond to public questions and requests. The purpose of the NOI is not to inform the public of all aspects of the proposed action or scope of the EIS but rather to provide enough information for stakeholders to determine whether they have potential interest and thus prompt them to seek additional information and, if warranted, initiate their active involvement. If an action under agency consideration is of national concern, the notice must be posted in the *Federal Register* and by mail to organizations (e.g., environmental advocacy groups and project proponents) that are reasonably expected to be stakeholders in the NEPA process. For actions of more local concern, notice must include direct communication to appropriate state agencies and through local media. Although not envisioned when the CEQ Regulations were promulgated, agency web pages and social media contacts to interest groups have become an important part of public notification.

As discussed above, coordination among federal agencies is a goal and element critical to the success of NEPA, and the CEQ Regulations (§1504) provide a mechanism to force and enforce this goal. As does the Act, the CEQ Regulations encourage the lead federal agency to involve other agencies with jurisdiction, interest, expertise, or responsibility over resources potentially affected (e.g., U.S. National Marine Fisheries Service if actions could affect marine mammals) early in the NEPA process. The CEQ Regulations go a step further and provide a procedure to resolve differences between the lead agency and another agency whenever voluntary cooperation is not successful. This process first requires the disagreeing agencies to exert a "concerted and timely effort" to resolve the differences, and if these are unsuccessful, the issue is referred to the CEQ for resolution. Other interested parties (e.g., environmental advocacy groups, states, or proponents of the proposed action) can add their views on the agency disagreement and CEQ has the option of holding public meetings or hearings on the issue. CEQ has the option of ruling in favor of one of the agencies or returning it to them for further negotiation. The CEQ referral process has rarely been used because with the threat of delays and bad publicity, a workable solution can usually be found.

2.3.5 Record of Decision

Once the environmental implications of a federal action under consideration have been analyzed, documented (including preparation of an EIS as described in Section 3.1.3), subjected to public comment and review, and considered by the responsible federal agency officials, a decision by the agency must be made. This decision is the final action under NEPA and no legal challenge to a federal action (see, Section 3.7, NEPA Enforcement, of the book) can be made under NEPA until the agency has concluded all deliberation and issued an ROD. The process for making the decision in compliance with

NEPA is described in CEQ Regulations §1505. The conclusion of the process is an agency's ROD and includes:

- A concise, simple, and clear statement of the agency's decision. Frequently the supporting NEPA document (e.g., EIS) is referenced for a more detailed discussion.
- A brief summary of alternatives considered and identification of the most environmentally preferable alternative and why.
- Factors affecting the selection of a proposed action over other the alternatives, including the rationale for rejecting the environmentally preferred alternative if it was not chosen as the proposed action. Factors discussed in making the decision should include economic considerations, technical implementability, agency statutory and mission goals, and consideration of national policy.
- How the above factors were balanced and entered into the agency's decision.
- A listing of all practicable means to mitigate the impacts of the proposed action identified during the NEPA process. A statement of commitment to implement these measures, or if not, why they will not be implemented.

CEQ Regulations also include a discussion on implementing the decision (§1505.3). The regulations make provisions for monitoring after implementing the decision, but monitoring is not required, which may be considered a shortcoming of NEPA and other environmental analysis legislation and regulations. NEPA concludes with the issuance of the ROD and does not apply during implementation of the proposed action (except for adherence to commitments made in the ROD), which could also be considered a NEPA shortcoming. However, other environmental requirements, approvals, and procedures come into play during implementation and can provide significant environmental protection (see Chapter 9).

CEQ requirements for making decisions are consistent with the overarching goal of NEPA and other CEQ Regulations with respect to transparency and "daylighting." The ROD must be made public and clearly explain the process; if it does not, it is in violation and subject to litigation. There is also an explicit requirement to make public the results of any postimplementation monitoring so that the historic record of NEPA effectiveness can continually be updated and hopefully improved.

2.3.6 Frequently Asked NEPA Questions

Following issuance of the CEQ Regulations, a series of public meetings were held to explain and discuss the regulations. During the meetings and as follow-up there were a number of frequently repeated inquiries and requests for clarification regarding the regulations. In response and in order to assist NEPA practitioners, CEQ prepared a memorandum entitled "Forty Most Asked Questions Concerning CEQ's National Environmental Policy Act Regulations" (*Federal Register*, Vol. 46, No. 55, March 17, 1981). The full text of the questions and CEQ's responses can be accessed both through the *Federal Register* and the CEQ web page (www.whitehouse.gov/administration/eop/ceq) and the questions are summarized in Table 2.3.

TABLE 2.3

Summary of 40 Most Asked NEPA Questions

1	What is the "range of alternatives" and how many alternatives must be considered in a NEPA document?
2	Must an EIS address alternatives that are outside an agency's jurisdiction or beyond the capability of an environmental permit applicant?
3	What is the "No Action" alternative and must it be addressed in all EISs?
4	What is meant by the "agencies preferred alternative," does it have to be identified in the draft and final EIS, and who selects this alternative?
5	What is the "proposed action," is it different from the "preferred alternative," and is it to be treated differently in the EIS from the analysis of alternatives?
6	What is meant by the "environmentally preferred alternative" and who selects this alternative?
7	What is the difference between the "environmental consequences" and the "alternatives" sections of an EIS?
8	What is expected of agencies with respect to "early application of NEPA" in cases of nonfederal entities' permit applications or other support by the agency?
9	What is an agency's responsibility for coordination in cases of issuing an environmental permit or funding if actions of other federal agencies are involved?
10	What actions can be taken by a federal agency or a delegated state or local agency during the preparation and review of an EIS?
11	If an agency learns that a nonfederal applicant is about to take action covered by NEPA prior to completion of a NEPA process, what is their responsibility?
12	Can proposed actions be "grandfathered" if they were conceived before CEQ Regulations?
13	Can the EIS scoping process begin before issuance of Notice of Intent, such as during preparation of an EA?
14	What are the roles, responsibilities, rights, and dispute resolution procedures among lead, supporting, and state agencies?
15	Are EPA's obligations under Section 309 of the Clean Air Act independent of their responsibility as a cooperating agency?
16	What is an EIS preparation "third party contract"?

(*Continued*)

TABLE 2.3 (*Continued*)

Summary of 40 Most Asked NEPA Questions

17	What constitutes a "conflict of interest" for a private firm in relation to preparing an EIS?
18	How should uncertainties about indirect impacts be addressed in an EIS?
19	What is the scope of mitigating measures that must be addressed and how should they be treated if they are outside the jurisdiction of the implementing agency?
20	Question 20 addressed "worst case" analysis but it was withdrawn when NEPA was amended to delete the requirement for worst case analysis.
21	If work on a project is done outside the formal NEPA process, to what degree can the NEPA document refer to prior or accompanying work?
22	May the lead agency role be shared with other state or federal agencies?
23	What procedures should be followed by an agency proposing an action if the action is in conflict with adopted land use plans?
24	When are EISs required on agency actions more comprehensive than an individual projects such as plans, policies, geographically broad actions, and follow-up activities?
25	When is the use of appendices appropriate in a NEPA document?
26	What are the guidelines for an EIS index?
27	Who should be included in the list of preparers?
28	Can reproduced copies of the EIA be filed prior to the public availability of printed versions?
29	What is an agency's responsibility to respond to comments regarding inadequacy and new alternatives?
30	Can a cooperating agency adopt only selected portions of an EIS to fulfill their responsibility under NEPA for their actions?
31	Do the CEQ Regulations apply to independent regulatory agencies and can an agency adopt a prior EIS prepared by an independent regulatory agency?
32	When and to what degree does an EIS require a Supplement before action can be taken?
33	When should an action be referred to CEQ if an agency has an issue with another agency's proposed action?
34	What is the required content and public availability of an ROD?
35	What is the expected time to complete the NEPA process?
36	What are the expected contents of an EA?
37	What is the required content of a FONSI?
38	What is the public availability requirement for EAs and FONSIs?
39	Can mitigation measures be included and required in EAs and FONSIs?
40	Can mitigation be used in an EA to eliminate otherwise significant impacts of a proposed action, thus eliminating the need for an EIS?

Source: Federal Register 46 (55), March 17, 1981.
Abbreviation: FONSI, Finding of No Significant Impact.

2.4 Clean Air Act as It Relates to NEPA

After a number of years and a somewhat rocky start, NEPA was well established and regulations to implement the Act had been promulgated. But who was to oversee and coordinate federal agencies' NEPA procedures and compliance? Congress had mandated that each agency was responsible for compliance and no external government enforcement or police force was authorized. But without some coordination, the inconsistencies among agencies would make it very difficult for the public to provide informed and constructive input because they would not know what to expect if each agency addressed NEPA requirements differently. Also inconsistency among federal agencies could be a litigation nightmare with the courts faced with determining what constituted an acceptable "hard look" at environmental implications if each of the dozens of federal agencies followed a totally different implementation of the Act and CEQ Regulations.

Some assumed CEQ would take the role of NEPA compliance coordination and oversight. However, CEQ had an extremely limited budget (restricted to $300,000 a year) and little or no internal technical expertise in all the disciplines required for comprehensive environmental analysis. Thus they were nowhere near equipped to coordinate the submission and review of approximately two EISs per day (see below for a discussion on EIS submission rate and related issues) plus perhaps 20 times as many other NEPA documents and agency regulations.

The U.S. EPA did have the authority under the Clean Air Act, Section 309 (42 U.S.C. 7609) to review and comment on the environmental impact of any federal action. EPA also had the technical expertise, staff support, and budget to take on this role, and they have taken seriously the responsibility, of coordination of NEPA, including serving as the clearinghouse for EISs and other environmental documents from all federal agencies. The 1984 promulgation of federal agency action impact review procedures (U.S. EPA 1984) made their role in NEPA compliance clear to the public and other federal agencies. Thus all agencies were required to officially file EIS documents with EPA and as of Fall 2012, filings were required to be electronic: http://yosemite.epa.gov/OEI/webguide.nsf/content/pdf_metadata.

Central components of the oversight procedures are the review, written comment, and rating of environmental analyses conducted by other federal agencies. The review is focused primarily on draft EISs but other documents such as EAs (i.e., mini EIS, Section 3.1.4 of the book) and CATEX (see Section 3.6 of the book) are frequently reviewed upon request. The review and rating are focused on the draft EIS because a proposed action requiring an EIS represents a major federal action with potentially significant environmental impact, and as a draft there is the opportunity to affect the analysis and agency-proposed action before irreversible action is taken. If the comments on the draft EIS and subsequent negotiations with the agency proposing the action do not address the concerns through the Clean Air Act Section 309 EPA Review procedures, EPA has the option of elevating the issue to the CEQ, as described above, at the final EIS stage.

EPA's rating system, as described in detail in U.S. EPA (1984) procedures for the review of federal actions and summarized below, addresses both the technical adequacy of the environmental analysis and the level of impact of the federal agency action. The rating categorizes are: 1 – adequate; 2 – insufficient information; or 3 – inadequate. A rating of 1 indicates no further data or analysis is necessary. A rating of 2 is applied if the draft EIS does not contain enough information to fully assess environmental impacts or a new reasonable alternative has been identified by EPA that could reduce the adverse environmental impacts of the proposal action. The EPA expects the federal agency preparing the EIS to address the 2 rating on the draft in the FINAL EIS. The lowest rating (3 – inadequate) indicates that the purposes of NEPA are not met in the draft EIS and a supplemental or revised draft EIS should be produced and made available for public comment.

The impact level of a proposed federal agency action is assigned one of the four ratings:

- **LO (Lack of Objections):** The predicted impacts of the proposed action are consistent with the goals of NEPA and other environmental regulations.

- **EC (Environmental Concerns):** The proposed action will result in adverse environmental impacts that should be avoided and there are changes available to the preferred alternative or mitigation measures that can reduce the environmental impact to levels consistent with the goals of NEPA and other environmental regulations.

- **EO (Environmental Objections):** The proposed agency action will result in significant environmental impacts and avoidance of the impacts may require substantial changes to the proposed action, selection of a different alternative, or implantation of substantial mitigation measures. Impacts forcing an Environmental Objection rating can include:
    - Violation of or inconsistency with an environmental standard.
    - Violation of the proposing federal agency-specific environmental requirements in the areas of EPA's jurisdiction or expertise.
    - Violation of an EPA policy declaration.
    - A precedent would be set by the proposed action that if left unchecked could cumulatively result in significant environmental impacts.

- **EU (Environmentally Unsatisfactory):** EPA believes the proposed action will result in adverse impacts to the degree that the action must not proceed as proposed because of one or more of the following conditions:

- A national environmental standard violation or inconsistency is substantive and/or will occur on a long-term basis or over a large geographic area.
- In the case where no applicable standard exists, the severity, duration, or geographical scope of the predicted impact warrants special attention.
- The predicted adverse impacts are a significant threat to national environmental resources or contradictory to environmental policies.

EPA coordination of NEPA compliance, and particularly the draft EIS rating system, has generally worked well in evaluating EISs but as discussed below there is much room for improvement in the quality of EISs, and the environmental sustainability of proposed federal actions. The EIS rating system has the potential to support development and implementation of a standard level of investigation and accepted impact prediction methodology consistent across agencies. The establishment of standard and accepted practices has been a benefit to agencies and the courts alike because the CEQ Regulations (§1502.22) require collection of the necessary information to conduct the impact analysis if the "costs of obtaining it are not exorbitant." Such a requirement is subject to differing interpretation and a history of comments from qualified environmental analysis practitioners on EISs for 25 years to determine whether the information and analysis methods are adequate makes it a much more objective determination. It is a wise NEPA practitioner who reviews EISs on similar projects to determine which approaches are rated adequate and which are not when they plan, scope, and execute an environmental analysis.

The EPA ratings and EPA comment letters authorized under Section 309 of the Clean Air Act are posted for each EIS submitted by a federal agency (http://yosemite.epa.gov/oeca/webeis.nsf/viEIS01?OpenView, Environmental Impact Statement Data Base). This database demonstrates how deeply NEPA has penetrated the functioning of several federal agencies. From July 1, 2007 to June 30, 2012, 2500 EISs were filed by all federal agencies (Table 2.4) and just over half of these were draft EISs. The number filed per year was relatively constant but during the period, 24 were the fewest filed in a month (about one a day), 65 were the most (three a day), and the average over the period was 42 per month. So NEPA practitioners and their environmental consultants filed approximately 10 EISs a week over the period and although a precise count is not available, the number of EAs and other NEPA documents has been estimated at 10–20 times that number.

There is a marked difference in the number of filings per agency over the five-year period (Figure 2.1 and Table 2.5). Of the 2500 EISs, four agencies accounted for over half of all the EISs during the period (U.S. Forest Service, 608; Federal Highway Administration, 296; Bureau of Land Management, 241; and U.S. Army Corps of Engineers, 203). On the other hand, there were nine

TABLE 2.4

Monthly Filings of Environmental Impact Statements, June 2007 to June 2012

	2007–2008	2008–2009	2009–2010	2010–2011	2011–2012
July	52	29	51	51	48
August	60	51	37	65	36
September	29	44	31	53	50
October	49	60	40	45	35
November	56	44	35	25	36
December	40	51	44	27	37
January	41	41	27	30	25
February	47	27	34	29	24
March	43	46	41	34	51
April	38	45	61	45	35
May	60	49	45	34	38
June	40	41	36	31	47
Monthly mean for the year	46	44	40	39	39
Maximum filings in a month	65				
Minimum filings in a month	24				
Monthly mean for the 5-year period	42				

agencies with 1% of the EISs filed (35 or 7 per year) and over a dozen with even fewer EISs. In 2006 as the agency with the highest number of EISs filed, the U.S. Forest Service had almost 8000 forest service employees engaged in nearly 6000 NEPA environmental analyses (including EAs) with a commitment of $365 million (Stern et al. 2010).

The ratings of the EISs also present an interesting picture (Figure 2.2). Based on a random sample, of a 100 draft EISs filed and rated by EPA, more than half (55%) received a rating of EC2 (Environmental Concern and insufficient information). With all the environmental analysis resources available, as discussed in this book and other NEPA reference books it is somewhat disheartening that half the environmental analyses conducted by federal agencies are viewed as based on insufficient information. It is a little more encouraging that LO (Lack of Objections) was the next highest rating (26% of the random sample of draft EISs) and only two were considered EU3 (Environmentally Unsatisfactory and inadequate).

Tzoumis and Finegold (2000) found that the EPA's rating of EISs for both the adequacy of information and the impact of the proposed action had

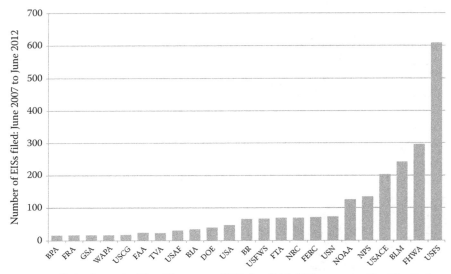

Federal agencies filing 1% or more of EISs (see Table 2.5 for acronym identification)

FIGURE 2.1
Environmental impact statements (EISs) filed by federal agencies: June 2007 to June 2012.

declined in the period 1970–1997 (i.e., the quality of EISs was worse and the degree of impact increased with time). The authors cite weak standardization in EIS guidelines among agencies and field offices within agencies as a major contributor to the lack of improvement in EIS quality. They also identified inconsistency in the rating system, possibly reflecting the lack of general availability of the rating results from around the country to all agencies and geographic regions, resulting in reduced coordination and sharing of lessons learned. Tzoumis (2007) updated the quality and degree of impact evaluation by evaluating the ratings of EISs since her earlier study. She anticipated an improvement in both the quality and degree of adverse impact of the proposed action because, among other factors, she and others have identified:

- Increased access to environmental information via the Internet
- An EIS and NEPA process that was more mature and standardized
- More sharing of lessons learned, particularly via the Internet
- An additional seven years of experience in both the methods of impact prediction and monitoring of past projects
- More environmental science and policy training. Beginning in the late 1990s, colleges and universities not only offered more environmental science and policy courses but also began offering BS and MS degrees in the field.

Summary of the National Environmental Policy Act

TABLE 2.5

Federal Agencies Filing Environmental Impact Statements from June 2007 to June 2012

Agency	Acronym	Number Filed	% of All Filings
Bonneville Power Administration	BPA	16	1%
Federal Railroad Administration	FRA	17	1%
General Services Administration	GSA	17	1%
Western Area Power Administration	WAPA	17	1%
United States Coast Guard	USCG	18	1%
Federal Aviation Administration	FAA	24	1%
Tennessee Valley Authority	TVA	23	1%
United States Air Force	USAF	31	1%
Bureau of Indian Affairs	BIA	35	1%
Department of Energy	DOE	40	2%
United States Army	USA	47	2%
Bureau of Reclamation	BR	66	3%
United States Fish and Wildlife Service	USFWS	67	3%
Federal Transit Administration	FTA	69	3%
Nuclear Regulatory Commission	NRC	69	3%
Federal Agency Regulatory Commission	FERC	71	3%
United States Navy	USN	73	3%
National Oceanic and Atmospheric Administration	NOAA	126	5%
National Parks Service	NPS	135	5%
Unites States Army Corps of Engineers	USACE	203	8%
Bureau of Land Management	BLM	241	10%
Federal Highway Administration	FHWA	296	12%
United States Forest Service	USFS	608	24%
Total agencies filing 1% or greater of all EISs during the period		2309	93%

In fact the evaluation of EIS ratings by Tzoumis (2007) concluded that for the agencies that produced the majority of the EISs (U.S. Forest Service, Federal Highway Commission, Bureau of Land Management, and U.S. Army Corps of Engineers, all combined accounting for over 57% of EISs filed) there was no overall improvement in either the quality or degree of impact from

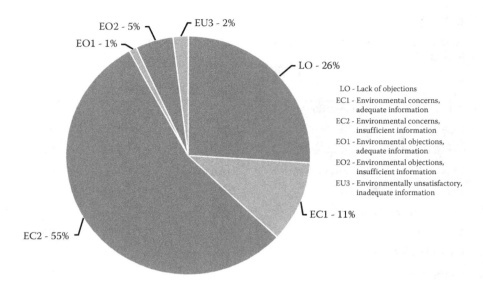

FIGURE 2.2
U.S. EPA ratings for a subset of federal agency submittals: June 2007 to June 2012.

1998 to 2004. The number of EISs rated inadequate was low and highly variable from 1970 to 1998 and this pattern continued in the more recent period. The relationship of adequate to inadequate rating remained similar in the two periods with the number of inadequately rated EISs high for most agencies in most years.

Based on her research and interviews with NEPA practitioners within agencies, Tzoumis commented that the increased availability of information was a mixed blessing. Because so much information was readily available they felt obligated, particularly in response to public comments, to consider all the information. This tended to dilute their focus on the critical issues. The interviews also revealed a general feeling among practitioners that the EISs and the entire NEPA process were somewhat compromised by the necessity to be both responsive and clear to the public and to provide decision makers with a solid and verifiable scientific/engineering/social analysis that was conclusive. The future challenge is to better balance these multiple objectives.

The evaluations of Tzoumis and Finegold (2000) and Tzoumis (2007) and the analysis presented above when viewed in combination depict a disconcerting trend that seems to be continuing. There is a committee of the broad based and highly respected professional society, National Association of Environmental Professionals that produces an annual report on NEPA (www.naep.org). These annual reports track EPA's EIS ratings and reflect the same and continuing trends described above. EISs appear to be declining in quality and projects proposed by federal agencies appear to be

Summary of the National Environmental Policy Act

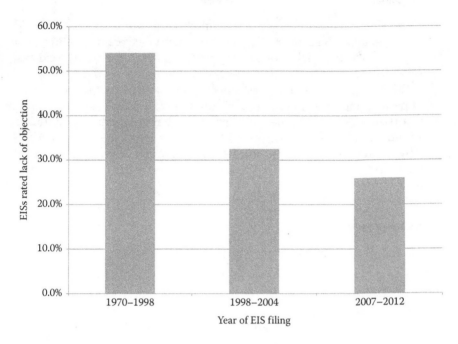

FIGURE 2.3
Environmental impact statements filed by federal agencies with a rating of Lack of Objection: 1970–1998 (Tzoumis and Finegold 2000); 1998–2004 (Tzoumis 2007); and 2007–2012 (see Figure 2.2).

creating greater environmental impacts. This is most clearly illustrated by the long-term record of EISs rated as Lack of Objection by EPA under the Clean Air Act Section 309 rating process (Figure 2.3). Other rating categories have changed over time (i.e., Environmental Concern and Environmental Objection since 1984 and a combined Environmental Restriction prior to 1984), so the trend in other environmental rating categories cannot be as easily depicted over time. There has been some maturation and change in the application of the rating system over time, which could account for some of the temporal variation. However, the drastic decline from 54% of EIS filings producing no environmental objection during the period 1970 to 1998 to only 26% in the most current period certainly indicates the lack of fulfillment of NEPA goals.

Equally concerning are the quality of EISs and trends over time in improvement. During 1970–2012, the percentage of EISs determined to have adequate information and analysis has been relatively constant at 15%–25% with some agencies showing a decline (Tzoumis 2007). During the same period, the percentage of EISs rated as inadequate or insufficient has typically been between 25% and 35%.

The quality of analysis and adequacy of information reported for NEPA EISs are similar to that reported under other environmental programs. The

European Union has a more prescribed method of evaluating environmental impact analyses with over 140 questions that reviewers must address on every study covering such topics as alternatives, description of the existing environment, mitigation, and nontechnical summary. Although the rating of the environmental analyses reflects substantial variation by country and type of projects, evaluation of 50 randomly selected studies in three EU countries consistently reflected over half as having inadequate information in most or all areas (Peterson 2010). Also similar to the rating of NEPA EISs, the evaluation of the randomly selected European environmental evaluations found "C" (on a scale of A to E with A being the highest rating) to be the most common rating.

2.5 NEPA Implementation

NEPA requirements as laid out generally by the Act and then fleshed out by both CEQ Regulations and CEQs follow-up of the Forty Most Asked NEPA Questions established the framework for implementing the national policy. Over the last 40 plus years, this framework has been filled in and expanded by separate and parallel processes including:

- Court rulings establishing what constituted NEPA compliance and specifically the required "hard look" at impacts before a federal agency took action.
- Each agency developing their own regulations consistent with CEQ Regulations and focused on their mission, goals, and common agency activities.
- Experience gained by agency personnel and the growing private sector of environmental analysis practitioners, particularly within a growing environmental consulting industry.
- General feedback from the public and more technical input from environmental, engineering, planning, and social science stakeholders.
- Development of state programs to add more specificity and fill gaps in NEPA jurisdiction.
- Ongoing higher education to produce BS and MS programs in environmental science and policy.

The framework and continuing development of environmental analysis processes and methods have resulted in an established approach to implement NEPA and expectations for the critical elements in the process. These overall and element-specific approaches are presented in the following chapter and illustrated by case studies. As described, there needs to be an overall plan and approach but as shown in the case studies, there is no single and universal plan.

Adaptation of the plan to each individual case and creativity from the environmental practitioner are critical to a successful environmental analysis. One of the goals of this book is that environmental analysis practitioners generally and NEPA practitioners specifically implement and enhance the methods presented in this and companion books and raise the quality of environmental analysis as reflected by the NEPA and similar rating systems. The improvement in environmental analysis should also reduce the degree of adverse environmental impact resulting from the proposed actions because a better understanding of the anticipated impacts is a first step in reducing the degree of impact.

References

Andrews, N.L. 1997. The unfinished business of national environmental policy. In: *Environmental Policy and NEPA*, Editor. Clark, R. and L. Canter, 85–98. St. Lucie Press, Boca Raton.

Bass, R.E. and Albert L.H. 1993. *Mastering NEPA: A Step-by-Step Approach*. Solano Press Books, Point Arena, CA.

Eccleston, C. 2008. *NEPA and Environmental Planning: Tools, Techniques, and Approaches for Practitioners*. Taylor & Francis, Boca Raton.

Eccleston, C. and J.P. Daub. 2012. *Preparing NEPA Environmental Assessments: A Users Guide to Best Professional Practices*. Taylor & Francis, Boca Raton.

Gustafson, P.F. 1993. The NEPA process: Its beginnings and its future. In: *Environmental Analysis: The NEPA Experience*, Editor. Hildebrand, S.G. and J.B. Cannon, 23–30. Lewis Publishers, Boca Raton.

March, F. 1998. *NEPA Effectiveness, Mastering the Process*. Rockville MD: Government Institutes.

Peterson, K. 2010. Quality of environmental impact statements and variability of scrutiny by reviewers. *Environmental Impact Assessment Review* 30: 169–176.

Reinke, D.C. and L. Low Swartz. 1999. *The NEPA Reference Guide*. Battelle Press, Columbus, OH.

Stern, M.J., S.A. Predomore, M.J. Motirmer, and D.N. Seesholtz. 2010. The meaning of the National Environmental Policy Act within the U.S. Forest Service. *Journal of Environmental Management* 90: 1371–1379.

Tzoumis, K. 2007. Comparing the quality of draft environmental impact statements by agencies in the United States since 1998 to 2004. *Environmental Impact Assessment Review* 27: 26–40.

Tzoumis, K. and L. Finegold. 2000. Looking at the quality of draft environmental impact statements over time in the United States: Have ratings improved? *Environmental Impact Assessment Review* 20: 557–578.

U.S. EPA. 1984. Policy and procedures for the review of federal actions impacting the environment. October 3, 1984.

USAID (U.S. Agency for International Development). 2005. Environmental compliance procedures. Available from: www.usaid.gov. (accessed August 6–23, 2012)

3

NEPA Process and Specific Requirements

The preceding chapter summarized the requirements of the National Environmental Policy Act (NEPA) and this chapter addresses meeting those requirements. The process of implementing NEPA mandates is first presented based on a flow diagram and is followed by a discussion of elements in the process that are NEPA specific. Although several of the NEPA-specific elements can be useful tools in any environmental impact analysis, they have unique application in NEPA and therefore are discussed in this chapter. The elements that are universal to any environmental impact analysis, including NEPA, are only briefly mentioned in this chapter but discussed in detail in the context of an overall impact analysis approach presented in Chapters 4 and 5. Following a discussion of the NEPA process, the critical elements central to NEPA compliance are discussed in detail. These include purpose and need for the proposed action, categorical exclusions, and NEPA enforcement. In some ways this chapter may seem out of sequence because many of the NEPA-specific aspects of analysis are best considered within the context of an overall environmental impact analysis (Chapters 4 and 5). However, it is important to point out that NEPA specifics immediately follow the history and requirements of NEPA (i.e., Chapter 2). Thus it may be beneficial to the reader to revisit some sections of this chapter after studying the overall environmental impact analysis and assessment approaches presented in the following chapters.

There is an important aspect of NEPA that is not discussed in either this or the preceding NEPA-focused chapters: multilevel environmental impact analysis. The concept of a phased approach to environmental impact analysis is embodied in both NEPA and the Council on Environmental Quality's (CEQ) implementing regulations, but the approach is not unique to NEPA. A phased approach, which can successfully increase efficiency, focus the analysis, and reduce duplication, is an important environmental impact analysis tool for any situation, not just NEPA. Thus this concept warrants separate discussion and is presented in Chapter 6 as it applies to NEPA and environmental impact analysis in general.

3.1 NEPA Process

NEPA requirements are summarized in Sections 2.2 and 2.3 of the book, but implementation of requirements is actually a process and best explained as a progression of steps. These steps can be depicted in a simplified form covering the major components of the process. Following this simplified process does not relieve the practitioner of complying with all the details of NEPA, CEQ Regulations, and all requirements of other laws and regulations. However, it provides an overview and a sense of direction when implementing NEPA. This process is depicted in Figure 3.1 and the components are discussed in the following sections.

3.1.1 Defining the Action

The first step in the process is to identify the "Major Federal Action." As discussed in Section 2.2, a federal action can be a project, plan, regulation, or policy implemented by a federal agency; a project funded by an agency; or a permit issued by a federal agency allowing another entity to act in compliance with a federal law or regulation. As discussed earlier, "but for" is a good test to determine NEPA jurisdiction. If the proposal could not occur "but for" the action of a federal agency, it is almost certainly subject to NEPA.

The next step in the NEPA process is to clearly understand and articulate the purpose and need for the proposed action. The purpose and need is a critical element of NEPA compliance and it is discussed in detail in Section 3.2. It should be immediately obvious to any potential stakeholder that the federal agency has a clear understanding of what it is trying to accomplish before it initiates the NEPA or any other process. Similarly, it should be obvious to stakeholders exactly what the agency is trying to achieve and what is not part of the agency action, but as discussed in Section 3.2, this is not always the case.

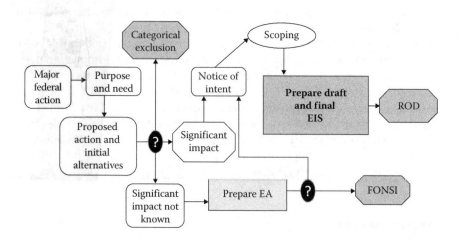

FIGURE 3.1
Summary of the NEPA compliance process. *Abbreviations:* EA, environmental assessment; EIS, environmental impact statement; FONSI, findings of no significant impact; ROD, Record of Decision.

Having defined the purpose and need for the action, the agency is in a position for the next step: describing the proposed action and an initial list of alternatives. This description should be conceptual, as more detail will be developed as the process progresses. But enough information has to be developed so the agency can make at least an initial judgment as to the environmental resources potentially involved and the general level of anticipated impact. Also, there must be enough detail developed so that a meaningful approach to evaluating the environmental impacts of the proposed action can be initiated (see Section 4.3 on the scoping process). In the case of a specific project, at a minimum, the description of the proposed action and alternatives at this stage of the process should include:

- General geographic area of the action, if there will be actual construction or other physical alteration of the landscape. The area potentially affected should also be defined when the action could result in plans or regulations affecting a specific area
- Magnitude of any construction, such as length of highway, size of a structure, etc.
- General qualitative description of any activity covered by a federal environmental regulation or act, such as air emissions, filling, of wetlands, or water discharges
- Acreage covered by the project
- General environmental setting
- Conceptual description of any alternatives, including the proposed action if one is known, identified to date

Once the purpose and need are established and the basic description of the proposed action and some initial alternatives have been developed, there is a major decision to be made by the agency based on the question: "what is the likelihood the action will cause an impact on natural or built environmental resources?" The federal agency proposing the action must answer this question in one of four possible ways:

1. There may or may not be an impact but we have dealt with this type of action numerous times. Therefore, we can fully comply with the NEPA requirement to incorporate environmental consequences and appropriate mitigation into our decision making without further environmental analysis for this specific action (see Section 3.1.2 if this answer applies).
2. There is the definite possibility of a significant impact (see Section 3.1.3 if this answer applies).
3. There will probably be some level of impact resulting from the action, but we don't know enough to assess the likelihood or severity of the impact (see Section 3.1.4 if this answer applies).
4. There will definitely *not* be a significant impact.

If the answer is number 4, "definitely no significant impact," the agency has met their NEPA obligation and can proceed with the action. If one of the other answers applies, there is a different course of action that must be followed to achieve NEPA compliance. Each of these is discussed below.

3.1.2 Categorical Exclusion (CATEX) as Part of the NEPA Process

If the proposed action is a common occurrence in the agency's normal course of doing business and fulfilling its mission, answer number 1 in the list likely applies: "There may or may not be an impact but we have dealt with this type of action numerous times." If the environmental impacts have previously been analyzed and found not to be significant, it can generally be categorically excluded from action-specific environmental analysis. CEQ instituted this procedure in an effort to streamline NEPA and focus environmental analysis on actions that are unique and the agency proposing the action does not know the likely consequences to the environment. The rationale is that the agencies perform these actions frequently and fully understand the consequences. Thus, if they decided to go forward they will incorporate the environmental concerns in their decision and the intent of NEPA will be satisfied. This is a major streamlining procedure and has been incorporated into many other environmental laws, regulations, and programs. The concept is discussed in detail in Section 3.6.

3.1.3 Environmental Impact Statement (EIS) and Record of Decision

3.1.3.1 *Planning and Structuring the EIS*

If the agency feels there is a strong likelihood of significant environmental impact (affirmative answer to question number 2 in Section 3.1.1) both NEPA and the implementing CEQ Regulations are quite clear that a "detailed statement" or EIS must be prepared. To assist in addressing this key question, many agencies maintain either formal or informal lists of the types of actions, or the environmental settings (e.g., wetland, endangered species habitat) that generally require a comprehensive EIS. More often than not, they will err on the side of preparing an EIS because if they turn out to be wrong, there can be considerable backtracking to comply with NEPA and an associated loss of time, agency energy, and resources. Once the decision to prepare an EIS has been made by an agency, the wheels have started running for a process that can take years, cost many millions of dollars, and if conducted consistently with NEPA intent, will have a major influence on the agency's decision.

It has been said that an EIS is a legal document and not simply a report of scientific and engineering evaluations. This is true, and in order to be successful, an EIS must be capable of withstanding legal challenges (see Section 3.7 which explains the role of legal defensibility in the enforcement of NEPA). But to successfully address the intent of NEPA, the document must: (1) inform stakeholders; (2) provide decision makers with a clear understanding of the environmental implications of their actions; and (3) objectively evaluate and compare alternatives and potential impact mitigation measures. The NEPA practitioner must meet the procedural requirements but cannot get so distracted by procedure that they lose sight of Congress' and CEQ's primary intent in requiring a "detailed statement." That intent is to conduct an unbiased, interdisciplinary, and technically defensible environmental impact analysis. A well-prepared and useful EIS is more than a disclosure document. The EIS and the entire NEPA process should be initiated at the earliest stage of planning the action, incorporated when making decisions, and integrated throughout the process. Managing the preparation of an environmental impact analysis, including an EIS is critical to its success and this book devotes Chapter 9 to describing management techniques and approaches.

The first step in the EIS preparation (Figure 3.1 for this and subsequent NEPA steps) is to inform the public, other federal agencies with jurisdiction, states that could be affected, and other parties potentially affected by the action (i.e., stakeholders) for which the agency is in the process of planning an EIS. As discussed in Chapter 2, if the action considered is of national concern the notice must be posted in the *Federal Register*. For actions more of local concern, notice must include direct communication to appropriate state agencies and through local media.

The agency's notice of starting the EIS process initiates the scoping process, and more often than not, the notice of intent includes the announcement of a scoping meeting and a link to a web page where comments can be recorded. Scoping is not unique to NEPA and is a key element in any environmental analysis because it establishes what should be covered in the analysis and frequently the methods that should be used. Because of its importance to any environmental impact analysis and its application beyond NEPA, scoping is discussed in detail in Chapter 4 of this book.

The EIS consists of several basic components: NEPA-specific procedural requirements, technical impact prediction, development and comparison of alternatives, and impact mitigation. The NEPA-specific requirements are covered by the CEQ Regulations and agency-specific guidance. They are also covered in detail by a number of classic and newer NEPA/EIS reference books (Bergman 1999, Hildebrand and Cannon 1993, Eccleston 2008, Bass and Herson 1993, Eccleston and Peyton Daub 2012). The reader is encouraged to consult the regulations, agency guidance, and appropriate EIS-specific reference books when preparing an environmental impact analysis subject to NEPA because this book does not provide the level of detail and nonsubstantive aspects covered by the other sources. A summary of the basic EIS requirements follows immediately below and critical EIS topics are addressed in the following section. The other aspects of NEPA (i.e., impact prediction, alternatives, and mitigation) are all critical aspects of any successful environmental impact analysis and these are discussed in subsequent chapters: Initiating (Chapter 4) and Conducting (Chapter 5) the environmental impact analysis and assessment.

An initial step in EIS planning should be the development of a draft outline to serve as a road map for the preparers. The outline should be developed in coordination with key stakeholders and periodically reviewed as more and new information on the project, environmental resources, and stakeholder concerns become available. As a starting point the CEQ Regulations include a recommended EIS format with the following substantive sections:

- A cover sheet that identifies the agency preparing the document, agency contact information, approvals, etc.
- Summary of the EIS contents
- Purpose and need for action
- Alternatives including the proposed action and the no-action alternative
- Affected environment (sometimes termed existing conditions)
- Environmental consequences or impacts
- List of preparers
- Appendices

Agencies have been reluctant to vary from the specific EIS format in the CEQ Regulations, primarily because strict adherence to CEQ's recommendations minimizes the scope of a legal/procedural challenge to an EIS. The CEQ's format has been used and accepted thousands of times; thus an agency is on "safe ground" if they strictly follow the format. However, this can sometimes make the document cumbersome and difficult for stakeholders to follow. As long as the required CEQ Regulations and NEPA requirements are satisfied, the format can be adapted for specific aspects of the project, program or plans addressed in the EIS. A project-specific format can be developed to focus on the key issues and highlight the environmental trade-offs and decisions to be made. This approach can greatly enhance the quality, readability, and usefulness of the EIS. If an original and project-specific format is developed, it should be compared with NEPA requirements and the CEQ Regulations (including their recommended format) to make sure every topic is covered somewhere in the document and if material has not been included, the EIS must document why it was not covered.

Another action critical to preparation of a successful EIS that should be started early in the process is the preparation of an environmental impact analysis plan and schedule (see Section 9.3.2). This plan and schedule lay out the EIA preparation efforts along a time line, identifying decision points and efforts on the critical path to EIS completion. This plan and schedule show where you are going and when you want to get there. Getting to the end successfully and on time are little more than random chance if you don't know where you are going before you start the EIS preparation journey. The environmental impact analysis plan and schedule are updated frequently and used throughout the EIS preparation to measure progress and redirect efforts as early as possible if necessary due to changed conditions or new information. It is essential to the successful completion of the analysis and EIS, and should be a tool and skill every NEPA practitioner has mastered. The analysis plan and schedule is so important and so often ignored, it is afforded an entire section of this book (Section 9.3.2).

3.1.3.2 Draft EIS Environmental Analyses

The draft EIS documents the analyses conducted to support the intent of NEPA, and thus must meet a high level of scientific and objective scrutiny. The critical technical analyses supporting impact prediction and appropriate methodologies to conduct them are generally identified during scoping. These technical evaluations are comprised of literature review, field data collection, laboratory testing, and modeling of existing and future conditions. These technical evaluations are typically the first activities initiated as part of the EIS preparation because they have the longest lead time. The detailed methods of any investigation should be reviewed and accepted by the appropriate stakeholders. This generally insures not only a superior investigation, but when the results are obtained they should be acceptable to the key technical stakeholders. Also they should be satisfactory to stakeholders in general because they have been blessed by independent, non-agency, non-EIS team technical experts.

There is also significant preparation work with respect to securing equipment, laboratory contracts, and logistical planning for detailed investigations; thus obtaining concurrence with the investigation methods from key stakeholders early in the process is critical. These detailed investigations into natural systems are required frequently to address conditions in multiple seasons or at least key seasons, such as breeding times for sensitive biological resources. It is a sad day (and experience shows it has happened more than once) when scoping identifies a critical sampling period for a biological resource or water quality and that period ended yesterday. Thus the sampling cannot be completed until the same time next year! Thus anticipation of such situations and initiation of critical time-sensitive investigation as soon as possible can be an essential element in maintaining the EIS schedule and maximizing efficiency.

The results of the investigations and the methods used are best documented in detail in appendices to the EIS and the results summarized in the body of the document. This approach satisfies the CEQ's brevity and readability requirements: the CEQ Regulations specify brevity and call for EISs to be 150 pages or less for most agency actions and not more than 300 pages for unusual or complex proposals. Use of appendices also satisfies a common sense approach because the majority of stakeholders do not want to read extensive, highly technical discussions of every environmental resource potential affected. Their interest is in the results, and the individual appendices can be delegated to experts in the specific discipline for review and comment.

Concurrent with initiating the technical investigation, the federal agency begins preparation of the EIS. Many sections can be gleaned from other documents, such as extracting the description of the proposed action and alternatives from previously prepared planning, and conceptual engineering documents. If these other documents are readily available to the public, they can be summarized in the EIS and then referenced, but if not they should be included as appendices. Also at this early stage of the EIS preparation, aspects of the action similar to other federal actions subject to NEPA and those that have been addressed in other EISs or environmental assessment (EA) should be identified. If they are relevant to the new proposed action they can be incorporated by reference and briefly summarized in the new EIS.

Recent NEPA documents prepared for the same geographic area, even if for a dissimilar type of action, can be helpful and improve efficiency. Relevant sections of the description of existing conditions (or Affected Environment, see Section 5.2) can often be referenced, used directly or minimally modified for the new EIS (see Chapter 6, for a discussion of incorporation by reference). Incorporation of existing conditions descriptions and relevant elements of the proposed action from other NEPA documents not only saves time and resources, it takes advantage of the review and acceptance of previous environmental analysis information and avoids the possibility of new information being challenged or criticized as not meeting the objectives of NEPA.

The meat of the draft EIS is the incorporation of the technical investigations into the prediction of impacts and evaluation of alternatives. As pointed out earlier, the details are best presented in appendices with the approach, rationale, and conclusions summarized in the body of the draft EIS using language clearly understandable by stakeholders at any level of technical knowledge and experience. The design, execution, and presentation of technical investigations are key to any successful environmental impact analysis, not just a NEPA draft EIS, and they are discussed in detail in Chapter 5.

3.1.3.3 Draft EIS Contents

The draft should meet all the procedural requirements for an EIS, clearly report the findings of the environmental analysis, and frame the critical issues and trade-offs the agency considered in developing and comparing alternatives. The clarity of an EIS is not a trivial issue, as courts have ruled on numerous occasions that, in essence, if the stakeholders potentially affected by the proposed action cannot be reasonably expected to understand the EIS, it does not meet NEPA or CEQ requirements. For example, a Bureau of Land Management (BLM) EIS that did not foster informed decisions and active public participation was invalidated. The court "considered not only its [EIS] content, but also its form...." (*National Parks & Conservation Association v. U.S. Bureau of Land Management*, 586 F.3d 735 [9th Cir. 2009]).

The EIS must also list all federal permits that are required to implement the proposed action. These permits and their specific requirements are critical to the development of impact significance criteria and comparison of alternatives (see Chapter 9). As discussed in Chapter 9, an understanding of environmental permit requirements of the proposed action and integrating the requirements into the environmental impact analysis process is key to efficient implementation of the proposed action. No one wants to progress through the entire NEPA or equivalent process only to learn that some aspect of the proposed action cannot be permitted or must be modified to the extent that a supplemental EIS is required.

The draft EIS can, but is not required to, identify the agency's preferred alternative and proposed action. There are advantages to not identifying the preferred alternative in the draft EIS, such as allowing the agency to consider public input, including identification of new information before they complete their alternative comparison and make a decision. This approach also provides a mechanism for stakeholders to be active and productive participants in the process and not just reactive to an agency decision. The stakeholder acceptance and support developed by facilitating active participation in the process can be important in successful implementation of the proposed project, program, or policy as it works its way through funding requests, permit issuance, litigation, and the court of public opinion.

Not identifying a preferred alternative in a draft EIS when the agency actually has one can have adverse consequences. Such an approach undermines the CEQ Regulations and NEPA's call for public disclosure and transparency. It can also affect the cooperation and trust with stakeholders, including other federal agencies that must issue environmental permits or approvals. The lack of full disclosure exhibited by an agency's failure to expressly identify its preferred alternative when it actually has one can also lead the agency to complacency and result in arbitrary dismissal of other alternatives, which can be an important arrow in the quill of a legal challenge of the EIS, Record of Decision (ROD), and agency's NEPA procedure.

3.1.3.4 Review and Comment on Draft EIS

The draft EIS, with or without identification of a preferred alternative, is released and noticed. As part of the release, the agency must establish a procedure to receive comments on the draft EIS. CEQ Regulations specify that the agency issuing the draft EIS must obtain comments from all other federal agencies with jurisdictional obligations or technical expertise related to a predicted impact of the proposed action. In addition, the issuing agency must request comments from other parties with interest in the action including:

- Appropriate state and local agencies
- Indian tribes with reservation land or other resources (e.g., fishing rights) that could be affected by the proposed action

- Any agency which has requested that it receive statements on actions of the kind proposed
- The applicant if the EIS addresses issuance of a permit or similar project proponent
- The public

The issuing agency must have a procedure in place to receive comments, and the procedure must be described concurrently with the release of the draft EIS. The comment period must be at least 45 days, but the lead agency can extend the period at its discretion.

Careful attention to the comments received on the draft EIS is important, not only to produce a better analysis but also to ensure adherence to NEPA procedure and thus produce a legally defensible document. The federal agency proposing the action and issuing the draft EIS must consider every comment and prepare a response. The response to most comments will be in one of three forms:

- State that the comment has already been addressed and cite the location in the draft EIS where it was considered.
- Objectively explain why the issue raised is not relevant to the NEPA process, such as documenting that it was beyond the purpose and need for the action (see Section 3.2).
- Perform the analysis, consider the alternative, incorporate the data, etc. referenced in the comment and reevaluate the impacts in a revision to the draft EIS (i.e., a final EIS or supplemental draft EIS; see Chapter 6).

The response to comments on the draft EIS has both administrative and efficiency elements. The administrative aspect is to keep track of the comments and ensure each is addressed by one of the responses discussed above. There are numerous software applications available to keep track of comments and they provide very useful techniques for tracking and the logistics of producing a response to comments.

Experience has shown that even if a large number of comments (sometimes in the thousands) are received, there are typically three to five concerns that cover 90% of the comments. Efficiency in addressing comments on the draft EIS begins by identifying those three to five concerns early. A common response can then be prepared and the appropriate prepared response noted as the comments are received. This approach not only reduces the effort (and associated cost) and time to respond to comments, it also minimizes the possibility of inconsistency in responding to comments. The unified response to common issues also makes for a much more readable document.

Experience has also shown there is considerable overlap between comments received during scoping and comments on the draft EIS. If stakeholders have a burning issue, they will raise it during scoping, and the scoping document must address the comment by explaining how the comment received during scoping will be addressed or why the comment is not relevant (see Chapter 4). Unfortunately, some narrowly focused stakeholders don't take the time to study the draft EIS in enough detail to determine how or even if their issue has been addressed; thus they will have the same or similar comment on the draft EIS. If the comment tracking system is designed and implemented as part of scoping, it can be continued as the draft EIS tracking system. This can increase the efficiency of the EIS development and also maintain consistency between scoping and EIS production.

3.1.3.5 Final EIS

The final EIS is produced by the federal agency proposing the action after the draft EIS has been reviewed and comments from stakeholders are received. Frequently the final EIS is a reproduction of the full draft EIS with additions or changes based on comments and some minor additions. This total reissuance of the draft is discouraged by CEQ Regulations as very inefficient and forces the stakeholders to reread substantial material they have already read. It also is a large distraction from the key issues raised during the comment period and the agency's consideration and response to the issues. The key issues, agency responses, and changes since release of the draft should be the focus of the final EIS and inclusion of other material which has not changed since the draft is frequently a distraction from this focus. A final EIS approach and contents proven successful and reader friendly include:

- A very brief summary of the draft EIS findings and preferred alternative if it was included in the draft
- Identification and response to the three to five general comments that, as discussed above most often encompass 90% of all comments
- Conclusions of the final EIS based on the draft and comments received by: confirmation of the decisions made in the draft including selected alternative and mitigation commitments; modification of the conclusions of the draft to reflect comments and/or new information; or a new conclusion and supporting analyses
- Any additional investigations or evaluations conducted for the final EIS (the investigations can often be included in an appendix but should be referenced and summarized in the body of the final EIS)
- Changes to any conclusions or decisions from those presented in the draft EIS
- New or replacement sections based on new or changed information from the draft EIS
- Accounting of all comments received and the agency's response or explanation of why a response is not warranted (this also can be

included in an appendix but should be referenced and summarized in the body of the final EIS)
- Documentation of the Public Outreach Program (this also can be an appendix and summarized in the body of the document)
- If necessary, an errata sheet correcting any mistakes in the draft EIS

The final EIS must complete the analysis by including stakeholder input and comparison of alternatives, but unlike the draft, the final must include identification of the preferred alternative. Although identification of the "environmentally preferred" alternative and why it was not selected (if it was not) is not required in the final EIS, it is required to be identified in the ROD. It can be efficient and useful to readers and decision makers to identify and discuss this alternative in the final EIS because the justification and rationale for the designation and choice are in the final EIS. If this process is followed, the ROD can simply state the conclusions and refer to the final EIS for the details. Similarly, the mitigation measures available and rationale for incorporating or rejecting the measures are required to be included in the ROD but not necessarily the final EIS. However, for the same reasons for inclusion of the environmentally preferred alternative in the final EIS, mitigation should be fully discussed in the document.

In contrast to the draft, there is no requirement for the agency proposing the action to request comments on the final EIS. The CEQ Regulations encourage solicitation of comments but leave it up to individual agencies to develop the policy and procedures best suited to the agency for comment on the final EIS. Each agency's guidance for compliance with CEQ Regulations documents the requirements and procedures, if any, for comments on the final EIS. The advantages of making the final EIS available for comment are that it can help build public and agency support in implementing the proposed action and also provide another chance to correct any deficiency in the EIS and minimize the potential for legal action rejecting the EIS. Similarly if a stakeholder is contemplating litigation against the EIS, the courts have commented that they have a stronger case if they raise the deficiency in comments on the EIS and the agency fails to address them, rather than raising them the first time in the legal challenge.

3.1.3.6 Record of Decision

Following release of the final EIS, the federal agency proposing the action prepares its ROD (see also Section 2.3.5). This should be a very concise document, frequently not more than a few pages and it represents an enforceable commitment by the agency. The ROD should be limited to summarizing the following:

- What the agency has decided to do
- Which alternatives were considered

- Which alternative was selected and why
- Which is the environmentally preferred alternative, why, and if not selected why not
- A listing of adverse impacts and all mitigation measures to address each adverse impact
- A statement of which mitigation measures were adopted and why the others were not adopted

Brevity and definitive statements are the key to a useful ROD. To achieve brevity, the explanation of alternatives, selection process, and mitigation should be accomplished in the final EIS and the ROD should reference the document. However, the ROD must be definitive. Both the draft and final EIS state impacts and actions as future possibilities such as, "the new U.S. Navy base *could* increase the school age population by up to 1,000 students which *could* exceed the existing municipal education capacity." These are stated as *could* because it would happen only if the particular alternative under consideration was implemented by the Navy. In contrast, ROD uses *would* and *shall* language because the agency has made the decision to go forward and fully expects the predicted impacts to reach fruition.

If monitoring or other commitments are included in the selected action, they must be included in the ROD, and this commitment is illustrated by the U.S. Air Force's development of the over-the-horizon radar system. This system consisted of eight proposed radar complexes that provided coverage and detection of potential security threats from aircraft, ships, and missiles up to 5500 km offshore of the continental United States. These facilities were strategically located around the continent in Maine, North Dakota, Minnesota, California, Oregon, and Alaska to provide full surveillance coverage of the U.S. borders. They were also very large (up to 1000 hectares each) in order to achieve the transmission and receiving capacity for the extensive areal coverage of the radars. As one might expect, construction on such a large scale had the potential for significant impact on the natural and built environmental resources and EISs were prepared for each system.

The final EIS and ROD for the radar system in northern California's Modoc National Forest identified the general location for the facilities and acknowledged potential impact on Native American cultural resources and migratory large mammals (prong horn antelope and mule deer). In response to comments from the California Department of Fish and Game and the U.S. Forest Service, the U.S. Air Force committed in the ROD to monitor and mitigate cultural and large mammal impacts during design and finalization of the specific layout of each radar. They fulfilled this commitment by working closely with the two commenting agencies to design and implement an extensive monitoring system prior to final design (see Chapter 5 on existing conditions for a detailed description of monitoring). The monitoring was a very large effort that exceeded in scope and duration all of the other biological and archaeological investigations that were done for the draft and final EIS and resulted in a final project design that mitigated cultural and natural resource impacts to the satisfaction of the commenting resource agencies.

The time between issuance of the final EIS and ROD gives the opportunity for federal agencies proposing the action to express and resolve any final concerns or conflicts among sister agencies. If the federal agency proposing the action has not responded in the final EIS to draft EIS comments by another federal agency with resource jurisdiction or expertise to their satisfaction, the responses can be included in the ROD to address the concerns. If an agreement cannot be worked out in the ROD, the issue can be elevated to the CEQ (see discussion above) which is the least desirable alternative to all involved. As in the above example, if the U.S. Air Force had not committed to mitigation in the ROD, the U.S. Forest Service in all likelihood would have elevated the issue to CEQ which could have required a supplemental EIS to work through the issue which could have delayed this fast-track project by a year or more.

3.1.4 Environmental Assessment and Finding of No Significant Impact

If the potential for "Significant Impact is Not Known" (Figure 3.1 and affirmative response to question number 3 "There probably will be some level of impact resulting from the action, but we don't know enough to assess the likelihood or severity of the impact") by the federal agency proposing the action, the CEQ Regulations include a procedure for determining the level of significance of anticipated impact without a full EIS. As discussed above, if the agency knows from implementing similar actions or by other means that the proposed action has a high likelihood of creating significant environmental impacts, they must prepare an EIS. Similarly if they know from past experience the action will not have a significant impact, a CATEX will satisfy NEPA requirements. If the proposed action falls between these two extremes and the level of impact is unknown, CEQ Regulations call for the agency to prepare an EA to make the decision.

Similar to an EIS, an EA is a decision document for the federal agency proposing the action. But in contrast to an EIS, where the decision is whether and how to proceed with the proposed action, the decision addressed in the EA is, "will the proposed action result in a significant environmental impact?" In order to address this question an environmental impact analysis must be conducted and the EA is the procedure prescribed by the CEQ Regulations for the analysis. The basic approach and structure for the analysis should be similar to that of an EIS and methods discussed in Chapters 4 and 5. An EA is similar to an EIS in that the primary goal of both documents is to further the intent and goals of NEPA; however, an EA has an additional goal of determining whether there would be significant impacts, and if so a full EIS is required. If not, the EA is the appropriate NEPA environmental impact analysis document. Both EAs and EISs must achieve NEPA goals and interest by transparency and objectivity (i.e., analyses, procedures, and decisions cannot be "arbitrary or capricious"). The two documents are also similar in their requirement to consider environmental permits in the evaluation and involve other agencies with jurisdiction or interest in resources potentially affected. However an EA is often referred to as a "mini EIS" and the environmental analysis in an EA differs from the one in an EIS in scope and intensity (Table 3.1).

Differences between an EIS and EA in treatment of alternatives have evolved as a major distinction in the two NEPA documents. Although CEQ Regulations and NEPA call for consideration of alternatives to the proposed action for any action subject to NEPA, including in an EA, the focus of an EA almost always is heavily on a single alternative—the proposed action. This is because the agency has identified a need, an action that addresses the need, and the agency has made an initial interpretation that the action has little or no adverse environmental impact. It is very rare that an agency would embark on an EA if they had not already identified a proposed action and could make a preliminary educated assessment of magnitude of environmental impact resulting from the action. If they had not identified the proposed action and developed at least some detail on how it would be implemented, it would be very difficult to assess the potential for significant impact and preparation of an EA could be a wasted effort.

The public participation component is also a differentiator between EIS and EA requirements. As discussed above, public input to scoping for an EIS is critical not only to the success of the document but also to procedural compliance. Circulation, comment, and response to a draft EIS are similarly required and critical parts of the EIS process. Litigation invalidating an EIS is frequently based on lack of sufficient compliance with public participation guidance and mandates, particularly addressing comments received during the process. However, in *California Trout v. Federal Energy Regulatory Commission*, 572 F.3d 1003 (9th Cir. 2009), the court ruled that while NEPA does not require the same level of public participation by federal agencies for EAs compared with EISs, "an agency must permit some public participation when it issues an EA." The courts have not stated what kind of public participation is required to meet NEPA standards, but their rulings make it clear that a complete failure to inform or involve the public as an integral part of the EA process would violate NEPA and CEQ Regulations. Such a failure to involve the public can also be a determinant to successful implementation of a proposed action and can sometimes miss an opportunity to enhance a project and minimize impacts by taking advantage of stakeholder knowledge.

In contrast to an EIS, public participation for EAs is not specified by NEPA or CEQ guidance but rather left to the discretion of each agency. For example, in *Theodore Roosevelt Conservation Partnership v. Salazar,*

TABLE 3.1

NEPA Environmental Assessment (EA) Compared with Environmental Impact Statement (EIS)

	EA	EIS
Length	Can be as few as 6–10 pages; typically around 50; or can approach the length of an EIS	150-page limit recommended; can be up to 300 for complex actions
Time to Prepare	Typically a few months	Rarely less than a year and typically 1.5–2 years
Data Requirements	Generally relies on existing data and field observations	Original data collection almost always conducted
Preparation Team	General NEPA practitioner prepares EA, frequently with input from technical experts in one or two critical areas	Multidiscipline team required with a senior NEPA practitioner coordinating. Also public participation, legal review, and other expertise involved.
Public Participation and Availability	Information available to public but review, comment, and participation procedures at the discretion of each agency	Mandatory review, comment, and public hearing/meeting requirements
Review and Comment	Largely at discretion of agency; draft document not required.	Specific requirement with public review and comment of draft EIS required.
Alternatives	Minimal treatment (see text)	Key component of EIS
Mitigation	Can be used and useful in certain cases (see text)	Must be included in the process and evaluated fully
Cost	Typically in the tens of thousands of US dollars	Rarely less than a million U.S. dollars and scientific investigations alone can run up to a million U.S. dollars
Final Outcome	Finding of no significant impact OR prepare an EIS	Record of Decision

605 F. Supp. 2d 263 (D.D.C. 2009), it was held that NEPA public participation requirements for the EAs were met even though the agency did not circulate the EAs for notice and comment. "[T]he agency has significant discretion in determining when public comment is required with respect to EAs."

In compliance with CEQ Regulations and similar to an EIS, the EA must include a no-action alternative, but some valid EAs have just the two alternatives, proposed action and no action. On the surface it appears that this subverts the NEPA intent of using environmental considerations and impact mitigation to compare alternative approaches to implementing the purpose and need. But the rationale is that the proposed action in an EA has no significant environmental impact, so minimizing impacts is not as great an issue as it is with a full EIS. As discussed below and illustrated in Figure 3.1, if there are significant impacts associated with the proposed action that cannot be mitigated, a full EIS must be prepared with a robust evaluation of alternatives and analysis of mitigating measures.

The completion of the EA results in one of two decisions: the proposed action will not result in a significant environmental impact or it will. If the EA concludes that a significant impact will not occur, the environmental analysis is completed and the agency can issue a finding of no significant impact (FONSI) which must describe why the agency has concluded there is no significant impact. The details, review procedure, and public dissemination of a FONSI are agency specific and should be consulted for each proposed action. It is not necessary for the FONSI to repeat all the information in the EA; the information can be summarized with emphasis and support for the conclusion of no significant impact.

If the agency reaches the other conclusion upon completion of the EA (i.e., there will be a significant impact from the proposed action), the agency has two options. They can initiate an EIS on the proposed action and draw on the EA for both environmental impact analysis and EIS scoping. Although preparation of an EIS following an EA can be more efficient than the one prepared from scratch, it will still involve close to a year-long process, full public outreach, and significant expenditure of money and agency effort. If the EA concludes that there will be a significant impact resulting from the action, the other option available to the agency is to develop measures to avoid or substantially reduce the level of impact. They can then modify the proposed action and issue a "mitigated FONSI" which either refers to a revised EA or documents the measures that have been incorporated in the agency's proposed action that reduces or eliminates any significant impacts previously identified.

3.2 Purpose and Need

3.2.1 Purpose and Need Overview

Purpose and need are most simply defined as the intended outcome of the agency's action, or what the agency is trying to accomplish by its action. As discussed earlier in Chapter 2, such a statement is required for each action subject to NEPA in the CEQ "Regulations for Implementing the Procedural Provisions of the National Environmental Policy Act" (40 CFR 1502.13). A well thought out and structured purpose and need statement should be a substantive element of any environmental analysis and used to simplify, guide, and increase the efficiency of the entire environmental analysis process. A wise and concise purpose and need statement can aid the environmental analysis practitioners because, "if you don't know where you are going, you will never know if you have arrived." Thus the statement gives not only the practitioners, but all other stakeholders a sense of where the process is headed, and at the end, whether it has been successful.

By simply and concisely stating the purpose and need upfront, not only in the draft and final environmental analysis documents, but also as part of all scoping and public involvement efforts, potential stakeholders are able to determine whether their interests could be affected. This then allows them to make informed and rational comments on alternatives to be considered, identify the environmental resources they feel could be affected, and provide informed input on the scope of the environmental analysis. If the agency or other entity proposing the action is not clear up front regarding what they hope to accomplish, they could receive voluminous input during scoping that is not relevant, yet they would be obligated in the draft EIS, or similar document, to respond to the comments and explain why they were not relevant. At the end of the process the purpose and need statement provides the stakeholders, including the courts in the case of a NEPA analysis, a meter stick to measure accomplishments and judge the success of the process.

The CEQ may have had in mind a distinction between "purpose" and "need" when they included both terms in their 1978 guidelines for implementing NEPA (discussed in Section 2.3) but since the concept is presented by the guidelines in only one sentence, they do not make the distinction apparent. Early NEPA practitioners attempted to make a distinction between purpose and need both by exploiting subtleties in the definition and using the two terms to formulate an alternative evaluation process consistent with the intent of NEPA. For example, Schmidt (1993) working with the Bonneville Power Authority on one of the first EISs prepared in accordance with the 1978 CEQ guidelines attempted to interpret CEQ guidance on purpose and need. His interpretation was that the underlying need represented a problem to be solved or an opportunity to be exploited by the agency and the purposes as other objectives related to the problem or opportunity. Following this interpretation, Schmidt (1993) proposed that the underlying need would define the range of alternatives for the EIS and if an alternative also met the purpose, it should be retained for a detailed analysis. Others have proposed satisfying the underlying need as the criterion for retaining an alternative for detailed analysis in the draft EIS and using the purposes, or agency goals, as criteria for comparing alternatives and designating a preferred alternative.

In practice, particularly in recent years, the distinction between purpose and need has generally been blurred and/or lost. A unified purpose and need statement is typically developed and used to determine the range of alternatives for the draft EIS. Achievement of the statement is frequently measured by identifying specific criteria that define the purpose and need. For example if the purpose and need of a proposed action is to relieve traffic congestion along the main street of a small town, the criteria established to measure conformance with the purpose and need might be that on average the projected traffic volume that could pass through each intersection without waiting for more than one change of the traffic light. Potential alternatives can then be compared with the criteria and those that meet the purpose and need are retained for a detailed analysis in the draft EIS.

Perhaps of greater concern than semantics and distinction between the terms purpose and need, is an ambiguous or otherwise unclear statement of purpose and need that fails because it did not initially attract the attention of stakeholders potentially affected. In such cases these parties would not be aware that their interests were potentially at risk until a preferred alternative is identified at the draft or even the final EIS stage. At this late juncture of the process, stakeholders could identify alternatives, areas of impact, methods to be used in the environmental analysis process, or conflicting future plans that were not addressed in the draft document. The proposing agency would then be obligated to evaluate each of these additional areas to determine whether they had merit and if so address them in the final or a supplemental draft EIS. Either of these options have the potential for significant time delays and unnecessary expenditure of funds.

Some background on why the agency is proposing to take action may in some cases be advantageous for inclusion in the purpose and need statement, but if included it must be brief and very general. There are much more appropriate sections of an EIS to discuss the history of the problem to be addressed by the action under consideration, including the section on existing conditions or affected environment. Lengthy discussion of the problem to be addressed does not belong in the purpose and need section and frequently detracts from a concise and easily understood statement of what the proposing agency is attempting to accomplish. It is important to keep in mind the objective of the purpose and need statement: establish the intended outcome of the agency's action and use this to determine the appropriate range of alternatives. A lengthy discussion of the problems that led to the need for action and the history of the problems can obscure a concise statement of purpose and need.

3.2.2 NEPA Compliance Related to Purpose and Need

As important as the statement of purpose and need is to stakeholder input, technical alternative selection, and focused preparation of the analysis, in a NEPA document the statement is often most important in relation to procedural compliance and litigation. Under virtually any interpretation, the purpose and need of an action determines what alternatives are appropriate for consideration and which can be eliminated at the screening stage or not considered at all. As discussed below (Section 3.7) a flawed alternative analysis is one of the most common defects identified in NEPA case law and a cause for rejecting an EIS. Thus the statement of purpose and need forming the basis for alternative analysis is often at the heart of NEPA litigation. There are several court challenges of NEPA EISs, as summarized in Schmidt (1993) which illustrate the importance of clearly defining the purpose and need. Several of these are presented in the following sections and these cases demonstrate how the purpose and need statement is used to narrow the universe of alternatives that must be considered, thus making preparation of the EIS more focused and efficient. Additionally, cases provide instruction on wisely and concisely constructing the statement, responding to EIS comments consistent with the statement, and producing an EIS that can survive legal challenges. Following the presentation of legal challenges to EISs based on purpose and need, case studies of successful purpose and need processes are discussed. Two NEPA EIS case studies are presented where the purpose and need statement was central to the development and screening of alternatives. As discussed, they supported the goals discussed earlier for a concise and wise purpose and need statement: solicit meaningful stakeholder input, focus and increase efficiency of EIS preparation, identify superior alternatives, and meet NEPA procedural requirements.

3.2.2.1 City of New York v. U.S. Department of Transportation. 715 F.2d 732 (2d Cir. 1982), cert. denied, 465 U.S. 1055 (1984)

The U.S. Department of Transportation (DOT) was charged with promulgating safety requirements for transporting radioactive materials on the nation's highways. Since this was a federal action with potentially significant impacts, the DOT as the lead agency prepared an EIS with the purpose and need of improving the safety of highway transportation of radioactive materials. DOT's resulting EIS and decision evaluated and established a set of regulations for highway transport of the material, which was challenged by the City of New York because DOT did not evaluate the alternative of barge transport of radioactive materials around population centers (e.g., New York City) to promote highway safety. The court upheld DOT's treatment of alternatives in the EIS because of the narrowly constructed purpose and need statement. The court concluded that based on the purpose and need statement (rules for safe highway transport of radioactive materials) appropriate alternatives might include various highway routes, transportation equipment, and driver qualifications, but alternatives means of transportation were not required. The court commented that if the purpose and need statement was broader, such as safe transport of radioactive materials, then inclusion of barge transport and many other alternatives should have been considered in the EIS.

3.2.2.2 Natural Resources Defense Council v. Morton 458 F. 2d 827 (D.C. Cir. 1972)

The U.S. Department of the Interior proposed to issue Gulf of Mexico oil and gas leases to private entities. The department prepared an EIS for this federal action, potentially affecting the environment, with a purpose and need statement of meeting the nation's energy crisis. The EIS was narrow in scope, and focused just on oil and gas leases in the Gulf of Mexico. The Natural Resources Defense Council, a national environmental advocacy group sued because the range of alternatives was too narrow. The court agreed with the Natural Resources Defense Council that alternatives such as developing oil shale, tar sands, eliminating oil import quotas, and geothermal resources were appropriate under the broad purpose and need statement. If the Department of the Interior had been more concise and identified developing fossil fuel resources in the Gulf of Mexico, such alternatives would not have been required and the EIS might have been adequate.

3.2.2.3 Isaak Walton League of America v. Marsh, 655 F.2d 346 (D.C. Cir 1981)

The U.S. Army Corps of Engineers proposed an underlying need at an existing lock and dam system, for: (1) more capacity at the system, (2) greater water transportation capacity in the waterways system containing the lock and dam, and (3) greater safety at the lock and dam. This was a federal action with potential environmental impacts and an EIS was prepared. The EIS was challenged by the Isaak Walton League, a national environmental advocacy organization, because the proposed action was to replace the lock and dam without consideration of the alternatives: a railroad system to carry cargo, rehabilitation of the existing system, or management of congestion at the existing system. The court ruled in the U.S. Army Corps' favor because the U.S. Army Corps had narrowly defined the purpose and need, which was their prerogative, and the alternatives identified by the intervener did not satisfy the stated needs of increased capacity and safety.

3.2.2.4 Methow Valley Citizens Council v. Regional Forester, 833 F.2d 810 (9th Cir. 1987); Reversed Robertson v. Methow Valley Citizens Council 490 U.S. 109 S.Ct. 1835, 104 L.Ed.2d 351 (1989)

A ski resort in Washington state's North Cascades was proposed by the U.S. Forest Service on Forest Service land. This was a federal action with potentially significant environmental impacts and thus required an EIS, which was challenged by a citizens' group. The EIS was upheld in the district court but the plaintiffs appealed and the Court of Appeals, Ninth Circuit agreed to hear the NEPA-related challenges.

Only one site for the ski resort was analyzed in the EIS, yet the stated purpose and need were for "winter sports opportunity." The Court of Appeals ruled that with such a broad purpose and need statement, evaluation of only one site and only as a downhill ski resort was not a reasonable range of alternatives. The challenging party offered evidence of other suitable sites that would readily satisfy the broad need for winter sports opportunities, yet these were not evaluated in the EIS. Also as suggested by stakeholders, the Forest Service neglected to consider expansion of existing ski areas, which the Appeals Court found was reasonable to assume would have less environmental impact than construction of an entirely new ski area. The Appeals Court did acknowledge that the U.S. Forest Service was not required to explore an unreasonably broad range of alternatives but it "should more clearly articulate its goal," thus providing limits to the breadth of alternatives that must be considered. Specifically the court stated that a clear and concise statement of purpose and need "will provide a clear standard by which it can determine which alternatives are appropriate for investigation and consideration in its EIS. In its present state, the EIS's discussion of alternatives to the proposed action is inadequate as a matter of law." Based on the failure to evaluate a reasonable range of alternatives satisfying the purpose and need statement, and other deficiencies in the EIS, the Court of Appeals judged the EIS as inadequate.

3.2.2.5 City of Angoon v. Hodel, 803 F.2d 1016 (9th Cir. 1986)

This example involves timber harvesting and a log transfer facility on Admiralty Island, near Juneau, Alaska, and the court's ruling not just of the relationship of purpose and need with alternatives, but also on formulation of the purpose and need statement. When the EIS for the project was challenged, the district court ruled that the U.S. Army Corps' purpose and need of "safe, cost effective means of transferring timber harvested on [permittee's] land to market" was too narrow, and the district court substituted a broader "commercial timber harvesting" as the underlying need. Under this revised, more general purpose and need statement, the district court ruled that the EIS was defective because it failed to consider a reasonable range of alternatives, including the alternative of a specific land exchange.

The district court's decision was reversed by the Ninth Circuit Court of Appeals based on the unjustified revision of the purpose and need statement by the district court. The appeals court ruled the purpose and need statement was the prerogative of the agency proposing the action, because they had the most complete knowledge of the problems and opportunities within their jurisdiction and the technical expertise to understand the issues. The court of appeals stated: "The preparation of [an EIS] necessarily calls for judgment, and that judgment is the agency's." Because the district court's broader statement of purpose and need was rejected, the specific land exchange alternatives did not meet the purpose and need and the EIS was deemed procedurally acceptable. Specifically the appeals court stated: "When the purpose is to accomplish one thing, it makes no sense to consider alternative ways by which another thing might be achieved." Thus in this case, the courts affirmed that the agency proposing the action has wide latitude and discretion in substantive matters, and the court will not substitute its judgment or otherwise interfere where there is no indication of an arbitrary or capricious decision and procedures are followed.

3.2.2.6 National Parks & Conservation Association v. U.S. Bureau of Land Management, 586 F.3d 735 (9th Cir. 2009)

This case involved a private developer who wanted to exchange privately held land for U.S. Bureau of Land Management (BLM) land in order to convert a former iron ore mine into a landfill. The bureau prepared an EIS because the exchange was a federal action that could have significant environmental impacts. The court ruled the EIS was invalid because among other reasons, the BLM unreasonably narrowed the purpose and need for the project and only included the private needs of the applicant. The BLM had initially identified several alternatives that would have been responsive to the need to meet long-term landfill demand, but the BLM did not consider these options in any detail. They were rejected because each of these alternatives failed to meet the narrowly drawn project objectives addressing only the applicant's private needs: "BLM cannot define its objectives in unreasonably narrow terms and may not circumvent this proscription by adopting private interests to draft a narrow purpose and need statement that excludes alternatives that fail to meet specific private objectives."

3.3 Purpose and Need Case Study: Washington Aqueduct Water Treatment Residuals

The Water Treatment Residuals Management Process for the Washington Aqueduct, Washington, D.C. (U.S. Army Corps of Engineers 2005) was ultimately a very successful project with stakeholder support, and a narrowly defined purpose and need statement contributed to the success. In particular, the narrow purpose and need statement at the initial stages of the project allowed a focus on practical and implementable alternatives addressing the substantial issues of the stakeholders. A broader statement of purpose and need could have obfuscated the real issues and delayed a project that was badly needed for water quality protection, safe drinking water for the nation's capital, and caused violations of the Clean Water Act. A summary of the project background and issues related to the purpose and need are presented below and additional project background detail is provided in Section 10.4.

3.3.1 Background

The U.S. Army was given the responsibility of providing potable drinking water for Washington, D.C. shortly after the founding of the nation's capital. Originally groundwater wells within the district were sufficient in quantity and quality to supply the initial inhabitants of Washington, D.C.

However as the population and demand for water grew, another water source was necessary, and the adjacent Potomac River was identified and put into service as the obvious and plentiful potable water source. Currently the Washington Aqueduct, a division of the U.S. Army Corps of Engineers, Baltimore District, operates two water treatment plants (WTPs) processing the Potomac River water. These facilities, the Dalecarlia and the McMillan WTPs, both in Washington, D.C., provide potable water in adequate quantity and quality for over 1 million people in the D.C. and Northern Virginia areas. When these facilities were first constructed at the beginning of the twentieth century, they represented the state of the art technology in water treatment and the processes were more advanced than virtually any in the country. The treatment removed solid particles (e.g., river silt) from the Potomac River raw water source, treated and disinfected the water, and distributed it to the metropolitan service area. The water treatment system historically consisted of a series of reservoirs and treatment facilities, with raw water diverted from the Potomac River and collected in the Dalecarlia Reservoir. River silt was naturally deposited in the forebay of the Dalecarlia Reservoir and the silt (forebay residuals) was periodically dredged and trucked offsite, or utilized onsite. The water treatment operations achieved an additional level of silt removal by adding aluminum sulfate (alum) prior to the sedimentation basins at the Dalecarlia Water Treatment Plant and the Georgetown Reservoir where the sediment was removed. Periodically flushing the settled residuals from the basins to the Potomac River was the historic residual management practice.

Thus the solids removed during the treatment process were returned to the Potomac River under the logic that they came from the river and it made sense to return them to the river. However, when the existing National Pollution Discharge Elimination System (NPDES) issued under the Clean Water Act expired a new permit was issued which effectively precluded the discharge of water treatment solids, or residuals, to the river. Since changing the treatment process to eliminate the discharge of solids to the river was a federal action with potentially significant impacts, NEPA applied and because of the magnitude of the project the U.S. Army Corps decided to prepare an EIS.

3.3.2 Washington Aqueduct Water Treatment Residuals Purpose and Need

Consistent with the discussion above regarding the relationship of the purpose and need statement and evaluation of alternatives, the Washington Aqueduct purpose and need statement was wisely and narrowly constructed. The forcing function for the action was compliance with the new NPDES permit requirement to minimize or eliminate discharge of solids to the river, and thus this was central to the purpose and need for the action. The purpose and need statement was included in the notice of intent, published in the *Federal Register* on January 12, 2004 and also in the "Project Introduction and Description of Proposed Action and Alternatives" (prepared in May 2004) which was a primary supporting document for the scoping process and ultimately preparation of the EIS. In summary, the purpose and need were stated as:

- Achieve complete compliance with the reissued NPDES permit and all other federal and local regulations.
- Do not compromise the current or future production of safe and sufficient drinking water for Washington Aqueduct customers.
- If possible, reduce the quantities of solids generated by the water treatment process.
- Minimize, if possible, impacts on various local and regional stakeholders and minimize impacts on the environment.
- Implement a solids management process that is cost-effective.

The purpose and need statement successfully precluded the need to consider alternatives in the EIS that were not reasonable or practical. Of particular concern was avoidance of the time- and resource-consuming effort of identification and evaluation of a new raw water source. Engineering analysis done over 100 years prior and periodically revisited conclusively demonstrated that the Potomac River was the only viable source of potable water for metropolitan Washington, D.C. Any attempt to find another source that would result in fewer water treatment residuals would not only be fruitless but it could engage numerous stakeholders (i.e., those associated with new possible but unrealistic raw water sources) and thus dilute and distract the focus of the EIS. Also, development of a different raw water source would likely require treatment facilities at additional or new locations and a significant reconfiguration of the distribution system. Consideration of a new raw water source could be immediately dismissed because it would not meet the purpose and need and these distracting issues could be avoided. Specifically because of the development of new intakes, protection of a new source, new treatment facilities, and modification of the distribution system, the cost-effective condition in the purpose and need would not be satisfied. Similarly, construction and operation of a new treatment facility and disruptions caused by major system modifications would not conform to the purpose and need to minimize stakeholder impacts. Also, any new unproven sources of water may not conform with the criterion of continued adequate and safe supply of water to Washington Aqueduct customers and thus would not be consistent with this provision in the purpose and need statement. For these reasons a new water supply and other unreasonable alternatives were dismissed without even a cursory examination and the EIS was able to focus on the critical issues, stakeholder concerns, and mitigation of impacts.

After excluding unrealistic alternatives based on obvious inability to meet the purpose and need, 26 potentially realistic alternatives were identified. These 26 alternatives were screened (see Chapter 5 for discussion of alternative screening) based on criteria developed to measure consistency with the purpose and need statement. Based on the screening, 3 of the 26 alternatives were found to meet the purpose and need:

- Process water treatment residuals at the Dalecarlia WTP, and dispose via contract hauling. Process forebay residuals by current methods and periodically haul off-site.
- Thicken water treatment residuals at Dalecarlia WTP and then pump via a new pipeline to the D.C. WASA Blue Plains Wastewater Treatment Plant. Process forebay residuals by current methods and periodically haul off-site.
- Process water treatment residuals at Dalecarlia WTP and dispose of the solids in Dalecarlia monofill (i.e., a dedicated landfill constructed to accept material from only one type of material from only one source). Process forebay residuals by current methods and periodically haul off-site.

The no-action alternative (i.e., continue the current practice of discharge to the Potomac River) did not meet the purpose and need because it did not satisfy the limits set by the reissued NPDES permit. However, as discussed above, the no-action alternative is required under NEPA as a basis of comparison of the action alternatives. Thus in addition to the three action alternatives, the no-action alternative was carried forward for a detailed analysis in the draft EIS.

Construction and operation of residuals treatment and/or disposal facilities at or adjacent to the Dalecarlia WTP was a common element of all the alternatives surviving screening and carried forward for a detailed analysis in the draft EIS. The addition of facilities at the Dalecarlia location and the operation of the facilities proved to be unpopular with the neighborhoods surrounding the facility because of the perceived noise, aesthetic, traffic, and other impacts. This issue was the most controversial aspect of the project and generated substantial and numerous comments. Many of the comments focused on carrying forward alternatives that did not require new facilities at Dalecarlia; however, since a thorough screening process demonstrated such alternatives were not consistent with the purpose and need statement, the project proponent was able to focus on the most controversial issue of facility location and design. By focusing attention on this issue, the U.S. Army Corps of Engineers was able to work with the stakeholders to develop facilities designs, layouts, locations, and operations that minimized impacts. If there had been a weaker or less definitive purpose and need statement, the focus would have shifted to an unproductive examination of less viable alternatives. This would have consumed time and resources, and an efficient and effective proposed action that was ultimately acceptable to the stakeholders may not have been developed and implemented. Thus the strong purpose and need statement served the intent of focusing the program on the real issues and allowed the U.S. Army Corps to achieve its goal of developing a residual management system that fully met reissued NPDES permit requirements while achieving cost-effectiveness and not compromising the quantity or quality of water supplied to the Washington, D.C. Metropolitan area. By focusing the U.S. Army Corps' efforts on the true issues, the purpose and need statement also facilitated development of a proposed action which minimized impacts on the neighbors and in the final analysis generally satisfied their concerns.

3.4 Purpose and Need Case Study: U.S. Coast Guard Rulemaking for Dry Cargo Residue Discharge in the Great Lakes

3.4.1 Background

The economic life blood of the Great Lakes region and in many ways the industrial revolution in the United States is the combined presence of plentiful and easily extractable natural resources (e.g., coal, iron ore, and limestone) in the region and easy transportation from the sources of these primary components required to steel manufacturing operations. There are significant coal deposits both at the western (e.g., Montana) and eastern (e.g., Pennsylvania and West Virginia) ends of the region, and iron ore is plentiful, particularly in Minnesota at the western end of Lake Superior. The other component required for steel manufacturing, limestone, is abundant in the center of the region. This distribution of resources is not unusual in North America but the ability to cheaply and quickly transport them to central locations for manufacture is unique. Since the 1800s, these materials have been loaded on Great Lake ships at or near their point of origin and transported to Ohio, Indiana, and other ports where they are converted to steel. The steel was then easily transported to Detroit and other industrial centers around the Great Lakes for the production of cars, appliances, and other goods that made U.S. manufacturing a dominant factor in the country's development and ultimately their place in history.

Transport by ship was easy, convenient, and quick. There was no need for roads or railroads, and many of the steel mills, such as those in Cleveland were on the lake's edge, and the raw materials for steel production could be unloaded directly to the mills. The cost of moving a ton by ship is a small fraction of that by any other means of transport, and although the ships might not travel at the speed of a train, they could navigate a direct line course from source to factory. Also, the ships never stop, so transport efficiency was also superior to other means of transportation.

Even with all these advantages, the Great Lakes shipment of coal, iron ore, and limestone can sometimes be messy, as the material ends up on the deck or as residue in the holds of the ship. This residue and spillage, termed dry cargo residue (DCR), created safety and maintenance issues and had to be addressed to maintain an efficient and safe shipping operation. Since the late 1800s the issue had been addressed by washing or sweeping the DCR overboard, either in port or more recently away from shore once the ship was underway. Around 2000, this practice started to generate interest and concern because of potential environmental impacts and inconsistencies with provisions of the Clean Water Act and Great Lake treaties with Canada. In recognition of the issue, the U.S. Congress mandated that the U.S. Coast Guard (USCG) develop regulations addressing the environmental and treaty/Clean Water Act concerns. Congress considered the USCG the appropriate entity to address the issue because they had a strong and longstanding presence on the Great Lakes; they had maritime safety, shipping, and environmental protection responsibilities on the Great Lakes; and they had the confidence and respect of most stakeholders who generally considered them unbiased.

The USCG (2008) promulgation of rules governing management of DCR constituted a federal action with potentially significant environmental impacts. Thus they were required to comply with NEPA and chose to prepare an EIS. The Coast Guard's DCR EIS provides another example of wisely and narrowly stating the purpose and need in order to facilitate alternative development and solicit appropriate and meaningful stakeholder input (see Chapter 10 for summary of the USCG DCR program and EIS). Having dealt with DCR for decades, the USCG was very familiar with the issues and current management practices, so they were well informed as they developed the purpose and need statement. They knew from their experience and interaction with environmental groups for over a decade that there was concern that the current practice had significantly degraded the Great Lakes ecosystem, and also the impression of various stakeholders was that the Great Lake carriers could easily contain the sweepings for discharge at port. They also knew that retaining the large volume of water associated with DCR sweeping would be impracticable at best and at worst a safety hazard to the stability of the ship because of the shifting weight when the carriers were underway.

3.4.2 USCG Rulemaking for DCR Discharge in the Great Lakes: Purpose and Need

The task of developing a purpose and need statement was simplified not only due to the USCG's extensive background and familiarity with the issues but also by their mandate. They had a very specific directive from Congress to develop the DCR management regulations summarized as balancing economic viability of the Great Lakes shipping industry and environmental protection. There was extensive NEPA and litigation history confirming a Congressional mandate as a legitimate and in fact a primary purpose and need for an EIS. In addition, the USCG had an overall organizational mission, and meeting the goals of the mission was an important purpose and need of any USCG action. Thus they developed a purpose and need statement that met their Congressional directive, was supportive of the organizational goals, and took into account their background knowledge to exclude, for detail, evaluation alternatives that were inconsistent with their directive and goals.

The purpose and need statement consisted of the following:

- Limitation of the regulations and EIS to nonhazardous, nontoxic DCR discharges from vessels in the Great Lakes that fall under the jurisdiction of the United States.
- Satisfying Congress's direction to develop a regulation that meets the mandate of the Coast Guard and Maritime Transportation Act (ACMTA) of 2004 Public Law 108-293 §623.
- Incorporating language from the ACMTA that gave the USCG the authority to develop regulations that "grant[s] the Commandant of the Coast Guard notwithstanding any other law the permanent authority to promulgate regulations governing the discharge of DCR on the Great Lakes." Thus alternatives which might be inconsistent with other laws (e.g., the Clean Water Act) could be considered in detail as long as they meet all the provisions of the purpose and need statement.
- Optimizing the DCR management to address the USCG organizational strategic goals, of maritime safety, protection of natural resources, maritime mobility, and maritime security.

This purpose and need statement formed the basis of screening criteria such that alternatives that met the criteria met the purpose and need and were evaluated in detail in the EIS. The resulting screening criteria and their link to the purpose and need statement are summarized in Table 3.2.

A comprehensive list of alternatives was developed based on: historic knowledge of the USCG; input from stakeholders, including Lake Carriers Association and environmental advocacy groups during scoping; and EIS team (including technical consultants) evaluation. The alternatives identified were measured against the criteria designed to determine the achievement of purpose and need (Table 3.2). The alternatives that did not meet the purpose and need were eliminated from detailed evaluation in the draft EIS with the caveat that insufficient information available at the time of screening would result in retaining the alternative for detailed evaluation. Three of the alternatives were eliminated from full evaluation in the draft EIS because

TABLE 3.2

USCG Rulemaking Dry Cargo Residue Discharge Alternative Screening Criteria to Meet Purpose and Need Statement

Alternative Screening Criteria	Link to Purpose and Need Statement
Prevent impacts that significantly degrade the Great Lakes aquatic resources	USCG's strategic goal of protecting natural resources
Regulate with only minimal additions to existing Coast Guard organizational structure and resources	Prevent compromising USCG's organizational strategic goals by diverting resources needed to meet the goals
Avoid regulating dry bulk carriers and related shoreside facilities in a way that threatens their continued economic viability	Congressional mandate to maintain the economic viability of the shipping industry
Avoid regulating dry bulk carriers in a way that threatens their safe operation	USCG's strategic goal of maritime safety and maritime mobility
Minimize additional energy use	USCG's strategic goal of protecting natural resources
Provide for an adequate and appropriate recordkeeping and compliance monitoring system	Provide a monitoring system so the USCG could demonstrate compliance with the regulations, Congressional mandate, and organizational strategic goals
Use proven dry cargo residue control measures	USCG's strategic goal of maritime safety and maritime mobility

Source: Adapted from USCG, *Final Environmental Impact Statement: U.S. Coast Guard Rulemaking for Dry Cargo Residue Discharges in the Great Lakes*, Washington, DC: Commandant USCG Headquarters, 2008.

of inconsistency with one or more of the criteria established to evaluate achieving the purpose and need as follows:

- The alternative to adopt the existing interim rule without modification was eliminated because it did not include a record keeping requirement, and the USCG would not be able to determine whether they met their Congressional mandate over the long term.
- A newly developed USCG permitting system was eliminated as an alternative for detailed evaluation because it would have required a significant modification with substantial expenditure of resources to the existing organizational structure, potentially compromising the USCG's ability to fully meet its organizational goals. There was no additional environmental or economic benefit to this alternative compared with others, and it risked other aspects of the USCG mission.
- Structural modification of the deck and ship's hold to retain DCR wash-water and prevent overboard discharge was also eliminated as an alternative. Based on the extensive knowledge of ship safety within the USCG (ship safety has been their primary mission for over 200 years) it was determined that the wash-water could weigh

as much as 400,000 kg and storing this mass of water on a moving ship would compromise the ship's stability and threaten crew safety. This important element of the purpose and need would not be met and the alternative was eliminated from detailed evaluation in the draft EIS.

3.5 Purpose and Need Conclusions

Even though it is typically the shortest section of an environmental document, the purpose and need statement plays a very important role in environmental analysis and NEPA. It focuses the efforts and energy of the analysis, and at the same time, encourages appropriate input from stakeholders truly affected by the proposed action or any alternative. It also can serve to discourage stakeholder input and anxiety over issues not related to actions under consideration. The purpose and need statement has the added importance in a NEPA investigation of being one of the most common factors in litigation and court's rejection of an EIS.

The breadth and scope of an EIS, particularly the range of alternatives considered, is set by the purpose and need statement. It must accurately reflect the project proponent's intentions, while establishing what an alternative must accomplish to warrant a detailed analysis in the environmental impact analysis document. The purpose and need must be narrow enough that only legitimate, practical, and implementable alternatives are considered in detail yet not so narrow that the purpose and need reflect only the proponent's needs or actions that are totally under the jurisdiction and authority of the federal entity proposing the action. As discussed earlier, NEPA court cases have provided guidance in establishing the breadth and scope of the purpose and need statement.

"Brainstorming" early in a project is a method that can frequently be employed to produce a superior purpose and need statement. The project team and unbiased stakeholders knowledgeable about the history, background, technical aspects, and public perception of the project can discuss the likely and unlikely successful alternatives. The reasons for success or anticipated failure can be discussed and documented. These reasons frequently can be built into the purpose and need statement and thus focus the environmental analysis. Objectivity and documentation of such a brainstorming exercise to develop the purpose and need statement is an appropriate and acceptable way of structuring the environmental analysis early in the process. Such a process is in contrast to the knowledgeable project team subjectively declaring during the final preparation of the draft environmental document that we are not going to evaluate that alternative in detail because it is "stupid and makes no sense." The better approach, and the one that will survive legal and procedural challenge, is to have that discussion early, document what doesn't make sense regarding the alternative, and build the reason into the purpose and need statement.

3.6 Categorical Exclusion as an Efficiency Approach to Environmental Analysis

A categorical exclusion (abbreviated CATEX, CATX, or Cat-X, by various agencies) is a common sense approach to NEPA compliance presented in the CEQ Regulations (see Section 3.1.2 for discussion of how CATEX fits into the NEPA Process). Section §1507.3 b 2 (ii) of the regulations require each agency to develop *"Specific criteria for and identification of those typical classes of action… which normally do not require either an environmental impact statement or an environmental assessment."* In general this process is limited to small-scale projects that result in a tangible product rather than a policy, program, or regulation development action. In accordance with this directive, each federal agency has reviewed the projects they conduct on a continuing basis (e.g., installing an electronic security surveillance system in a facility, data collection, payroll processing, purchasing standard office equipment, or conducting land surveys) and determined whether they singly or collectively have a significant environmental impact. If they conclude they do not, they document the finding so that the next time they take the action they can simply refer to the documentation and no further NEPA action or other environmental analysis is required. This categorical exclusion of actions is not an exception from NEPA but rather another and simpler way to comply with the Act.

CEQ has recognized CATEX importance and opportunities, noting that it represents the most frequent method of NEPA compliance. In recognition of the importance, opportunity for efficiency, and common use, CEQ has recently issued guidance to assist agencies in maximizing these attributes of the CATEX process (75628 *Federal Register*/Vol. 75, No. 233/Monday, December 6, 2010). This guidance also adds consistency among agencies and reflects a 40-year history of agencies employing (and in some cases failing to employ) categorical exclusions. Among other innovations the guidance encourages agencies to include the public at some level in various aspects of the CATEX process, including development of criteria for listing as a CATEX action, the listing of specific actions, and announcement that an action has been "CATEXed." CEQ leaves the nature and extent of public involvement up to the discretion of the agencies but suggests factors to consider such as the anticipated public concern over the action and the degree of mitigation commitments factored into eligibility of the action for exclusion. The CEQ CATEX guidelines also recommend the agencies have some level of tracking of CATEXed projects and periodically review their list of eligible actions.

The guidance recognizes the varied nature of "typical" actions and importance of environmental setting in the process. The guidance encourages agencies to take these into account in executing their CATEX procedures. They recommend that an agency consider potential "extraordinary circumstances" that might apply to an action in an atypical situation or environmental settings that would otherwise be standard for the agency and previously shown not to have an impact. When an action may fall into this category, the agency should make a case-by-case determination whether it differs substantively from the actions considered when the CATEX list was developed and if so an EA or EIS may be warranted. Several agencies establish criteria in determine extraordinary circumstances such as the potential presence of endangered species' habitat or hazardous materials.

The CEQ CATEX guidance also establishes proven and acceptable methods for establishing an action on the CATEX list. Although an agency can use other approaches specific to their circumstances in developing the list, the CEQ guidance cites the four methods discussed below. The first method for qualifying an action for the list is experience from previously implemented actions, which is probably the most common method for listing an action. It can take the form of a previous action that was the subject of an EA (any action requiring an EIS would not qualify because of the uniqueness and potential for significant impact) and FONSI. However the guidance specifies that a FONSI alone does not qualify similar actions for the CATEX list, and there must be some form of confirmation that the predictions of the EA and conclusions of the FONSI were generally accurate and impacts did not occur. In the extreme case, confirmation would be a formal monitoring program, but for most actions more casual observation and documentation would suffice. For example, if several storm water control systems on an Air Force base had been proposed, an EA was prepared, and FONSI issued, these activities could qualify for inclusion in a CATEX list if the base civil engineer or environmental officer observed and documented activities during construction and during a storm to confirm that the EA predictions and FONSI were accurate. Even if no EA was prepared, a similar informal monitoring and documentation of no impact for common agency practices could qualify the action for the CATEX list.

An agency can also conduct impact demonstration projects to determine eligibility for CATEX listing by the agency (second method identified in the CATEX guidance). This process would be useful if an agency anticipates several similar projects that are new to the agency and they target the first one as the demonstration project. As described in the CEQ guidance, use of this approach for CATEX listing consists of preparing an EA and FONSI; implementing the action; and a subsequent evaluation of the environmental effects in accordance with a predetermined monitoring plan. If a postimplementation evaluation confirms the FONSI, such actions can be included on the agency's CATEX list.

The third approach cited in the CEQ guidance to CATEX listing is "Professional Staff and Expert Opinions, and Scientific Analyses." This approach acknowledges that agency personnel who deal with agency actions on a day-to-day basis can apply their technical knowledge and experience in a common sense fashion and reasonably conclude that "when we have done this in the past, it has had no unacceptable environmental consequences, so we can make reasonable predictions about similar actions." This provision also acknowledges that scientific investigation of an action showing no impact can apply to a similar class of actions typically taken by an agency. The guidelines do recommend that the qualifications of the expert and details of the scientific investigation supporting the no-impact conclusion be documented and made part of the record.

The final method cited by CEQ guidance for CATEX listing is "Benchmarking Public and Private Entities' Experiences." Thus if another agency (federal, state, or other) has qualified an action for their CATEX list, the listing can be used by other agencies. CEQ points out that transfers from another agency are not automatic, and the listing agency must consider and document the similarity of their action to that of the original listing agency.

Each agency or other federal entity has developed a list of actions common to executing their mission and fulfilling their responsibility that have been shown to have no significant impact. In some cases, the actions were originally the subject of an EA or even an EIS, and mitigating measures developed, perfected, and incorporated in the standard practice for implementing the measure. Thus the expected impacts, if any, are fully understood prior to a decision, and a NEPA EA or EIS review of the same or very similar action would not provide any additional information; and thus a CATEX is appropriate. In other cases, the actions have been conducted by the agency so many times and no adverse environmental impacts have been observed that a detailed NEPA analysis is not necessary to understand environmental implications prior to a decision and they can be CATEXed. Some federal entities, such as the U.S. Navy, have included in their CATEX list conditions that must be met if no further NEPA analysis is required (Table 3.3). Other agencies have just listed the actions eligible but then specified that certain actions must be compared to a check list and a CATEX determination be prepared before the action is eligible to be CATEXed. The USCG has taken this approach (Table 3.4).

The concept of categorically excluding a routine action from a specific and detailed environmental impact analysis is not unique to NEPA. As discussed in Chapter 8, many international and individual states embrace this concept. In many cases the jurisdiction of these entities is much more limited than that covered by NEPA (i.e., the entire U.S. and many overseas actions) thus a universal list of actions not warranting individual analysis is included in the general environmental regulations of many entities. In some cases, the approach goes beyond a binomial decision of an action requiring an individual environmental analysis versus the action not requiring an individual evaluation analysis. In these cases a hierarchical list is developed where classes of actions are defined which require no environmental evaluation, limited and well-defined level of analysis (similar to a NEPA EA), or comprehensive evaluation (similar to a NEPA EIS).

TABLE 3.3
Department of the Navy's List of Categorical Exclusions

1 Routine fiscal and administrative activities, including administration of contracts	11 Routine movement of mobile assets (such as ships and aircraft) for homeport reassignments, for repair/overhaul, or to train/perform as operational groups where no new support facilities are required
2 Routine law and order activities performed by military	12 Routine procurement, management, storage, handling, installation, and disposal of commercial items, where the items are used and handled in accordance with applicable regulations (e.g., consumable, electronic components, computer equipment, pumps)
3 Routine use and operation of existing facilities, laboratories, and equipment	
4 Administrative studies, surveys, and data collection	13 Routine recreational/welfare activities
5 Issuance or modification of administrative procedures, regulations, directives, manuals, or policy	14 Alteration of and additions to existing buildings, facilities, structures, vessels, aircraft, and equipment to conform or provide conforming use specifically required by new or existing applicable legislation or regulations (e.g., hush houses for aircraft engines, scrubbers for air emissions, improvements to storm water, and sanitary and industrial wastewater collection and treatment systems, and installation of firefighting equipment)
6 Military ceremonies	
7 Routine procurement of goods and services conducted in accordance with applicable procurement regulations, executive orders, and policies	
8 Routine repair and maintenance of buildings, facilities, vessels, aircraft, and equipment associated with existing operations and activities (e.g., localized pest management activities, minor erosion control measures, painting, refitting)	15 The modification of existing systems or equipment when the environmental effects will remain substantially the same and the use is consistent with applicable regulations
9 Training of an administrative or classroom nature	16 Routine movement, handling and distribution of materials, including hazardous materials/wastes that are moved, handled, or distributed in accordance with applicable regulations
10 Routine personnel actions	

(Continued)

TABLE 3.3 (Continued)

Department of the Navy's List of Categorical Exclusions

17. New activities conducted at established laboratories and plants (including contractor-operated laboratories and plants) where all airborne emissions, waterborne effluent, external ionizing and non-ionizing radiation levels, outdoor noise, and solid and bulk waste disposal practices are in compliance with existing applicable federal, state, and local laws and regulations

18. Studies, data, and information gathering that involve no permanent physical change to the environment (e.g., topographic surveys, wetlands mapping, surveys for evaluating environmental damage

19. Temporary placement and use of simulated target fields (e.g., inert mines, simulated mines, or passive hydrophones) in fresh, estuarine, and marine waters for the purpose of non-explosive military training exercises or research, development, test and evaluation

20. Installation and operation of passive scientific measurement devices (e.g., antennae, tide gauges, weighted hydrophones, salinity measurement devices, and water quality measurement devices) where use will not result in changes in operations tempo and is consistent with applicable regulations

21. Short-term increases in air operations up to 50% of the typical operation rate, or increases of 50% operations per day, whichever is greater. Frequent use of this CATEX at an installation requires further analysis to determine there are no cumulative impacts

22. Decommissioning, disposal, or transfer of navy vessels, aircraft, vehicles, and equipment when conducted in accordance with applicable regulations, including those regulations applying to removal of hazardous materials

23. Nonroutine repair and renovation, and donation or other transfer of structures, vessels, aircraft, vehicles, landscapes, or other contributing elements of facilities listed or eligible for listing in the National Register of Historic Places which will result in no adverse effect

24. Hosting or participating in public events (e.g., air shows, open houses, Earth Day events, and athletic events) where no permanent changes to existing infrastructure (e.g., road systems, parking and sanitation systems) are required to accommodate all aspects of the event

25. Military training conducted on or over nonmilitary land or water areas, where such training is consistent with the type and tempo of existing nonmilitary airspace, land, and water use (e.g., night compass training, forced marches along trails, roads and highways, use of permanently established ranges, use of public waterways, or use of civilian airfields)

26. Transfer of real property from the Department of the Navy to another military department or to another federal agency

27 Receipt of property from another federal agency when there is no anticipated or proposed substantial change in land use

28 Minor land acquisitions or disposals where anticipated or proposed land use is similar to existing land use and zoning, both in type and intensity

29 Disposal of excess easement interests to the underlying fee owner

30 Renewals and minor amendments to existing real estate grants for use of government-owned real property where no significant change in land use is indicated

31 Land withdrawal continuances or extensions that merely establish time periods and where there is no significant change in land use

32 Renewals and/or initial real estate in grants and out grants involving existing facilities and land wherein use does not change significantly (e.g., leasing of federally owned or privately owned housing or office space, and agricultural out leases)

33 Grants of license, easement, or similar arrangements for the use of existing rights-of-way or incidental easements complementing the use of existing rights-of-way for use by vehicles (not to include significant increases in vehicle loading) electrical, telephone, and other transmission and communication lines; water, wastewater, storm water, and irrigation pipelines, pumping stations, and facilities; and for similar utility and transportation uses

34 New construction that is similar to existing land use and, when completed, the use or operation of which complies with existing regulatory requirements (e.g., a building within a cantonment area with associated discharges/runoff within existing handling capacities)

35 Demolition, disposal, or improvements involving buildings or structures when done in accordance with applicable regulations including those regulations applying to removal of asbestos, PCBs, and other hazardous materials

36 Acquisition, installation, and operation of utility (e.g., water, sewer, electrical) and communication systems (e.g., data processing cable and similar electronic equipment) that use existing rights-of-way, easements, distribution systems, and/or facilities

37 Decisions to close facilities, decommission equipment, and/or temporarily discontinue use of facilities or equipment, where the facility or equipment is not used to prevent/control environmental impacts

38 Maintenance dredging and debris disposal where no new depths are required, applicable permits are secured, and disposal will be at an approved disposal site

39 Relocation of personnel into existing federally owned or commercially leased space that does not involve a substantial change affecting the supporting infrastructure (e.g., no increase in vehicular traffic beyond the capacity of the supporting road network to accommodate such an increase)

(Continued)

TABLE 3.3 (*Continued*)

Department of the Navy's List of Categorical Exclusions

40 Prelease upland exploration activities for oil, gas, or geothermal reserves, (e.g., geophysical surveys)	44 Routine testing and evaluation of military equipment on a military reservation or an established range, restricted area, or operating area; similar in type, intensity, and setting, including physical location and time of year, to other actions for which it has been determined, through NEPA analysis where the Department of the Navy was a lead or cooperating agency, that there are no significant impacts; and conducted in accordance with all applicable standard operating procedures protective of the environment
41 Installation of devices to protect human or animal life (e.g., raptor electrocution prevention devices, fencing to restrict wildlife movement onto airfields, and fencing and grating to prevent accidental entry to hazardous areas)	
42 Reintroduction of endemic or native species (other than endangered or threatened species) into their historic habitat when no substantial site preparation is involved	45 Routine military training associated with transits, maneuvering, safety, and engineering drills, replenishments, flight operations, and weapons systems conducted at the unit or minor exercise level; similar in type, intensity and setting, including physical location and time of year, to other actions for which it has been determined, through NEPA analysis where the Department of the Navy was a lead or cooperating agency, that there are no significant impacts; and conducted in accordance with all applicable standard operating procedures protective of the environment.
43 Temporary closure of public access to Department of the Navy property in order to protect human or animal life	

Abbreviation: PCB, polychlorinated biphenyls.

TABLE 3.4

Summary of USCG's List of CATEXs[a]

1. Administrative Actions	a.	Certain personnel actions
	b.	Recreational activates at designated locations
2. Real and Personal Property Actions	a.	Lease of easement on USCG property if use is similar to existing
	b.	License on USCG property if use is similar to existing
	c.	Permit on USCG property if use is similar to existing
	d.	Lease of USCG historic property if in accordance with memorandum of agreement with historic authorities
	e.	Acquisition of property if use is similar to existing
	f.	Acquisition of property from another federal entity if use is similar to existing
	g.	USCG use of property if use is similar to existing
	h.	Construction or improvement on land if a strict set of conditions are met
	i.	Property inspections
	j.	Transfer of property to another federal entity that results in no immediate change in use
	k.	Determination of excess property
	l.	Congressionally mandated conveyance of property to another entity
	m.	Relocation of USCG personnel
	n.	Decommission or close shore facilities until no long under USCG control
	o.	Demolition of structures
	p.	Transfer of USCG-controlled personal property to another federal entity1
	q.	Minor renovations of buildings and infrastructure
	r.	Installation of devices to protect human or other animal life.

(Continued)

TABLE 3.4 (Continued)

Summary of USCG's List of CATEXs[a]

3. Training	a. *Defense preparedness training and exercises on USCG property that don't involve undeveloped property*
4. Operation Actions	a. *Realignment of mobile assets to existing facilities*
5. Special Studies	a. Site environmental characterization.
6. Regulatory Actions	a. Regulations addressing vessel operation safety
	b. *Congressionally mandated regulations*
(1) [from previous list, gaps in numbers indicate items deleted]	Routine personnel, fiscal and administrative activates
(2)	Routine procurement activates
(3)	Maintenance dredging if depths not increased with disposal at approved site
(4)	Routine repair of vessels
(5)	Routine repair of buildings and infrastructure
(7)	Routine repair of waterfront facilities
(8)	*Minor renovation and additions to waterfront facilities*
(9)	Routine grounds maintenance
(18)	Defense preparedness training on other than USCG property where lead agency has completed the NEPA analysis
(20)	Simulated exercises involving small numbers of personnel
(21)	Classroom or administrative training

(22) Safety, law enforcement, search, rescue, ice breaking, and oil removal programs
(23) Operations of aids to navigation
(24) Routine movement of personnel and equipment
(25) Disaster relief under leadership of another federal entity
(27) Natural and cultural resource management and research
(28) *Contracts for activities at established laboratories and facilities*
(30) Review of documents
(31) Planning and technical studies
(32) Bridge administration program for selected activates
(33) Preparation of documents implementing instructions, regulations, procedures, and guidance
(34) Promulgation of selected regulations not involving physical alteration
(35) Approvals of regatta and parade event permits not in environmentally sensitive areas

Abbreviations: CATEX, Categorical Exclusion; USCG, U.S. Coast Guard.
[a] For this CATEX and other italicized entries in the table, a checklist and CATEX determination document are required.

The advantages and experience of CATEXs came to light during implementation of the American Recovery and Reinvestment Act of 2009 (Recovery Act, or Economic Stimulus). This act was passed in response to the 2008 global economic crisis as a method to create jobs and at the same time strengthen the country's infrastructure. The goal was to create jobs as rapidly as possible and thus fund the construction, rather than the planning or engineering aspects of public works projects, because the construction phase typically creates 10 to 20 times the number of jobs and economic stimulus compared to the planning and design phases. Thus projects that were "shovel ready" and could be constructed immediately were given preference.

As important as the economic stimulus and construction job creation were, Congress and President Obama made the conscious decision not to ignore NEPA and environmental considerations. In fact they incorporated the concept of CATEX in development of the act by specifying in Section 1609 of the Act that all federal agencies devote adequate resources to ensuring that applicable environmental reviews are carried out under NEPA. However they are to be completed in an expeditious fashion using the shortest existing, applicable process allowed by the law, which as discussed above maximizes the use of CATEX.

This approach worked in expediting economic stimulus while maintaining consideration of the environment. From January to December 2009, there were approximately 171,000 projects under the Recovery Act. Of these projects NEPA was found not to be applicable for 4141, EAs were performed for 7600, and EISs conducted for about 800. The vast majority of projects (158,316 representing 93%) qualified as CATEXs and could be implemented immediately (Bass 2010). The lessons learned from the Recovery Act for streamlining NEPA and applying CATEXs has had an effect on the broader NEPA practice. Following the frantic activity of the Recovery Act, some agencies initiated changes to their NEPA procedures to incorporate techniques and approaches, including maximizing the use of CATEX for making NEPA compliance more efficient and expedient (Bass 2010).

3.7 NEPA Enforcement

NEPA is a law with associated implementing regulations, but no one has ever been arrested, charged, imprisoned, or fined for violating any NEPA provisions or regulations. The philosophy behind NEPA enforcement is that NEPA applies only to federal agencies, and if an agency can be trusted to implement policies critical to the defense, diplomacy, safety, education, and health of the nation, without threat of incrimination, they can be trusted to comply with the national environmental policy. Thus NEPA enforcement is based on trusting the federal agencies to carry out the national policy, but as President Ronald Reagan once stated in a now famous comment on the nuclear weapons nonproliferation treaty with the Soviet Union, "trust but verify." There are two avenues of verification for NEPA compliance: litigation challenging NEPA procedural requirements through the U.S. federal court system; and through the court of public opinion forced by the public transparency requirements.

As discussed earlier and in preceding chapters, NEPA does not require federal agencies to achieve environmental protection or enhancement, but it does require that they follow a specified procedure in making decisions that could impact the environment. An agency's compliance with the procedural requirements can be challenged in the courts and subjected to judicial review by any entity that has "standing" in the case and the courts have interpreted standing broadly. For example, an environmental advocacy group with concerns regarding wildlife as part of their charter, or a hunter who utilizes the resource is generally judged to have standing regarding a proposed federal action that could affect wildlife habitat. Litigation regarding an agency's failure to comply with NEPA's required procedures is actually brought under the Administrative Procedure Act. This act, passed in 1946, mandates that federal agencies observe established procedural requirements set forth in a law and specifies that agency actions cannot be arbitrary, capricious, or an abuse of discretion.

The courts have consistently deferred to federal agencies regarding substantive matters addressed as part of the NEPA process. The rationale is that the federal agency proposing the actions has or has access through sister agencies the technical expertise necessary to understand the environmental implications of their actions, and the proposing agency best understands how the action relates to the agency's mission. Thus they are in the best position to weigh the consequences and make a decision consistent with NEPA and the agency's mission. However, the court has ruled on numerous occasions that agencies must take NEPA seriously and take a "hard look" at the environmental impact of their action (see discussion of invasive mussels in Chapter 5 for an example of what does and does not constitute a hard look). Similarly, courts have been asked to rule on the objectivity of decisions and have consistently ruled that if an agency seeks and considers available technical input and then either accepts the input or clearly explains why they have not, their decision is not arbitrary or capricious.

Although the courts are very reluctant to overrule an agency decision regarding interpretation of data, use of technical information, or other substantive issues, the courts have given little latitude to the federal agency proposing the action regarding NEPA procedures. They have held the agencies to the CEQ Regulations implementing NEPA, and if the agencies stray from the regulations, they must clearly explain why and how the variation does not contravene the intent of CEQ. Some of the common nonconformances with NEPA procedures causing the courts to reject an EIS or other aspects of an agency's compliance with NEPA are:

- Failure to address a comment on a draft EIS or a comment submitted during scoping. The courts have ruled that the agency does not have to accept the comment at face value or comply with the request,

but they must consider it and either address the comment or clearly explain why it does not have to be addressed.
- Treatment of alternatives. NEPA and CEQ Regulations are clear that consideration of alternatives is at the heart of NEPA and if the agency doesn't consider a "reasonable range of alternative[s]" (see Chapter 5) or rejects an alternative suggested by a stakeholder without explaining why it is not appropriate, the courts have been very clear that the NEPA analysis is flawed. Similarly as discussed above in detail, the alternative evaluation must be consistent with the purpose and need statement.
- An agency's failure to fully set forth the scientific basis for all its conclusions or give adequate consideration of the relevant and pertinent facts. However, to challenge an EIS or NEPA procedure in general, there must be a demonstration that the relevant and pertinent facts were known to the agency (e.g., they had been used elsewhere within the agency) or their existence made known to the agency (e.g., by a scoping or draft EIS comment).
- Failure to acknowledge an environmental impact in an ROD or address mitigation of the impact. As discussed above an agency must identify adverse impacts and measures to mitigate the impact. They are not required to mitigate adverse impacts but if they decide not to, they must explain why.

So if the court rules that NEPA procedures have not been followed and there are no penalties imposed, what are the repercussions for the agency? The most immediate is an injunction prohibiting implementation of the proposed action. Thus if the agency wants to proceed, they must rectify the procedural flaw, which most often means preparing a new or supplemental EIS (see Chapter 6 for a discussion of supplemental and other multilevel environmental analyses). In practice, a court ruling for a new or supplemental EIS frequently results in such delays that the agency frequently loses either public support or funding for the project and they abandon the proposal. The result of an adverse judicial decision in many cases is the agency modifying the proposed action to include significant mitigation or selecting an alternative with less environmental impact in the revised document to avoid additional litigation and the accompanying delays.

In each of these numerous court cases discussed under purpose and need (Section 3.2.2) in which the court ruled against the federal agency, the ruling was directed at the EIS and not the project under consideration. In some cases the EIS was deemed inadequate or deficient, but in other cases the courts ruled that the agency made an arbitrary or capricious judgment or decision which violated the NEPA process. In no case did the court rule that the proposed action "had too great an impact." The courts have universally treated NEPA as procedural and allowed the proposing federal entity to exercise its authority to fulfill its mission, regardless of environmental impact as

NEPA Process and Specific Requirements 99

long as procedures are followed, they are not arbitrary or capricious, and that they have taken a "hard look" at the impacts before making decisions.

Although in the above and similar examples the courts did not prohibit implementation of the proposed action as long as NEPA procedures were followed, their decisions frequently can have that effect. It generally takes one or more years for a challenged EIS to make its way through the district court, and if the decision goes to appeal it can be another year or more. During this time the proposing agency often loses funding or refocuses its energies on other more implementable projects. If there is an immediate need to implement a solution to a problem or exploit an opportunity, the federal agency will often abandon their proposed alternative which is unpopular (which is generally the true reason for the challenge as the plaintiff frequently could care less if the agency strictly followed NEPA procedure) and choose an alternative and/or mitigating measures that satisfies both their legitimate purpose and need as well as the primary objections of the plaintiffs and other stakeholders. This approach by the proposing agency has often been taken to avoid the lengthy delays and expenditure associated with litigation.

The court of public opinion can play just as large a part in NEPA enforcement as the judicial court. Almost any dedicated and self-respecting senior federal agency official will only very reluctantly confirm in a public forum that they have decided to proceed with a project even though the agency acknowledged it will have serious adverse environmental effects that could be mitigated. Thus they strive through the process to find an environmentally acceptable alternative, gain support from environmental advocates, achieve maximum practicable impact mitigation, be thoroughly convinced there is not a more environmentally acceptable approach to meeting the purpose and need, or some combination of these measures to avoid being forced to make such a confession and live with the repercussions. This is illustrated by the urban legend of a U.S. Air Force colonel at a NEPA public meeting to site a radar installation (may or may not be true but it is repeated as part of environmental training). When asked whether the Air Force had considered bird strikes (i.e., measures to prevent birds from flying into the radar facility) and the potential impact to the endangered bald eagle, which is the national bird of the United States, the colonel allegedly replied, "Yeah, but we are talking about national defense here, who cares about a frigging bird." The full-bird colonel was moved to another project and the proposed radar installation site was moved much farther from the bald eagle nesting habitat.

3.8 NEPA Conclusion

At its core NEPA is a simple concept designed to provide a mechanism for environmental impact analysis and protection while maintaining a balance between economic development and the missions of individual federal agencies. Sometimes posing questions regarding what NEPA requires and does not require is useful in understanding this simple concept and the intent of the Act and implementing regulations. Under NEPA:

- What is mandated for all federal agencies?
  - Take a "hard look" at the impacts of their proposed action and acknowledge the impacts when making decisions.
  - Consider alternatives to the proposed action that might have fewer or different environmental impacts.
  - Evaluate mitigation of all impacts.
  - Employ technically defensible and multidiscipline environmental impact analysis methods and avoid arbitrary and capricious decisions.
  - Conduct a transparent process that engages stakeholders and the public at large.
- Must an agency of the federal government achieve absolute environmental protection and enhancement? No, but it should be a goal and the environment must be considered in every decision. Specifically each agency must:
  - Promote efforts to prevent or eliminate damage to the environment and stimulate health and welfare (note there is no absolute mandate, just a philosophy).
  - Use all practicable means, consistent with other essential considerations of the national policy to improve and coordinate federal plans, functions, programs, and resources to:
    – Fulfill responsibilities of each generation as trustees of the environment for future generations.
    – Assure safe, healthful, productive, and aesthetically and culturally pleasing surroundings.
    – Attain the widest range of beneficial uses of the environment without degradation, risk to health or safety or other undesirable and unintended consequences.
- Is an agency required to select the alternative with the least impact? No, but:
  - The federal agency proposing the action must identify the alternative with the least environmental impact.
  - They must explain to all stakeholders in a clear and transparent manner the selection they made and if they do not select the environmentally preferable alternative they must explain why not.

- Does a federal agency make a decision to or not to implement a project based on its environmental impact? The simple answer is no, but they must understand the environmental implications prior to making the decision. They are also publicly accountable for their decision and the courts have determined that it cannot be arbitrary or capricious.
- Is the agency required to select the alternative with the most public support? Absolutely not, the decision is at the discretion of the agency proposing the action. But they must inform the public, solicit their input, and respond to all submitted questions and comments. Fostering public and other stakeholder support can also be an effective tactic to successfully implementing the proposed action.
- Must an agency follow codified procedures? This is one of the few absolute requirements of NEPA: the agency must follow established procedures. Failure to do so is the primary factor in court decisions regarding the violation of NEPA.
- Must agencies make a smart decision? No, unfortunately not even Congress can force wisdom in decisions. But NEPA does require open decisions and daylight generally increases the wisdom of a decision. Also in theory, the input from knowledgeable, affected, and involved stakeholders should make for a better and more sustainable decision.
- Is mitigation of all environmental impacts required? No, but an agency must fully consider mitigation of identified impacts, and if they have not been incorporated in the decision and implementation of the proposed action, they have to explain why not.
- Are there required scientific and engineering methods to be used in environmental analysis and impact prediction? No, there are no mandated methods. Every project, program, or policy is different and the methods must be developed and chosen for each individual case. However, the selected methods must utilize a systematic, interdisciplinary approach that will ensure the integrated use of the natural and social sciences and the environmental design arts in planning and decision making that may have an impact on the human environment. This promotes the use of objective, quantitative, and accepted methods and discourages approaches that are purely qualitative (if other methods are available), subjective, arbitrary, and have a large "waving of hands" component. The courts have interpreted NEPA and the implementing CEQ regulations as giving the agency proposing the action great latitude in selecting and implementing the environmental analysis methods to be used. However, they have ruled that if another credible source identifies a particular approach or a technical issue to be considered and the agency does not use the suggested methods or address the issue raised, they must explain why not.

- What are the penalties for violating NEPA and implementing regulations? There are no penalties in the classic sense (e.g., fines, sanctions, retribution, incarceration). The federal agencies have been entrusted to carry out NEPA at the same level of commitment and responsibility that they address their respective missions. If coordination with other federal agencies or litigation following an ROD identifies a lapse in NEPA compliance, the agency is required to "fix" the flaw or omission before they can implement the action.

The purpose of NEPA as viewed by the U.S. Forest Service, which prepares the largest number of EISs annually, was also recently evaluated (Stern et al. 2010). The evaluation included a survey of over 3000 U.S. Forest Service employees involved with NEPA and interviews with focus groups. The study concluded that the two primary purposes of NEPA identified by all categories of NEPA practitioners (e.g., advisor, line officer, and implementer) were (1) public disclosure of environmental analyses and (2) public disclosure of decision-making process. The ranking of the other nine potential purposes of NEPA in the U.S. Forest Service diverged based on the respondent's role in NEPA but the purpose "to increase environmental sensitivity of agency actions" was ranked seventh or eighth (out of 11) by every group and "protection from litigation" was ranked 10th or 11th. Similarly when asked which of 18 possible measures of NEPA successes respondents consistently listed, "full disclosure of environmental analyses has taken place" and "well documented rationale for decision is developed" were identified as the two most important success factors. "The final decision minimizes adverse impact" ranked fourteenth and "lack of litigation" was last.

Thus within the federal agency conducting the most environmental analyses, and likely a prevalent view across the board, NEPA is viewed as a process more than a tool to achieve environmental protection or enhancement. Some, particularly line officers, viewed it as a process to "get through" rather than addressing environmental protection or improving decision making (Stern et al. 2010). It is encouraging that the public disclosure aspect of NEPA is so broadly recognized, which sets an important foundation for environmental protection and better decisions. It also emphasizes the importance of NEPA and similar environmental impact analyses as a foundation for integrating environmental permits and other processes that are specifically intended to achieve defined levels of environmental protection (see Chapter 9 on coordinated analysis with other environmental protection measures).

Even after 40 years, NEPA compliance and practice are evolving and maturing with new methods, approaches, and interpretations by court cases. One way to keep abreast of recent actions and changes is through the National Association of Environmental Professionals' web site, www.naep.org, which issues biweekly newsletters and an annual NEPA report. There are also several peer-reviewed journals that report research and comment on NEPA, with the *Journal of Environmental Management* (www.elsevier.com/locate/jenvman) and *Environmental Impact Assessment Review* (www.elsevier.com/locate/eiar) being two of the primary journals.

References

Bass, R. 2010. Commentary. In: *Annual NEPA Report 2009 NAEP* (National Association of Environmental Professionals) *Working Group*, April 2010.

Bass, R.E. and A.I. Herson. 1993. *Mastering NEPA: A Step by Step Approach*. Point Arena, CA: Solano Press Books.

Bergman, J.I. 1999. *Environmental Impact Statements*. Boca Raton: Lewis Publishers.

Eccleston, C. and J.P. Daub. 2012. *Preparing NEPA Environmental Assessments: A Users Guide to Best Professional Practices*. Boca Raton: Taylor & Francis.

Eccleston, C. 2008. *NEPA and Environmental Planning: Tools, Techniques, and Approaches for Practitioners*. Boca Raton: Taylor & Francis.

Hildebrand, S.G. and J.B. Cannon. 1993. *Environmental Analysis: The NEPA Experience*. Boca Raton: Lewis Publishers.

Schmidt, O.L. 1993. The statement of underlying need determines the range of alternatives in an environmental document. In: *Environmental Analysis: The NEPA Experience*, ed. Hildebrand, S.G. and J.B. Cannon. 42–66. Boca Raton: Lewis Publishers.

Stern, M.J., S.A. Predomore, M.J. Motirmer, and D.N. Seesholtz. 2010. The meaning of the National Environmental Policy Act within the U.S. Forest Service. *Journal of Environmental Management* 90: 1371–1379.

U.S. Army Corps of Engineers. 2005. *Baltimore District Washington Aqueduct. Draft Environmental Impact Statement for a Proposed Water Treatment Residuals Management Process for the Washington Aqueduct*. Washington, DC: U.S. Army.

U.S. Coast Guard. 2008. *Final Environmental Impact Statement: U.S. Coast Guard Rulemaking for Dry Cargo Residue Discharges in the Great Lakes*. DOT Document Number: USCG-2004-19621. Washington, DC: Commandant U.S. Coast Guard Headquarters.

4

Overview and Initiating the Environmental Impact Analysis and Assessment

The preceding two chapters dealt with the regulatory, institutional, administrative, and procedural aspects of environmental impact analysis and assessment with a particular focus on NEPA, the first government-legislated environmental analysis program. These two chapters laid the foundation for environmental analysis, documenting the history, need, and requirements. The discussion in Chapters 2 and 3 conveyed information on why an analysis is necessary, the objectives of an analysis, and what can be accomplished through environmental impact analysis, but there was little discussion on how to actually conduct the analysis or use the results to address the overall objectives and goals.

This chapter and the next address the technical approach to environmental analysis with a focus on how to conduct the analysis and then incorporate the results to create a better and more environmentally sustainable project, plan, or policy. Chapters 4 and 5 address the technical approach not in the sense of scientific/engineering/planning disciplines, such as air-quality modeling, toxicity testing, wildlife population studies, or comparable techniques in the social sciences. The approach described in these chapters is more focused on how to use the results of such truly technical disciplines and methods to predict impacts and incorporate the results.

The information presented in this chapter and the next is somewhat analogous to a description of the scientific method. The scientific method informs the research scientist of a prescribed set of logical and rational steps to better understand phenomena and explore the causes and implications of the phenomena; it does not inform on how to conduct specific scientific tests or experiments. For example, James D. Watson and Francis Crick used the scientific method when they uncovered the structure of DNA in 1953. They followed the prescribed procedures they knew intimately from a lifetime of scientific research and they made observations, took measurements, formulated a hypothesis, tested the hypothesis, and then modified the hypothesis and retested until it explained the phenomena. The scientific method taught them to follow these steps, but it told them absolutely nothing about physical chemistry, biochemistry, genetics, or laboratory techniques that were critical to making the discovery. It simply and directly helped them use the results of sophisticated science to understand and explain the phenomena. The intent of this and the next chapter is to convey a similar understanding of the environmental impact analysis process to environmental practitioners. It would be a bit presumptuous to compare environmental analysis practitioners to Watson and Crick. But we can learn from their use of the scientific method, and this chapter and Chapter 5 present methods for environment impact analysis and assessment analogous to the scientific method for understanding and explaining phenomena. Similar to the scientific method, the focus is on the process and not the specific tests and disciplines that are so critical to the analyses.

As discussed in Chapter 2, environmental analysis is an interdisciplinary endeavor requiring numerous and often sophisticated scientific, planning, and engineering expertise. These two chapters, and in fact this book, make no attempt to teach such expertise to the environmental practitioner. Experience shows that a single book or academic course cannot make an engineer out of a planner or an anthropologist out of a hydrologist, nor would it be desirable to do so. Every practitioner comes to the environmental analysis team with her or his own expertise, and the goal of this book is to present methods and approaches to use and incorporate each discipline into an integrated, productive, and efficient approach to environmental impact analysis.

This chapter presents the environmental impact analysis process used in planning and establishing the groundwork for the more technical aspects of the analysis. Specifically, determining the need for environmental analysis, developing the scope of the analysis, the alternative consideration approach and the public outreach program are described in this chapter following an overview of impact analysis approaches. The subsequent Chapter 5 addresses approaches to predicting; evaluating; and mitigating environmental impacts for proposed projects, plans, and policies. The major elements of the environmental impact analysis process, presented under separate headings in Chapter 5 are existing or baseline condition, impact prediction, and mitigation.

4.1 Environmental Impact Analysis Approach

Environmental impact analysis has been defined as identifying and understanding the environmental consequences of human activities before they occur (Morgan 2001). This certainly is accurate and is the core of the analysis, but it can be much more. It can be used to formulate "better" projects, plans, and policies and achieve environmental protection and sustainability. An effective environmental impact analysis can also be used to develop and enhance stakeholder support for a proposed action. But before these benefits can be realized, an effective understanding of the consequences of the activities must be developed, which is the "meat and potatoes" of environmental impact analysis.

Another analogy that is sometimes helpful in understanding environmental impact evaluation is economic analysis. If an entrepreneur is contemplating starting a retail, manufacturing, software, or some other business venture, to be successful she or he must first conduct some form of economic analysis. This entails understanding what is already in place related to the proposed business, such as the competition, market, and availability of resources (e.g., real estate, raw materials, and labor). A next step in the economic analysis might be to develop the costs associated with producing the envisioned goods or services and based on the understanding of what is already in place, an expected return from the proposed business can be predicted. Foolish and slated for failure is the entity that starts a business without understanding the financial consequences of their investment before they launch, and as described below, an environmental impact analysis has many similarities. First there is a need to know what is currently present, then what is proposed to be implemented, what are the implications of the implementation, and finally how will existing conditions be altered by the direct action and its implications? But as opposed to a business endeavor, in the case of environmental analysis it is generally not the entity taking action that experiences the consequences, but the environment and society.

Understanding the consequences and their environmental impacts is all about predicting change. A consequence or an impact is a change from the existing situation or the future conditions that will develop if not for the proposed action under analysis. Similar to the economic analysis analogy mentioned earlier, the first step is to develop a preliminary understanding of what currently exists with respect to each environmental resource potentially affected, which is commonly termed the affected environment, existing conditions, or baseline conditions. Then each alternative for accomplishing the purpose and need of the project, plan, or policy can be projected on the existing conditions, and the intersection of these two is the area of impact (Figure 4.1). This intersection should be the area of focus both for investigation and analysis. It is also the critical area used in the comparison of alternatives and development of mitigation measures.

This simplistic depiction of impact analysis is obviously not a detailed guide to predicting impacts, but it should be kept in mind while planning and conducting the analysis and used to stay on track. If the evaluation approaches the fringes of the "credit card" diagram (Figure 4.1) and details of the proposed action are being developed which have no interaction with the existing conditions of relevant environmental resources, the analysis is off course. Similarly, if biological investigations are proposed that detail breeding activities of a species, which migrates far from the project area to breed, the goal of environmental impact prediction will not be advanced. In both cases not only do the investigations not advance the process, they consume valuable resources and time and give fodder to critics who view environmental impact analysis as a useless process which delays important actions.

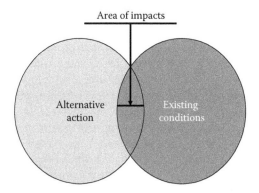

FIGURE 4.1
Simplicity of an environmental impact prediction.

This concept is expanded in the following presentations of the components of environmental impact analysis.

4.2 Need for Environmental Impact Analysis

A first step in the environmental analysis is to determine the need for and general extent of the analysis. Some authors (Morgan 2001) have termed this process "screening," but in this book the term screening is applied to an initial selection of alternatives (Section 4.5.3) and is not used here to avoid confusion. The need for analysis can be determined based on several factors including magnitude and type of proposed action, environmental setting, environmental resource potentially at risk, and preliminary assessment. Each of these is discussed subsequently.

4.2.1 Magnitude and Type of the Proposed Action

The size and breadth of the proposed action, be it a project, plan, or policy, are the most obvious and common factors that determine the need for environmental analysis. Many organizations have established threshold criteria defining the need for environmental analysis, and in some cases there are additional criteria to define the level of analysis required. The presence of significant impact used to determine the need for a full environmental impact statement (EIS) as opposed to an environmental assessment (EA) and the listing of actions for categorical exclusion (CATEX) (as discussed in Chapter 3) are examples of threshold criteria. There are other examples in state and international programs as described in Chapter 8. Also individual organizations, such as the World Bank and many U.S. federal agencies have established lists of actions commonly implemented within their organizations requiring various levels of environmental analysis and assessment. If the proposed action requires a permit related to an environmental resource, there is a strong likelihood that at least some level of environmental impact analysis is necessary. Thus, a first step in determining the need for environmental analysis is to identify environmental resources (e.g., historic properties, regional transportation, and water supply) potentially at risk and the regulatory mandates and procedures related to such resources. If a permit based on environmental criteria could be necessary for the action, then some level of environmental analysis is beneficial if not required.

If the organization or entity proposing the action does not have threshold criteria and there are no apparent regulatory mandates, the practitioner must make a case-by-case determination of the need for analysis based on the magnitude and intensity of the proposed action. If the area affected is large, the construction duration is long, the number of people/properties involved is large, it is highly visible, or there is a high cost of the action, there is likely a need for environmental analysis. The relative magnitude of "large, long, great, highly, and high" can be determined by comparison to other similar actions by the proposing or similar entities. If other historic actions of similar or less magnitude have conducted environmental analysis, it will likely be required and beneficial to the success of the action under consideration.

There are certain types of actions, regardless of size or intensity, where environmental impact analysis is typically warranted. These types of actions have historically generated substantial public interest, opposition, and controversy and in some cases avoidable environmental impacts. Examples of such actions are:

- Changes in land use
- New or substantially modified transportation resources (e.g., roads or changes in mass transit systems)
- Power generation
- Water supply
- Air emissions and wastewater discharges

If the need or level of environmental analysis is uncertain based on the type and size of the proposed action there are simple steps that can be taken to make the determination. One is to study the list of recent EISs reviewed by EPA under the Clean Air Act, Section 309, regarding level of environmental analysis (see Section 2.4 of this book) if the action is within the U.S. or similar inventories in another jurisdiction. If there have been numerous similar projects, plans, or policies requiring environmental analysis, chances are that the action under consideration would benefit from a full EIS or comparable analysis. Another approach is to inquire within government resource or regulatory agencies and environmental advocacy groups to determine whether actions similar to that proposed are generally subject to environmental analysis, but keep in mind there will be inherent biases in these groups.

4.2.2 Environmental Setting

Substantial environmental impacts, and thus the need for environmental analysis, can occur in certain sensitive settings regardless of the type or magnitude of the proposed action. These are typically related to ecological resources and have been categorized as valued ecosystem components (VECs) by several authors (Noble 2006 and Morgan 2001). Publicly or privately (e.g., Nature Conservancy Trustees, wildlife refuges, and land banks) held conservation lands or habitats for special status species are strong candidates as VECs, and thus proximity to such resources can necessitate environmental analysis.

The Commonwealth of Massachusetts and some other states have comprehensive programs to identify VECs. Areas of Critical Environmental Concern (ACECs) are geographic resources in Massachusetts that receive special recognition because of the quality, uniqueness, and significance of their natural and cultural value. ACECs represent a broad range of resources and include: sensitive drinking water sources (e.g., Canoe River Aquifer and associated areas); unique natural features (e.g., Sandy Neck Barrier Beach System); and wildlife habitats and sensitive natural areas threatened by urbanization (e.g., Rumney Marsh). These areas are identified and nominated at the community level, and the enabling legislation requires state reviewing agencies to apply much closer scrutiny for proposed actions within the general vicinity of an ACEC. The purpose of the intense scrutiny is to consider uncertainty and other aspects of the proposed actions that might threaten these areas of special value and vulnerability. This closer scrutiny is interpreted as the need for a full Environmental Impact Report (equivalent to a full EIS under NEPA, see Chapter 8 for a discussion of the Massachusetts environmental analysis and assessment program).

Similar to projects potentially affecting sensitive natural environmental resources, proposed actions in proximity to sensitive built land uses can necessitate environmental analysis. Land uses such as hospitals, transportation hubs, first responder facilities (e.g., fire stations and ambulance dispatch facilities), economically disadvantaged neighborhoods, hazardous waste sites, schools, and housing for vulnerable populations can be sensitive to adverse impacts from even seemingly minor actions. Thus, a combination of minor actions and sensitive environmental resources can prompt the need for environmental impact analysis.

4.2.3 Preliminary Assessment

If the magnitude or location of a proposed action does not definitely indicate the need or lack of need for environmental impact analysis, a preliminary assessment can be a last resort. This mini environmental impact analysis can frequently be accomplished by site reconnaissance; discussions with natural resource, local government, and environmental regulatory entities; and/or some preliminary calculations regarding possible impact-producing activities (e.g., actions generating noise, traffic, or pollutants). It is important to make these calculations conservative to represent a realistic worse case and also to compensate for uncertainty. If there are environmental impact standards or criteria (e.g., air-quality standards) to compare with the calculated results, the decision can be simplified and validated. If the calculations representing a realistic worst case show that expected levels of impact or environmental characteristics (e.g., water-quality concentrations) are well below applicable standards (e.g., half of allowable noise levels or water-quality standards), then a full environmental impact analysis may not be necessary. If on the other hand, predicted levels exceed or even approach the values considered acceptable, more analysis is probably necessary. This progressive approach to determine the need for additional analysis was developed, refined, and is now standard practice in Ecological Risk Assessment (see Chapter 7).

Caution must be exerted when applying preliminary assessment to determine the need for full environmental impact analysis and all areas of potential impact must be considered. In the early 2000s when the U.S. Coast Guard (USCG) was under the umbrella of the newly formed Department of Homeland Security, they were required to arm even their smaller vessels. The USCG was concerned that firing from small vessels (as small as 7 meters long) was substantially different from firing from much more stable platforms offshore or larger ships which was the more common USCG practice. Thus they felt extensive training and practice were necessary to maintain safety and effectiveness for the program of fixed fire arms on small vessels.

As the USCG prepared to institute the training and practice procedures in the sensitive environment of the Great Lakes, there was concern about "bullets in the water" resulting in contamination of the aquatic ecosystem, particularly the sediments from lead and other potentially toxic materials in the ammunition. In order to determine whether a full environmental impact analysis was necessary, the USCG commissioned a preliminary assessment of what they considered to be the primary, if not the only, environmental concern; the impact of toxic metals originating from the bullets on aquatic organisms, particularly those associated with sediment. In order to make the determination, the area of bullet deposition from the live firing was conservatively estimated by tracking the line of fire, the course of the boat, and firing accuracy (Figure 4.2). For every input parameter, the value which would result in the smallest deposition area (with the smallest area for a given mass of material, the concentration, and thus the toxicity of the material in the bullets would be the greatest) was used. Similarly, the greatest realistic mass of bullets was used and the sediment concentration calculated. The concentration was then compared with criteria considered to pose no impact to aquatic organisms, and the values were found to be well below concentrations of concern (Figure 4.3). The calculated risk quotients (i.e., the calculated concentration divided by the safe concentration so that a quotient of 1.0 represents a calculated concentration equal to the safe level and less than 1.0 is a concentration below the safe level) were generally 0.5 or less (Figure 4.3).

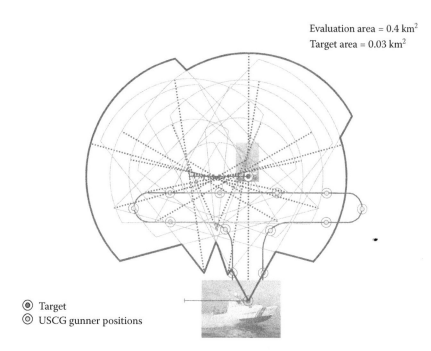

FIGURE 4.2
Area potentially impacted by USCG live fire training.

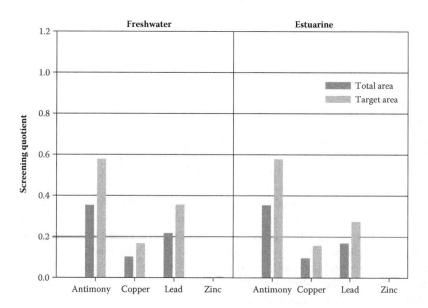

FIGURE 4.3
Comparison of safe levels to calculated sediment concentration resulting from USCG live fire training.

Based on these results and given the conservative nature of the calculations the USCG felt that no additional environmental impact analysis was required. However, they misread the concern expressed by some critical stakeholders when they discounted the temporary impact of restricting the target area to recreational boating during the short and infrequent live fire training activities. These stakeholders put pressure on the USCG through the Great Lakes' Congressional delegation, and the Coast Guard acknowledged the oversight and agreed that a full and comprehensive environmental impact analysis was warranted. Ultimately they abandoned the program in large part due to the concerns, impacts, and opposition uncovered through public input. Thus although the planning and execution of the preliminary assessment were flawed, it did serve a purpose in that the USCG identified the issues and could make a decision incorporating environmental concerns (although the process was neither painless or efficient).

4.3 Scoping

Scoping is the procedure used to determine what should be included in an environmental impact analysis and the resulting documentation of the analysis process. It is the road map and directions for the process, but unlike a literal road map, the environmental practitioner has to develop the map before the analysis trip can begin. With a well-planned and executed scoping process, the practitioners and the stakeholders alike will not only know where they are going, but how to get there, and even when and if they have arrived. As discussed above, a primary objective of environmental analysis is to identify and understand the environmental consequence of human activities before they occur. The intent of scoping is to support this objective by pointing out early in the analysis process the areas where the consequences could occur and how to understand them. Properly executed scoping can minimize the pitfalls associated with discovering late in the environmental impact analysis process, when the time is gone and the budget has been spent, what the issues are, where there are possible impacts, and where it is clear there are no adverse impacts or other concerns. In a more abstract sense, scoping can be considered a process of defining and constraining the area of impact as shown in Figure 4.1 (the intersection of existing conditions and alternatives).

The environment is a big place and the possible methods of investigating the environment can approach infinity, so the potential content of an environmental impact analysis is virtually unlimited. Scoping has evolved into a process, which when used appropriately, can not only streamline the analysis but also ensure that critical factors are included and emphasized. These are sometimes competing objectives because the project proponent wants to make the environmental analysis process as efficient as possible and thus not address, particularly in great detail, the nonissues. But on the other hand they would not want to ignore an area of impact during the preparation of the draft environmental impact analysis only to have to come back at a later and inopportune time. It is the environmental analysis team's and each practitioner's challenge to understand the stakeholders concerns, evaluate the potential for impact, and digest the scoping input. They then must apply what they have learned, their wisdom, and experience to first identify each issue, concern, and each environmental resource at risk. Next they must locate it on the continuum of exclusion to detailed analysis to determine what should be addressed in the environmental analysis and to what level of detail.

If approached with the wrong attitude and/or ulterior motives, scoping can be counterproductive by delaying, convoluting, and unnecessarily complicating the environmental analysis process (Snell and Cowell 2006). Problems can arise from either side of the table: the proponent of the proposed action or the potentially affected stakeholders. Fox and Murphy (2012) identify several breaches in sincerity that can inhibit successful scoping, particularly social scoping (see Section 4.3.2 for a discussion of social scoping):

- Primarily by the public stakeholders including
  - Lying regarding existing conditions or impacts from similar activities
  - Manipulation of alternatives ranking through public input to identify the preferred alternative
- Primarily by the project, plan, and policy proponent including
  - Breaking promises
  - Inviting public participation with little or no intention of incorporating the input received
  - Soliciting input that is not meaningful to create the false impression of cooperation

If the entity proposing the action is not committed to the scoping process and unwilling to legitimately consider stakeholder input in exchange for their commitment to participation, the stakeholders' retribution can be detrimental (Mulvihill 2003). Similarly if the potentially affected stakeholders use the scoping process to delay or add layer upon layer and issue upon issue to the analysis process just because of their opposition, the effort can be counterproductive. Because of the fear of these potential pitfalls, there is frequently reluctance from both parties to engage in more than the minimum required scoping.

However, experience has shown that in many cases if there are counterproductive forces at play, such as extreme and self-serving opposition or commitment by project proponents to move forward regardless of the facts, it will surface at some point in the process. Ignoring them during scoping does not necessarily make them go away but more likely just delays their exposure. If the disruption occurs during the comment period on the draft analysis or worse at the litigation phase, the delays and complications can be equal to or greater than the delays and complications uncovered during the scoping phase. At least if the issues and biases become apparent during scoping, the extreme surprise ending can be avoided. Also, when the biases, opposition, and intransigence are identified early in the process, there is at least the possibility of addressing these obstacles. If the obstacles of pacifying the extreme and self-centered opposition or changing the culture of the project proponent cannot be overcome, at least the issues can be exposed to the more rational stakeholders and their support used to forge a compromise.

Another potential trap of scoping is to substitute popular issues for real and potentially serious detriment to sensitive or nonrenewable environmental resources. The public input and concerns must not overshadow the environmental analysis practitioners need for a rational, objective, technically based, and rigorous appraisal of the likely impacts (Morgan 2001). Similarly, a survey and evaluation of USFS environmental practitioners and managers' concerns with environmental evaluation identified attention to public concerns as consuming time and resources that could be much better put to use in objective impact evaluation (Stern et al. 2010). The study even cited practitioners who felt all too often that addressing the public's subjective concerns became the primary focus of the environmental evaluation to the detriment of understanding, incorporating, and mitigating valid environmental impacts.

Successful scoping is a transparent and public process, thus there is much overlap between scoping and public outreach. There is more discussion of the relationship of scoping to public interaction in the following section (4.4), but the subsections below describing the elements of scoping include some of the interaction between scoping and public outreach.

4.3.1 Scoping Topics

The information generated as part of scoping can advance environmental analysis objectives including: streamlining; maximizing efficiency; focusing; and early identification of critical issues and concerns. Some areas of information that can advance these objectives include:

- Potential areas of impacts and concern, including relative importance and level of significance
- Environmental setting, including: relevant sources of information, VECs, and sensitive receptors
- Alternatives for consideration
- Methods available for environmental impact analyses of various components (e.g., impact prediction, baseline investigation, and comparison of alternatives)
- Environmental approvals that may be required

Each of these topics is addressed in the subsequent sections.

4.3.1.1 Potential Areas of Impacts and Concern

The areas of impact and concern are the classic and primary areas of information that should be identified during a scoping process. It is incumbent on the environmental analysis practitioner to identify the stakeholders for inclusion in the identification of impacts and concerns (see following sections for discussion of targeted audiences). Similarly the environmental analysis team must develop and disseminate sufficient information to the stakeholders so that they are able to generate meaningful and complete input for each issue and area of concern. Providing sufficient information is not as simple as it might seem. Overdevelopment of information on the proposed action and alternatives for dissemination during scoping can delay the process and result in significant backtracking if issues, concerns, and alternatives are identified during scoping that were not considered when the initial extensive and detailed information was prepared. Presenting too much information during scoping can also create the perceived and sometimes valid impression that the entity proposing the action has already made a decision and proceeded with a detailed development of the proposed project, plan, or policy without public input and thus subverted the critical full disclosure and transparency objectives of environmental analysis.

At the other extreme is too little information: there must be sufficient information presented for stakeholder to first determine whether they are in fact stakeholders and if so, whether they can provide meaningful, timely, and useful input. The information provided during the scoping process should also have enough detail and specifics to generate ideas and discussion among stakeholders. At a minimum the following should be developed by the environmental impact analysis team and made available during the scoping process (see also Section 3.1.1, "Defining the Action"):

- Clear and concise purpose and need statement (see Chapter 3, Section 3.5)
- Possible geographic locations of the project, area covered by plans/policies, and area potentially affected
- A brief project history
- Size range of any structures or infrastructure components that could be part of the proposed action (e.g., up to 10 km of new highway, a building up to 5000 m^2, power generation up to 1200 mW)
- Any other metrics that could be helpful in understanding the magnitude of the proposed action (e.g., quantity of water involved; number of jobs created; anticipated traffic volume)
- Any known critical existing conditions such as environmentally sensitive areas or resources
- A listing and description of any impacts, issues, or concerns that have already been identified by the environmental analysis team or through the project history
- Description of any alternatives already identified

- A listing and explanation as to why any alternatives have already been eliminated
- A summary and reference to any relevant previously prepared documents such as planning studies, conceptual engineering, needs analyses, etc.

Information on the proposed action and alternatives can be presented in a number of venues, including web pages, social media, meetings, newsletters, and presentations to interest or advocacy groups. All of these are useful tools for scoping, but scoping is more than just a set of tools. It is a process initiated at the beginning of an environmental analysis and integrated into the entire environmental impact analysis. Following presentation of the relevant information, the stakeholders should be encouraged to express their concerns and identify the issues they feel should be included in the environmental impact analysis and used as input to decision making. Once the topics and their relative importance are identified, the environmental impact analysis can be structured to address the concerns and predict the magnitude and intensity of the potential impacts.

Successful determination of the potential areas of impact and areas of concern during scoping can be measured by comments received on the draft environmental impact analysis document: the more comments and the more surprises, the less successful the scoping process. The objective should be to identify all areas of concern during scoping and then address them to the appropriate level of detail in the draft analysis document. If comments on the draft are voluminous and substantial, one major reason frequently is a scoping process that was less than successful in anticipating and addressing the issues. A common flaw causing an unsuccessful scoping process is not identifying and including the stakeholders with the most at stake early in the process. Another oversight that can limit the success of the scoping process is the failure to provide sufficient information so stakeholders can recognize their issues and concerns during initial scoping.

The other primary cause of voluminous and substantial comments on the draft is strong opposition to the proposed action on environmental or other grounds. Even a well-planned and well-executed scoping process cannot avert such opposition and the resulting comments. However, it can allow the environmental analysis practitioner to anticipate the comments and have them "preaddressed" in the document so that the response to the comments can simply be a reference to the appropriate section of the environmental analysis where the issue was thoroughly addressed and evaluated. This can take the wind out of the sails of the commenter, expose that they did not read the document, and reveal that the comment is just opposition and they are not interested in the answer, but just creating obstructions.

Identifying issues and concerns during the scoping process can have substantial benefits beyond simply identifying the scope of the analysis. Actively soliciting input from stakeholders early in the process can build a foundation of trust and cooperation as the analysis progresses. An environment of trust and cooperation can produce benefits throughout the analysis as well as supporting funding for the proposed action, and also community support during implementation. If the stakeholders and, even project opponents feel they are part of the process, they are more likely to work toward a solution rather than try to derail it. Scoping can also be useful to control expectations and counteract incorrect perceptions by stakeholders early in the process before they take on a life of their own. Finally, the scoping process can be useful in building the alternative comparison and selection of the proposed action as demonstrated in the example that follows.

A water supply project in south central Connecticut by the Regional Water Authority (RWA) illustrates some of the benefits of a thorough environmental impact analysis scoping process. The RWA and its predecessors have provided potable water to the City of New Haven and the surrounding communities since the 1800s. A key feature in their system is Lake Whitney, named after the Eli Whitney family of cotton gin fame because they built the dam to power manufacturing at the site in the 1800s. The lake was used for water supply from the late 1800s and gradually declined in use until about 1990 when it was finally taken completely out of service because the water treatment plant built well over a hundred years before had reached its useful life.

In the mid 1990s, the RWA began planning the construction of a new state-of-the-art water treatment plant to bring Lake Whitney back to full service as a potable water supply. This created substantial concern among Lake Whitney abutters and other stakeholder groups, such as the watershed association, neighborhood associations, and friends of the park adjacent to the lake and Mill River which supplies Lake Whitney. A major concern was lake level drawdown created by withdrawing up to 5000 m^3 a day for water supply. The stakeholders had become accustomed to the benefits of a lake unaffected by changes in aesthetics caused by the historic withdrawals for water supply. There was the perception by many stakeholders that this drawdown would create adverse effects such as: reduced downstream releases; aesthetics of the lake water view; odors from exposed mud flats resulting from extreme low lake levels; and wildlife habitat effects. There was also the feeling that the RWA should mitigate the perceived impacts by dredging portions of the lake to eliminate unsightly and odoriferous mud flats in compensation for the water level changes associated with withdrawal for drinking water supply.

The RWA committed to a lake management study and associated environmental impact analysis to address these concerns as part of their plans to bring Lake Whitney back online. In preparation for the environmental analysis, the authority instituted a scoping process and actively sought the participation of stakeholders they had been dealing with in some cases for decades as they operated the lake for water supply and protected the tributary watershed. Consistent with the earlier description on scoping, a first step was to identify the stakeholders' issues, concerns, and perceived areas of impact. This was accomplished through a multistep process with the first being a series of meetings to explain the history, purpose and need, and potential lake management alternatives. This was followed by workshops in which stakeholders were encouraged to raise their issues or alternatively submit them in writing, which produced a list of approximately 20 issues and concerns (Table 4.1).

After the stakeholders had an opportunity to review the list of issues/concerns, another workshop was held. The workshop included a discussion of issues, first by the RWA and then the stakeholders who had raised the issues were encouraged to elaborate and ask questions. Toward the end of the workshop, each stakeholder was asked to "rank" the concerns from their perspective with a score of 1 to 100 assigned to each issue. The process was well publicized by the RWA, the media, and the workshop participants, which created a larger list of stakeholders. The additional stakeholders were then encouraged to participate in a survey, similarly scoring the concerns and issues. The mean score from all stakeholders was calculated with the rank and relative importance of all 20 issues/concerns established (last column of Table 4.1). The environmental analysis then proceeded by addressing each issue with the level of importance indicated by the stakeholders, representing an important factor in establishing the level and intensity of investigation and analysis for each issue and concern. Thus more effort was expended on the issues and concerns that ranked the highest.

TABLE 4.1

Issues, Concerns, and Their Relative Importance (with 100 being the most important) for Lake Whitney Management

Number	Issue or Concern	Relative Importance
1	Bird Habitat	76.7
2	Downstream Environment	75.2
3	Wildlife Habitat	74.5
4	Water-Quality Improvement	73.1
5	Offsite Wetlands	71.9
6	Mosquitoes	68.6
7	Perceived Odors	66.4
8	Potable Water Quantity and Quality	65.1
9	Duration of Objectionable Odors	64.6
10	Open Water View	64.3
11	Diversity of View	63.2
12	Sediment Quality	60.6
13	Area of Objectionable Odors	60.4
14	Human Disturbance	57.0
15	Odor During Construction	56.3
16	Water-Quality Degradation	44.1
17	Construction Nuisance	43.1
18	Construction Duration	39.4
19	Recreational Use	39.2
20	Costs	39.0

In addition to identifying the issues and level of investigation for the analysis, the scoping had other important benefits for the process. First it established a strong working relationship between the project proponent, RWA, and the stakeholders, some of whom were initially in strong opposition. Working through the issues in an open forum and mutually addressing some hard questions created a sense of trust among all parties: between RWA and individual stakeholders and among various stakeholder groups with competing and conflicting interests. It also created an atmosphere of team effort to address mutual problems.

The scoping process also created a forum to dispel misconceptions, such as the initial desire to dredge the lake. During scoping, the dredging process was explained, including the adverse and disruptive attributes of the process and the stakeholders immediately adjacent to the dredging, dredged material processing sites, and dredged material transport routes were then allowed to publicly express their concerns, and the attractiveness of dredging as a lake management alternative was substantially diminished. It also allowed RWA technical experts to explain in a nonadversarial forum that dredging was just a temporary solution and the dredged areas would return to current conditions within a short time due to watershed contributions of sand and eroded silt from untreated storm water. The final result was that dredging was not even an alternative carried forward for detailed evaluation in the environmental analysis. The time and money saved by not addressing the dredging alternative in detail could then be used productively to develop alternatives, evaluate impact mitigation, and address stakeholder issues.

The integration of the scoping input into the alternative comparison and selection was another benefit of the scoping process. As part of the environmental analysis, the impact of each alternative related to each of the 20 issues/concerns raised during scoping was determined and assigned a relative level of significance. When the level of significance was combined with the relative importance defined by the stakeholder survey and then summed for each alternative in a decision algorithm, each alternative had an impact score. The alternatives could then be ranked based on the impact score, which was the major factor in the selection of an alternative as the proposed action. The result was dredging was least preferred, primarily because it ranked very poorly on the top issues such as bird habitats, wildlife habitats, and impacts during construction. In contrast, the alternative of minimizing lake sedimentation by controlling and treating storm water runoff in the watershed ranked highly because of the advantages to the top ranked issues and concerns. Since the stakeholders had been the primary participants in determining the ranking, they fully understood and supported the selection process and the proposed action. As a result of the study, storm water management facilities were constructed over time, which addressed the major concerns of the stakeholders, such as:

- Creating wildlife habitats because they are natural treatment facilities utilizing wetland plants

- Retained solids which controlled expansion of unsightly and odoriferous mud flats in the lake
- Improved water quality by removing pollutants before they reached the lake
- Retained more storm water in the watershed, and thus maintained more consistent lake discharge volume and improved downstream river flow and aquatic habitat

4.3.1.2 Environmental Setting

The environmental analysis is played on the field or stage of the environmental setting, as represented by the right side of the simplified environmental impact evaluation model (Figure 4.1). The more that is known about the conditions that host the proposed action, the more complete, accurate, and efficient is the impact prediction and the entire environmental analysis process. During scoping, the stakeholders closely associated with the location can provide important information related to both the existing and historic environmental setting not always apparent to the more technically focused environmental analysis team. The local stakeholders are also perhaps the ones most likely to be aware of previous environmental analyses or other investigations that have been conducted in the area, which can be important sources of environmental resource description and data (see Section 5.2.2). Both the general public and more technical stakeholders play a role in this aspect of scoping, with the public more focused on local conditions and sometimes more anecdotal information. Of course, the information provided by stakeholders can be biased and only selectively revealed to support specific interests. However, if all sides are represented in the scoping process the bias can be at least partially balanced. In any case scoping could be a starting place for information gathering and supply material to be evaluated and verified by the environmental analysis team.

Stakeholder input can also be useful in identifying and describing VECs and other sensitive resources. As discussed above, these include lands set aside as natural areas, habitats for special status (e.g., rare and endangered) species, culturally significant resources (e.g., historic properties and locations), economically disadvantaged neighborhoods, and institutional land uses vulnerable to disturbance. Once such resources are identified during scoping and confirmed as fragile or strongly susceptible to change, they generally warrant additional and special consideration in the environmental impact assessment investigations.

Soils are a typical example of a natural resource where technical stakeholder (see Section 4.3.2 for a discussion of technical vs. social scoping) input during scoping can be important and foster efficiency. If soils within the project area are known to be stable, not contaminated, and erosion resistant, they most likely will require little attention during the environmental impact analysis process. In contrast, if they are known to be highly erodible, with potential to create silt and sediment problems in downgradient water bodies, they might require specific and detailed analysis. Also if scoping has identified the potential for soil contamination, the need for detailed and thorough environmental analysis of soils may be required. Thus a goal of scoping is to pick the brains of stakeholders to learn which resources might be at risk and which should receive less attention in the environmental analysis.

The existing condition of a resource can be evaluated at a level of detail commensurate with the magnitude of the proposed action and susceptibility of the resource to impact, and scoping can provide valuable input in making this determination. In some cases, qualitative evaluations may be adequate and in other cases review of existing literature and other relevant environmental impact analyses from the same area or type of action might be necessary to evaluate a resource's vulnerability. When there is substantial overlap in the existing conditions and the proposed action (the impact area in the "credit card" depiction of environmental impact, Figure 4.1) there can be high impact potential to the resource and detailed vulnerability analysis, as that described by Kvaerner et al. (2006), may be appropriate. This will frequently involve substantial original data collection and advanced methods of impact prediction (see Chapter 5).

4.3.1.3 Alternatives for Consideration

As discussed throughout this book, alternatives are a critical component of any environmental impact analysis because they provide a basis for decisions and are critical in evaluating environmental trade-offs. For example, one alternative highway alignment might avoid all wetlands and critical wildlife resources but infringe on the setting of an important cultural/ historic resource while another alignment would have the opposite effects. Thus the consideration of alternatives and identification of issues and concerns during scoping is an interactive process. Once issues are brought out by stakeholders, alternatives should be developed to address the issues and similarly once alternatives are developed the impacts they could potentially generate are identified for inclusion in the environmental analysis.

It is the role of the environmental impact analysis team to begin the alternatives development process. In virtually every case the team will have considered at least conceptual approaches to addressing the purpose and need and these must be presented to the other stakeholders for a number of reasons. First, it is an early and powerful way of opening a transparent and cooperative dialogue with the stakeholders. An understanding of potential alternatives will also stimulate identification of issues, concerns, and potential impacts among the stakeholders and it is a productive approach to initiating the conversation and generating stakeholder identification of alternatives. Finally, there is little that can destroy stakeholder trust, support, and cooperative participation more than learning late in the process that the project proponent and environmental analysis team had alternatives in mind, particularly a preferred alternative, and they had not been revealed at the outset of the scoping process.

Frequently stakeholders have lived with the existing situation for some time and have brainstormed on possible solutions. These alternative solutions can either be related to possible sites that are frequently identified by local residents or organizations, or technical, planning, or policy approaches to addressing the purpose and need that can originate from more technically oriented stakeholders. In either case the environmental analysis practitioner should take advantage of a broader range of potential alternatives identified through local knowledge and technical expertise. Frequently, alternatives identified in such a manner require refinement, but they can be a starting point. Also it can initiate a reexamination and brainstorming by the environmental analysis team, which can result in better or refined alternatives.

Identification of alternatives as part of scoping for the Washington Aqueduct Water Treatment EIS (see Chapter 10 for background on the project) broke a logjam in the project and ultimately became part of the proposed action. The residents of the neighborhoods adjacent to the Dalecarlia water treatment plant were concerned about the noise, aesthetic quality, traffic, and other impacts resulting from installation of water treatment residual (i.e., the solids removed from the water during treatment, which creates a "sludge") management facilities on the site. They were also very familiar with the characteristics of the site (i.e., environmental setting or existing condition, right side of Figure 4.1). After a less than totally successful initial scoping effort, the local stakeholders were able to work with the project team to identify truck access points and routes that minimized impacts to existing traffic patterns. Also through their familiarity with the site and neighborhood characteristics, again working cooperatively with the project team, they identified a location for the facilities that cut into the side of a hill such that it was only minimally visible to the neighborhood and the hill significantly mitigated noise impacts. The project engineers and environmental analysis practitioners working alone, before stakeholder scoping input, had not identified these alternatives which ultimately became components of the proposed action.

4.3.1.4 Methods for Environmental Impact Analyses

A primary role of the environmental practitioners on the impact assessment team is to consider the concerns and potential impacts raised during scoping and select methods to address these issues. The interdisciplinary team should have the technical expertise to understand these issues and an appreciation of methods and approaches available that have worked for previous analyses to satisfactorily identify impacts. The team can then select the methods to use in the analysis, adapting to project-specific conditions as warranted. They can then be presented to stakeholders following the initial scoping where the issues were identified and supplemented by stakeholder input.

Environmental impact analysis is not always an exact science, and almost always there is more than one way to predict an impact. The environmental analysis team will most often select the methods familiar to them and within the allocated funding for the analysis. Frequently, they can convince the stakeholders that the selected methods are appropriate, but the presentation of the selected methods should absolutely be made before the investigation is concluded, or even started. If the team presents the methods and approaches used in impact prediction or any other aspect of the analysis after the results are in, particularly if the results are contrary to some stakeholders' previously held views and interests, the potential for stakeholder constructive support is diminished.

A preferable and more productive approach is to present the proposed methods before the investigation is initiated. This has multiple benefits, including giving stakeholders the opportunity to understand and inquire about the methods before they know and can take exception to the results. Also once they have accepted the methods, the stakeholders are much more likely to accept the results of the investigations and thus the conclusions of the environmental impact analysis even if they are not in line with their specific interests or preconceived conclusions. Advance presentation of methods is another opportunity to foster stakeholder trust and participation that is positive and active.

Probably the greatest benefit of opening the discussion of impact analysis methods to technical stakeholders is the identification of methods frequently used and accepted in the local setting. Often the technical stakeholders are just as familiar with the technical aspects of the project (e.g., wastewater treatment, power generation, bridge construction) as the project team. They may also be even more familiar with the local setting and thus can suggest alternative methods of impact analysis or description of existing conditions that are just as technically sound as those selected by the environmental analysis team but better adapted to the local situation.

Methods identified by stakeholders have the added advantage of generating even more stakeholder support and cooperation. If a stakeholder's method is used, the results will obviously have that stakeholder's support, but also typically the results will be viewed by other stakeholders as unbiased and have their support also. It is not an uncommon practice to use multiple methods for impact prediction and analysis (see Chapter 5, particularly the USCG dry cargo residue EIS example) in a Weight-of-Evidence approach (Maughan 1993 and Menzie et al. 1996) to address uncertainty in predictions, and the approach can be used to accommodate multiple methods suggested by stakeholders. The multimethod approach can require more funds, but often the input data for several methods is the same so the added cost can be minimal. Also the inclusion of key stakeholder identified methods can substantially increase the acceptance of the results so the environmental analysis team must weigh the cost of adding the analysis at the outset against the benefits and likelihood of having to conduct the analysis later in response to comments. In any event, it is incumbent on the environmental analysis team to consider any methods suggested during scoping and if the suggested method is not used in the detailed impact analysis, the scoping statement (see Section 4.3.3) should explain why it is not used as part of the full analysis.

4.3.1.5 Environmental Approvals

The environmental impact analysis is just the first step in comprehensive environmental protection and enhancement. As discussed in detail in Chapter 9, following the environmental impact analysis process and selection of a proposed action, environmental permits, and other approvals are required and the environmental analysis must lay the groundwork for the approvals. The first step in laying this groundwork is to identify the environmental approvals and permits required for each alternative under consideration. This can and should be done first by the environmental analysis team through research of regulations and past similar actions. But actively including the permitting and approval agencies in the scoping process is not only an efficient way to confirm the determination of approvals needed, it can ensure that the particular requirements of the approval and the approving agencies are incorporated into the environmental analysis. As with other aspects of scoping, it can help build a positive working relationship with the entities that must ultimately support and approve the proposed action.

If the appropriate regulatory agencies are not included in the scoping process and the analysis does not include consideration of environmental approvals, there are two possible repercussions, both of which can adversely influence the efficiency and effectiveness of the environmental impact analysis. The first is that the alternative selected at the conclusion of the environmental analysis is not "permittable." In other words it would violate one or more statutory requirements and could not receive the required environmental permits or other approvals. Thus the proposed action would not be implemented and the environmental analysis would have to be reopened, causing delays and inefficiencies. The other possible repercussion is that the selected alternative could be permitted, but there were specific requirements for an individual environmental approval, such as a prescribed analysis or public outreach procedure that could have been part of the environmental analysis procedure. But since the requirements were not revealed during scoping they would have to be repeated after the environmental analysis process was completed to obtain the permit, producing time delays and inefficiencies.

A hypothetical example (but one constructed from multiple actual situations) of adequately including the permitting and other environmental approval agencies in the scoping process involves a project with the purpose and need of providing wastewater management for a new multiuse development. Following the scoping and completion of the environmental analysis including technology screening and prediction of water-quality impacts, the selected alternative was a wastewater treatment facility with advanced treatment to reduce the phosphorus in the effluent to a state-of-the-art level of 0.1 mg/l. The project proponent and stakeholders alike were excited about the environmental sustainability of the proposed action and had proceeded with the implementation. During postenvironmental analysis discussions with the regulatory agency responsible for permitting, it became apparent that the proposed receiving water for the effluent from the state-of-the-art wastewater treatment plant was an impaired water body due to phosphorus loading (designation under the U.S. EPA Total Daily Maximum Load Program) and it was an antidegradation stream under the state water-quality classification. Thus no pollutants could be discharged, particularly phosphorus, unless a greater mass of phosphorus input to the stream from other sources was removed. In order to address these constraints, the environmental analysis would have to be reopened, and/or a separate and subsequent process would have to be initiated to meet the permit requirements. In either case, including the effluent discharge permitting agency in the scoping process would have avoided this substantial delay and unnecessary duplication of effort.

4.3.2 Scoping Targets

There are two types of scoping employed to address the objectives discussed in the preceding section: social scoping and technical scoping (Morgan 2001; Snell and Cowell 2006). As discussed below each has specific stakeholder and input targets.

4.3.2.1 Social Scoping

Social scoping is the more common type designed to reach a large audience with a wide range of interests, experience, and concerns. Typically this type of scoping addresses issues, concerns, alternatives, and environmental setting that can potentially affect, on a day-to-day basis, the perceived quality of life for nearby residents. The majority of participants in social scoping are generally only involved for a specific project, plan, or policy and sometimes only for a limited specific aspect of the action where they could be directly affected. They lose interest and cease involvement when they do not see a direct link to their specific interests or if the alternative that could directly affect them is eliminated (Carnes 1993). The extreme case of issues raised by social scoping is NIMBY, or not in my back yard. A well-known example is the Cape Wind project to harness wind energy to generate electricity by constructing a wind farm in the relatively shallow waters of Nantucket Sound offshore of Cape Cod, Massachusetts. The environmental analysis team had identified a number of adverse impacts, such as bird strikes, bringing electricity to shore, and distributing the electricity. They also identified potential positive impacts such as reduced greenhouse gases, less dependence on oil, and the creation of a more diverse marine habitat. During scoping, stakeholders raised issues of aesthetics, noise, and visual impact as social issues. The more technical issues were addressed and mitigated through scientific studies and engineering solutions in a relatively short time frame; however, the issues raised as part of social scoping delayed the project for more than 10 years. There are many similar examples of social scoping identifying, early in the process, local, sensitive, environmental resources and receptors frequently unknown to the environmental impact analysis team, which can be addressed as part of the overall environmental analysis in an integrated fashion.

The primary target participants for this type of scoping include:

- Citizens living, working, or playing in the immediate area of the proposed action. The neighbors abutting the site proposed for the action are the most common example.
- Other stakeholders directly affected by the action. This could include the labor pool for a proposed action generating jobs or institutions, such as hospital or school, in the vicinity of the proposed action.
- Environmental or economic advocacy groups and organizations. These can be at the local level, even organized in direct response to the proposed action or on a national level depending on the nature and size of the proposed action.
- Elected officials and citizen organizations. This would include neighborhood associations, watershed associations, and elected officials representing the geographic or interest areas potentially affected.
- Interest groups. Hunters, fishers, local historic organizations, and enthusiasts of other hobbies fall into this category.
- Chambers of commerce, local businesses, and other business interests.

With such an extensive list of potential stakeholders, for a large, complex, or controversial project, plan, or policy, the possible participation in scoping could be overwhelming. In such situations it is not unusual to form a representative citizens advisory committee (CAC). The CAC is typically established by the project proponent, or if it is a nongovernmental project, an agency with responsibility for approving the proposed action. In some cases the local or state elected government official or body can form the CAC, but this is generally done in collaboration with the project proponent. To be effective a CAC must represent all interests and most, if not all, members must have expressed some concern (either positive or negative) over implementation of the proposed action as envisioned by the proponent. But it is most important for members to demonstrate a willingness to work cooperatively. For the concept of a CAC to work, the project proponents must be genuinely willing and able to demonstrate their commitment to offer the members something in return for their donation of time and energy to the committee, which is most commonly an opportunity to influence the decision.

Social scoping generally provides the most input in the areas of potential impact and concern, environmental setting, and occasionally alternatives for consideration. Social scoping can be very productive at generating a list of concerns and in some cases fears; however, it is generally less useful in providing the environmental analysis team with input on how to address the concerns and fears. The "how" is sometimes more productively addressed as part of technical scoping.

4.3.2.2 Technical Scoping

There are three primary objectives of technical scoping to achieve the ultimate goal of adequately assessing the nature and extent of impacts to each potentially affected environmental resource (Morgan 2001):

- From a technical perspective (i.e., without the bias of potential direct impacts on the stakeholder), identify the areas of potential impact.
- Identify the scientific studies necessary to pose and test environmental impact hypotheses.
- Select the methods to conduct the studies that are necessary.

Two types of stakeholders are typically the dominant participants in technical scoping: academic and natural resource or regulatory agency personnel. In contrast to participants in social scoping, the technical scoping community is frequently involved with scoping for many projects, in many different areas, and over long periods. Experts in a type of action (e.g., power generation, wastewater treatment, housing) will participate in the scoping for numerous similar projects, frequently on a national scale. Similarly, technical scoping participants with extensive local knowledge or responsibility (e.g., member of planning boards, watershed association members, agency personnel) will likely participate in scoping in any and all projects within their geographic area of interest regardless of the type of proposed action (Carnes 1993). It is often useful to organize these stakeholders in a technical advisory committee (TAC) as a parallel to the CAC discussed earlier. The input from these technical stakeholders is based on their experience and background knowledge of the environmental conditions in the study area, the type of action proposed, and the basic science related to the issues.

The members of the TAC can provide important inputs based on their combined experience. Sometimes the interaction of the "experts" can generate ideas and information that would not surface in a one-on-one consultation with the experts. However inclusion of even a well-qualified, committed, and unbiased TAC does not relieve the environmental analysis interdisciplinary team of their responsibility. They must work through their internal scoping, including development of an impact prediction conceptual model (see Section 5.3.1 for description and discussion of the model) to understand the interaction of existing conditions and alternatives (Figure 4.1) and formulate impact hypotheses. It is this process by the analysis team that identifies the most obvious and potentially most severe impacts and generally represents the bulk of the environmental impact analysis. The initial identification by the internal team can also stimulate the thought process for the TAC so that the potential for more subtle impacts can be considered.

Another benefit technical scoping provides the environmental impact analysis process is identification of the types of investigation and associated methods that are appropriate based on the magnitude and intensity of the potential impact. Certainly, the TAC's combined knowledge and experience can help identify the appropriate and efficient investigations and methods to describe existing conditions, predict impacts, or formulate mitigation. Also, TAC input often results in an overall better understanding of the issues and impacts, and a better decision.

Proactive involvement of the TAC can also have other more subtle influences on the process. Conducting studies that the TAC supports in advance, or better yet suggested and helped develop, all but ensures their support of the results. When the relatively unbiased TAC supports conclusions, it goes a long way in generating broad stakeholder support for results of the individual studies and ultimately the proposed action.

The over-the-horizon (OTH) U.S. Air Force (USAF) radar EIS demonstrates the advantage of engaging experts in the technical scoping process. The OTH radar in northern California required approximately 1000 hectares of USFS lands and the Forest Service was concerned with potential impacts to native American artifacts and cultural sites. As part of technical scoping, the USFS's regional archaeologist suggested a method for mapping sites and identifying artifacts based on his extensive local knowledge and experience (see Section 5.2.5 for discussion of the methods). The USAF implemented the suggested method, which was successful, relatively efficient, and was able to develop a layout for the radar that was acceptable to the USFS. It did not hurt that the method employed by the USAF also made a major contribution to the USFS unfunded mandate to map the cultural resources and sites on their land.

The USCG DCR EIS experience includes another example of productive and effective use of technical scoping (see Section 10.2 for the background and description of DCR EIS). The environmental impact analysis team had identified physical alteration of the aquatic habitat resulting from the discharge of DCR to the waters of the Great Lakes as a potential area of impact. However, the identification was only of a very general nature based on the observation that the presence of coal, iron ore, and limestone on an otherwise soft mud lake bottom would alter the physical characteristics of the benthic habitat.

As a result of interactions with the DCR EIS TAC and the internal environmental impact analysis team, specific impacts resulting from the physical alteration of the sediment were identified. The combined knowledge and experience of the internal team and the TAC identified an increase in the density and distribution of the invasive and nonnative *Dreissena* spp. mussels as a potentially serious adverse impact of DCR discharge to the Great Lakes. As discussed in detail in Chapter 5 (Section 5.3.4) the presence of the DCR in the lake's soft mud could create an optimum habitat for the invasive and destructive mussels. Working with the TAC studies, associated methods were designed to test the hypothesis of increase in mussel density and distribution. As discussed in Chapter 5 (Section 5.3.4) the studies provided the information needed to predict impacts and for the USCG to make an informed decision based on a hard look at the impacts. Because of their active involvement throughout the process, the TAC understood the impacts and supported the USCG's decision.

The Boston Harbor Cleanup EIS (Sections 5.3.5 and 10.1) represents a much more complex technical scoping and TAC example that is not strictly technical. At a projected $3 billion, the cleanup was one of the largest single-site public works projects in the United States, and as such, it drew significant public attention, including creating an issue in the 1988 presidential election between Democratic nominee Massachusetts Governor Dukakis and Republican nominee Vice President George H.W. Bush. Because of all the attention, magnitude of the project (up to 3.5 million cubic meters of waste water a day) and the court ordered and monitored process, all of the prestigious academic institutions in New England felt they had to weigh in and provide the benefit of all their wisdom. Thus periodic TAC meetings were held with the environmental impact analysis team and frequently with the press in attendance. These sessions were held at critical junctions of the investigation, with top scientists and engineers from Massachusetts Institute of Technology, Harvard University, Rhode Island Graduate School of Oceanography, Woods Hole Oceanographic Institute, and other renowned institutions sitting around the table. Meetings were held during impact hypothesis formulation, study design, method selection, results interpretation, and recommendations. Each representative felt they had to make a contribution, and one that demonstrated their extensive knowledge and experience. The technical scoping process produced an extensive investigation program utilizing sometimes the most advanced methodology, even if common sense sometimes indicated the magnitude and probability of the impact did not warrant extensive investigation. The result was a wealth of information and data, some more useful than others. But the most important result was a final EIS and selected proposed action that had the full support and endorsement of a respected scientific and engineering assemblage that fostered public support, ultimately approved funding, and implementation of what proved to be a very effective and expedient Boston Harbor cleanup.

Both the technical and social scoping contingents are important in the public outreach process. After a lengthy and detailed discussion of the multimillion dollar scientific investigation proposed for the AJ Mine EIS (see Section 10.3 for a description of the AJ Mine EIS), one of the indigenous people, with cultural resources and subsistence requirements potentially affected, inquired at a public meeting, "What are you doing for those of us who don't worship at the altar of science?" The commenter was right on with his assessment of the process driven largely by scientifically oriented academics and natural resource agency personnel. The environmental analysis team reassessed the approach and incorporated studies to understand and predict impacts not only on the cultural resources potentially affected but also the current way of life for the local indigenous populations. As it turned out, most of the scientific investigations into marine resources were directly applicable to potential impact to indigenous subsistence reliant on seafood. But the study objectives, methods, results, and interpretations had to be presented in a manner meaningful to multiple publics: commercial fishers; environmental advocates; recreational fishers; and indigenous people.

4.3.3 Scoping Logistics and Statement

Environmental impact analysis scoping and public involvement in general vary in level and intensity and have been described as a spectrum (Council on Environmental Quality 2007). At one end of the spectrum the process is "informational" with dialogue in one direction and the proponent informing the public of their intentions. The next step in the spectrum is "Consultation," in which the dialogue is in both directions, but there is no commitment from the proponent. The "Involvement" step includes a commitment by the project proponent to give the stakeholders a seat at the table and address their concerns. "Collaboration," is at the other end of the spectrum and the parties are committed to work together toward agreement on issues. The classic scoping process can be considered near the informational end of the spectrum. At this end of the spectrum the process is centered on a scoping meeting. This basic scoping meeting approach as outlined in this section can be considered the bare minimum for an effective scoping process. Enhancements to the basic approach are presented in the following section.

4.3.3.1 Basic Scoping Approach

Under the basic approach, in preparation for the meeting or meetings, the environmental impact analysis team generally prepares a brief fact sheet consisting of the following:

- Purpose and need statement, and some background information
- Issues already identified
- Potential locations of any action or at least the general geographic area under consideration
- The proposed action, if determined, and any alternatives under consideration
- Project proponent contact information
- Date, time, location, and format of scoping meeting(s).

The meeting is noticed by publishing the time, place, date, and fact sheet. Typically the information is disseminated through local newspapers, other news media, publicly posted fliers (e.g., at the local library, town/county offices, local store bulletin boards), public and proponents' web pages, and sometimes mailers to stakeholders who have expressed an interest. If the action is of national interest or significance, notification of the scoping meeting is frequently through the *Federal Register*, sometimes as part of the Notice of Intent (see Section 3.1). Government agencies with potential jurisdiction over some aspect of the proposed action (e.g., wetlands permit, historic or archeological review, local planning, and zoning) or particular expertise in an environmental resource potentially affected by the proposed action or an alternative are also generally notified directly of the scoping meeting. Similarly, nongovernment organizations (watershed associations, chambers of commerce, and environmental advocacy groups) that have historically expressed an interest in the action or the environmental setting are notified directly. The notice also generally includes a website or snail mail address for comment submittal as an alternative or supplemental avenue to the scoping meeting.

The typical and minimal scoping meeting resembles a college course lecture. The proponent makes a half to one hour presentation elaborating on information in the fact sheet, emphasizing the need for action. It is then opened for questions and solicitation of issues, concerns, and suggestions for the environmental analysis. It is also common to circulate forms to submit written comments and a signup sheet to receive periodic information and notices regarding the environmental analysis process.

The basic scoping process typically concludes by documenting the input from the scoping meeting and any other comments or suggestions received as a result of the scoping notice. The documentation can be a scoping statement (see Section 4.3.3.2), which is required under the agency-specific NEPA regulations for many U.S. government agencies, or simply an appendix in the draft environmental impact analysis report. It can range from a simple listing of the comments received to an expanded discussion of the scoping process and explanation of how each of the comments will be addressed. Even if only the most basic scoping process is implemented, it can be successful if the input received is legitimately considered and incorporated in the environmental analysis.

4.3.3.2 Enhanced Scoping Approach

The basic process outlined above can be expanded, adapted, and enhanced to meet the needs of specific actions, stakeholders, environmental settings, and interests for a project, plan, or policy under consideration. The enhancements can be applied to any aspect of scoping from the notice of intent to addressing scoping in the environmental impact analysis document. If the proponent and environmental impact analysis team conclude that an enhanced scoping process is warranted, the process should start with the scoping fact sheet.

A simplified scoping fact sheet can be expanded to actually begin the environmental analysis. Frequently, substantial work is conducted to investigate critical aspects of the action before a decision is made to pursue approval of the action or determine that an environmental impact analysis is warranted or required under applicable regulations (see for example Preliminary Assessment, Section 4.2.3). If such investigations have been conducted, including a summary of the investigations and the conclusion reached in the initial investigations as part of the material disseminated at the initiation of scoping, it can be extremely productive and beneficial to the scoping process. Inclusion of the information has multiple benefits: it gives stakeholders more information to stimulate feedback; if there is a fatal flaw or extreme opposition to the conclusions in the preliminary work, it can be addressed in the draft environmental analysis rather than later in the process; sharing of the early evaluations can promote a feeling of inclusion among stakeholders; and withholding of conclusions and decisions violates the spirit of an open scoping process and can generate distrust. Some of the information that is frequently generated as part of early project planning that could be included in the prescoping is discussed below.

A needs analysis is an example of work that is frequently done before initiation of the environmental analysis and associated scoping process. For example prior to embarking on a water supply study, a detailed evaluation must be done to determine the volume of water needed. Similarly, before initiating planning on a highway bypass project, the existing traffic problems are typically documented through relatively detailed studies. A summary and the results of these planning level investigations can be included in an expanded scoping fact sheet or brochure to give stakeholders a better appreciation of the purpose and need and the project proponent's vision.

Preliminary consideration of alternatives is another early activity that could be included in publicly distributed scoping information. Before a project proponent embarks on detailed planning or environmental analysis, they frequently want to be sure there is at least one feasible way to achieve the purpose and need for the action. Thus they engage in a feasibility study, which typically produces a list of possible alternatives. During the process, a number of possibilities are identified and then evaluated conceptually, usually from an engineering perspective (for projects), but sometimes obvious environmental constraints or other issues are included in the conceptual evaluation. The feasibility study can even include a screening of alternatives by comparison to preestablished criteria (see Section 4.5 for discussion of alternative screening). The results of the feasibility study and any alternative screening can also be very useful information to stakeholders and generate constructive and well thought out input from stakeholders at the very beginning of the environmental analysis.

NEPA regulations developed by many federal agencies have institutionalized the early consideration of alternatives as a Description of Proposed Actions and Alternatives (DOPAA). Such a process was followed for the Washington Aqueduct Water Treatment Residuals Management EIS (see Section 10.4 for a description of the background on the EIS) and the DOPAA was included in the scoping material. The original 26 alternatives were screened down to 4 slated for detailed evaluation in the draft EIS based on comparison with evaluation criteria. Unfortunately, the environmental criteria resulted primarily from technical scoping, such as traffic patterns, engineering feasibility, cost, water quality, and air quality and the initial criteria were developed prior to social scoping, neglecting some issues very important to the neighborhood stakeholders (some of whom were Washington insiders, influential and experienced in affecting government decisions). Once the DOPAA was disseminated as part of the scoping material, the lack of social scoping input to the alternative development and screening process became apparent and the alternative process was reopened to include the issues raised as part of social scoping. But by releasing the DOPAA as part of scoping, the issues were identified early and could be addressed during the draft environmental impact analysis process. This was far superior to revising the alternative identification, initial evaluation, and screening for the first time in the final document based on comments from the draft, thus potentiality requiring a supplemental Draft EIS. Even worse the concerns could have been brought up in litigation following the ROD.

If conditions such as the complexity, controversy, broad scale, and potential for significant environmental impacts warrant an enhanced scoping process, the notification of the environmental impact analysis can also be expanded. A common enhancement is to proactively identify stakeholders that could be affected by any of the alternatives or affect the decisions and encourage them to participate in the process. This can include contacting government entities directly and even requesting they be the participating agencies and similarly directly engage nongovernmental organizations such as neighborhood or watershed associations and interest groups. The noticing can also be enhanced by expanded use of websites and social media. As productive and efficient as social media can be, it must be used with forethought as part of the environmental analysis public outreach. If the key or majority of the stakeholders are not experienced and comfortable with the tools, it can be a wasted effort and lead to invalid assumptions as to the concerns of the stakeholders and the dissemination of information.

The solicitation of stakeholder input is the area of greatest potential scoping enhancement. Basic scoping is predominantly a static and one-way exercise with the proponent presenting their purpose and need to a poorly attended scoping meeting, then taking the feedback they hear and working in internal isolation to produce a draft environmental analysis document. Making the process more dynamic and stimulating greater stakeholder participation can begin during the scoping meeting. The standard format of lecture and questions can be expanded by preceding the formal meeting with a workshop or poster session with a table or station set up for each theme of anticipated interest such as background, purpose and need, alternatives, proposed investigations, regulatory compliance, sensitive environmental resources, investigation methods, and public outreach. The stations are staffed typically by members of the environmental impact analysis team or project proponent's technical staff and include informational displays. The stakeholders are free to visit the stations that interest them, talk one-on-one with environmental analysis team members, and ignore topics of little or no interest. This format can simultaneously accommodate both social and technical scoping. The displays should be designed to answer anticipated stakeholder questions such as:

- Why do we need this action?
- What are the environmental analysis requirements that must be met and what is your process to fulfill these requirements?
- How is the proposed action going to work?
- Why didn't you consider, x, y, and z?
- Where is it going to be located?
- Will you consider how it will affect me?

It is also helpful to make the displays interesting to attack attention. For example, during a poster session for the USCG Live Fire Training program, there were waiting lines at several of the stations. The most popular had the automatic firearm proposed for use on small USCG vessels in the Great Lakes on display. Others, almost as popular, showed videos of training exercises, examples of ammunition, actual targets used in the training, and navigation charts showing the proposed training locations. Less popular, but critical to technical scoping, was a station addressing the aquatic toxicity of the proposed ammunition.

The poster/workshop approach can provide a number of benefits for successful scoping including:

- Providing adequate detail for full understanding to stakeholders interested in a topic without boring others to distraction.
- Encouraging one-on-one two-way discussions in an informal setting which not only conveys information but can foster trust and positive working relationships.
- Discourages non productive "grandstanding" by stakeholders with the primary objective of public exposure by making inflammatory comments in the more formal setting of a structured scoping meeting with a large audience and the press in attendance.
- Presents environmental impact analysis team members as people and not just talking heads on a stage.
- Presents stakeholders as people not just adversaries.

Scoping can also be enhanced through continuing activities following an initial scoping meeting. Formation of a CAC and TAC as discussed above is an often-used approach for progressive and proactive scoping. The committees meet at critical junctions during the environmental impact analysis, such as screening of alternatives, receiving results of existing condition investigations, evaluating impact prediction methodologies, assigning significance levels to predicted impacts, and comparison of alternatives. At each juncture they provide input and assistance in determining the next steps. This approach allows the stakeholders to remain actively involved and understand the process, so when the analysis is concluded they are not presented with a *fait accompli* but understand all the steps, decisions, and compromises that went into the analysis. This often leads to the CAC and TAC support of the proposed action, which can be important unbiased input to the general public as they review the project. It can also foster well-informed, objective, and rational input early in the environmental evaluation process so that issues can be addressed, resolved, and incorporated into decisions prior to issuance of the draft environmental analysis.

The AJ Mine EIS (see Section 10.3 for a description of the project and EIS) relied heavily on a dynamic and continuous scoping process as part of the environmental impact analysis. Not only was a TAC formed, there were subtopic TACs for each critical environmental resource. The seabird subtopic TAC had concerns regarding the impact on seabirds, and particularly their food source within the potential marine areas of mine tailings disposal. One alternative disposal area was approximately 15 km offshore and the ornithological experts were not sure which species frequented the area. In order to accommodate their concern, the project proponent (mining company) and the lead agency regulating the activity (U.S. EPA) arranged a trip to the area. When the boat was on station, the experts had a field day identifying the birds and their concerns were satisfied. In fact, they had an added benefit because many of the birds were actively feeding, but the food source was not readily apparent. The experts felt it was important to know what they were feeding on so that the environmental impact analysis team could assess the impact on the food source from the marine disposal of the mine tailings. Unfortunately, the trip was organized for observation only and no equipment for collection of biological specimens was on board. But one of the environmental team members sacrificed her pantyhose, which were towed behind the boat and collected specimens from the high population density of the euphausiids (shrimp-like organisms) which was provoking the frenzied feeding behavior of the birds. The event was an analysis team/TAC bonding experience that supported inclusion of euphausiids in the impact prediction conceptual model.

4.3.3.3 Scoping Statement

The outcome of a scoping process is a scoping statement document made available to the public, which is a guide instructing how the analysis will be conducted. In theory, if an entirely new environmental impact analysis team

Environmental Impact Analysis and Assessment

is formed to conduct the analysis, they could pick up the scoping statement and use it as a set of instructions to conduct the analysis without missing a beat. The statement typically includes:

- Description of the scoping process
- Summary of the information disseminated during the process, including purpose and need statement, project location, etc.
- Identification of the issues raised
- Explanation of how each issue raised will be addressed in the environmental impact analysis; if it will not be addressed, why
- Identification of the alternatives that will be addressed in the environmental analysis, and if some identified during scoping are not included, why
- A summary description (sometimes with details provided in an appendix) of the evaluations that will be conducted to address the issues, male impact predictions, compare alternatives and the methods proposed for the evaluations
- A description of the public outreach program
- Contact information for the designated environmental analysis team liaison with the public and how additions, comments, and questions can be submitted
- Identification of where relevant information is available
- A schedule for the analysis including target dates for public outreach events such as meetings and site visits.
- A list of comments received during scoping and statement of how each will be addressed, or if not why.

The AJ Mine EIS (see Section 10.3 for a description of the AJ Mine environmental analysis) included a broad CAC, full TAC, subtopic TACs, periodic scoping meetings at critical junctures, and thus represents a significantly enhanced scoping process. Because of the enhanced process the resulting scoping statement is longer, broader, and more detailed than many full environmental impact analysis documents. The total length of the scoping statement for the AJ Mine EIS was 276 pages. Approximately 150 addressed the tasks to be conducted to complete the analysis and most of the rest of the document was a table, listing the issues raised (1031 comments assigned to 76 categories) during the scoping process and how each would be addressed. The scoping document for AJ Mine was a bit over the top, but it was perhaps a reaction to the rejection of a first attempt at an EIS conducted by the mining company and rejected largely due to a lack of openness and response to the public.

A more typical scoping statement would be a few dozen pages at most. It might include a table, frequently as an appendix, with the issues raised and the intended approach to addressing the issues in the environmental analysis. But the body of the scoping statement would include a synthesis discussion, addressing the 6 to 12 overarching issues that typically cover 90% or more of the individual comments and an integrated discussion of how these will be comprehensively addressed in the analysis. The studies to be conducted, proposed investigation methods, and analysis of results can also frequently be addressed in summary fashion in the scoping statement. More details may be provided by reference, or if the methods are uniquely developed for the subject analysis, they can be included as appendices.

Whether the comments are addressed in text, table, or appendix, management of comments throughout the entire process is critical to a successful environmental impact analysis, and the management begins with the scoping statement. It is important to keep a dynamic record of comments received that can serve as a checklist at the completion of the draft environmental analysis to ensure each has been addressed or to confirm that it is not relevant to the purpose and need. Also, when comments are received on the draft analysis, experience has shown that a large portion, and frequently most, of the comments on the draft were expressed in one form or another during scoping. Thus thorough tracking, consideration, and documentation of response to scoping comments as part of the draft analysis provides a headstart in addressing comments on the draft during preparation of the final environmental impact analysis document. Also, if a dynamic comment management system, either software or manual, is established during scoping, the comments received on the draft can easily be added to the system.

4.4 Public Outreach

Virtually all EIA regulations in democratic countries have a requirement for some form of public participation (Noble 2006). The requirements vary from country to country and state to state within the United States with a strong trend since the 1970s of increased public involvement. An example of the increasing focus on public outreach is the U.S. Agency for International Development (USAID), where during the 1980s and into the 1990s, there was strong resistance to any form of environmental analysis and even more reluctance to public outreach in the environmental analysis process. The resistance partially stemmed from the potential strain on international relations because many of the countries receiving assistance were authoritarian, and acknowledgment to the public of plans and impacts potentially affecting the environment was contrary to the workings of the government in all areas of the society. Progressing beyond acknowledgment to active public participation was not even within the lexicon of many authoritarian countries receiving aid. In fact, it took a presidential commission and executive order to implement environmental analysis for international projects, and now USAID requires active and extensive public participation in almost every case. In their guidance, they even acknowledge how public outreach as part of environmental analysis can have benefits in the broader society: "Through this process [environmental analysis], together we reinforce practical civil society and democracy through transparency and public participation" (USAID 2005).

The benefits of public participation as an integral and important part of environmental analysis are largely the same and stem from the same concepts as full citizen participation in the governance of a democratic society. When information, options, and the decision process are open and transparent, those in a position to implement decisions are held accountable. Also when there is an avenue for public input, a combination of wisdom, knowledge, and experience from multiple sources is fully available and there is a strong potential for a better product compared with a process held closely by a small group with similar experiences and objectives. Of course the opposite is also true. If the process is not open, the agenda of those in positions of influence and authority will be addressed first and they will not be held accountable. Also, the concerns of others and ideas for a superior action are never considered, much less incorporated, in the environmental impact analysis or implementation of the action.

In the early years of environmental impact analysis, public outreach was figuratively and literally only an afterthought and was not an integrated component of the process. In most cases public participation was largely limited to a public meeting or hearing after the draft analysis document was made public. The results of the analysis would be presented, and then limited questions and comments would be received. However in the typical case, the proponent of the proposed action with obvious biases presented only the supporting and positive information that led to the selection of the proposed action, thus limiting the range and scope of comments that the public could contribute: if the public is not presented with the positive and negative aspects of all options, their input is limited.

As discussed in the preceding section, public outreach at the initial stages of the environmental impact analysis process, particularly during scoping, has become the largest and most productive avenue for public input. The public has the potential for influence while the paper is still blank and they become a part of the process rather than an audience or often unwanted appendage. The public outreach aspects of scoping are covered in detail in Section 4.3 and are not repeated here. Other aspects are discussed under separate headings.

4.4.1 Public Outreach Commitment and Extent of Involvement

An agreement by stakeholders to be an active part of an environmental impact analysis represents a significant commitment of time and energy. Only in the rarest of cases is there any monetary compensation for the commitment, thus there must be some reason for stakeholders to volunteer their time and energy.

The most common reason for stakeholders' commitment is the opportunity to influence decisions made as part of the environmental impact analysis such as inclusion of issues to be addressed as part of the analysis, selection of alternatives for detailed analysis, designation of preferred alternative, and adoption of mitigation measures. In order for the public outreach program to be successful and provide benefit to the environmental analysis process, the project, plan, or policy proponent must balance the stakeholders' commitment with one of its own and allow stakeholders to influence these important decisions. Stakeholders also expect the proponent to acknowledge their commitment by being open, honest, and timely in all dealings with the public. If the proponent does not make provisions for stakeholders to have a seat at the decision table and honor comments of honesty and openness, the process will ultimately collapse and there will be no benefit to the proponent or parties potentially impacted by the proposed action.

In fact, if the proponent does not reciprocate by offering and fulfilling a commitment to stakeholders, public outreach can end up as a detriment to all parties. The proponent will not gain the benefits of information and implementation support discussed above and the disenfranchised stakeholders will likely be forced into other, often counter productive, avenues of involvement such as:

- A vocal constituency group lobbying for political intervention
- Letter writing and other forms of publicity to organize opposition
- Independent analyses conducted by opposition groups, applying the results in negotiations or litigation
- In the case of an "ignored" CAC, an official report and statement released in opposition to the proponent's proposed action
- Litigation, even if it is ultimately unsuccessful will cost the proponent resources and delays.

Thus in almost every case it behooves both sides to engage in a cooperative and committed public outreach program. But this is easier said than done, for several reasons: (1) the proponents do not want the delay, exposure, and complication of public involvement; (2) the proponent wants to "do things their own way" and not deal with explaining; (3) stakeholders do not trust the proponent; and (4) the stakeholders are not willing to commit the time and energy necessary to make the process work. It is the responsibility of the environmental analysis practitioner for each individual analysis to determine benefits of an expanded public outreach program, the hurdles to implementing the program, and the effort and chance of success in overcoming the hurdles. Taking all of these factors into consideration, a decision must be made as to the scope of the program and then a plan of action developed to ensure the program is successfully implemented.

Once the decision is made regarding the scope and nature of public outreach, a draft plan is developed that details the elements of the outreach program (meetings, web page, newsletters, CAC, TAC, etc.). Frequently a draft plan is presented prior to the initial scoping meeting and then public feedback is solicited on the structure and organization of public outreach as part of the scoping process. If a CAC and/or TAC is to be included in the outreach program, this first meeting, and the publicity leading up to the meeting, is a prime opportunity to solicit membership in the committees. But before stakeholders are expected to commit to a role in the process, they must understand what they are getting into.

The first step in establishing a public outreach program is to determine the "rules of engagement." In other words, what is the role of the public (including subsets such as a CAC and TAC), what are their responsibilities, and perhaps most importantly what if any authority do they have? Making this determination is the shared responsibility of the major players including proponent and funding entity of the proposed action, government agency with oversight or permitting authority, local government, and sometimes an environmental or citizen advocacy organization which has a history with the issues. Also it is important to ask the public what they want out of outreach and consider their desires when structuring the program and establishing roles, responsibilities, and authority. The role of the public can evolve with time and can vary for different aspects of the environmental evaluation, but certain aspects of the outreach program must be firmly established from the beginning and be held sacrosanct throughout the process. The most important is the commitment, clearly expressed by the proponent (and if appropriate, government oversight agency) to hold firm to the role the proponent assigns to the public.

The role of the public can range over a broad spectrum. At one extreme, the public is assigned the role of a passive audience. In this role they are simply kept informed so that they can be aware and prepared for implementation of the proposed action. At the other extreme, the CAC or some other body representing the public can be assigned equal input with the project proponent or regulatory agency to major decisions, such as selection of the proposed action among alternatives. Such delegation of authority is unusual, but it can happen. When the South Central Connecticut Regional Water Authority began the planning for the new Whitney Water Treatment Plant, they delegated (within defined bounds) the selection of the architectural design firm and approval of the design to a citizen group. Major responsibility can also be assigned to the CAC, such as "veto" power over major decisions, although this is also rare.

More common is the example discussed above for the Lake Whitney Management Plan. The public was afforded the opportunity to actively rank various alternatives, and the RWA committed upfront to seriously take into account the public's ranking in making a decision. A common variant on this assignment of authority is to accept the authorized public group's (e.g., CAC or TAC) recommendation as long as it falls within certain bounds and meets certain criteria. The critical consideration in determining roles and responsibilities is to assign a role and authority to the public that is within their area of interest and expertise, clearly state the commitment, and then live up to it.

For the basic public outreach program, participation is very simple, the events are publicized, and whoever shows up participates. For expanded programs, who participates and in what role is much more complicated. Expanded programs almost always have a CAC or a similar steering group as the focal point of the outreach program. Composition of the group is critical to the success of the program and members must:

- Be willing to commit to extensive time and energy demands.
- Collectively represent all interests and actively communicate with their constituents.
- Be unbiased or must acknowledge that they represent a particular point of view but be willing to listen to other perspectives and open to compromise.
- Be conversant (not necessarily an expert but some background knowledge and willingness to learn) on most of the aspects, issues, local setting, and components of potential proposed actions.
- Be willing to accept and support the role and authority assigned to the group.
- Be willing and have a demonstrated record of working productively in a team or group setting.

So how is a group with such attributes selected and assembled? Similar to establishing the roles and authority of the CAC, the selection is the shared responsibility of the major players including the groups identified above responsible for establishing the rules of engagement. These organizations negotiate the structure of the CAC or a similar group, including its size, interests to be represented, and member selection process. The agreed upon structure often includes assigned seats on the CAC designated by major stakeholders (e.g., specific elected or appointed government official from the most affected municipality or from a heavily involved advocacy group). Once the structure of oversight or advisory groups is established there are several options for populating the group (Bregman 1999):

- Self-identification: interested parties volunteer and if the available pool exceeds the available positions, a selection is made based on a predetermined and publicized process.
- Group identification: groups (e.g., neighborhood association, watershed association, environmental or economic advocacy group) with

interest and expertise are defined as part of the CAC structure and each group selects a representative.
- Third party identification: a public participation specialist, volunteer citizens committee, or a recognized nonbiased organization does research and conducts interviews with stakeholders to identify and prequalify participants.
- Direct search: this is similar to third party identification except it is typically conducted by the environmental analysis team and resembles a search to fill a paid position.

Other stakeholder groups, such as a TAC, can be formed in a similar manner with the CAC (or equivalent) participating with other major players to structure the group. Alternatively, the CAC can accept full responsibility for identifying, structuring, and populating other groups to provide input. Under any scenario, it is generally a wise approach to involve the CAC in any public outreach activity or decision.

4.4.2 Benefits of a Public Outreach Process

The benefits of involving the public discussed at length in the previous section on scoping (Section 4.3) including identification of issues, improved understanding of existing conditions, ultimate support for proposed action, a better decision, and impact mitigation apply to all aspects of public outreach. However, when viewed in a larger perspective there are different benefits to different parties. At one end of the spectrum, the proponent of the proposed action benefits from increased knowledge of existing conditions, which can support a more informed and better decision. This translates into a conclusion and decision at the end of the environmental analysis that acknowledges all the facts and is at a reduced risk to successful challenge and the inherent delays at the end of the process. At the other end of the spectrum, the stakeholders have an opportunity to influence the decision in favor of their objectives and incorporate avoidance or other forms of impact mitigation to environmental resources important to the stakeholder (O'Faircheallaigh 2010).

Both the proponent and the stakeholders can benefit from using public outreach to assist in alternative comparison and selection (Schoepfle et al. 1993). As described in detail for the Lake Whitney Management environmental analysis, a key element in alternative comparison and selection was the magnitude and extent of impact of each alternative on each key environmental resource. Impacts to different resources can be expressed on a common basis by establishing levels of impact significance (see Section 5.3.3), and the public can play a key role in establishing the level of significance. The public input to the RWA Lake Whitney Management Plan (Section 4.3.1 and Table 4.1) established the level of significance of each resource that formed the basis for comparing the impacts of each alternative and ultimately for the proponent (RWA) deciding which alternative approach would constitute the selected lake management plan. Similar processes can be used on a variety of projects, plans, and policies with adaptation to specific situations and composition of the public participants.

The CEQ recognizes the extensive benefits that can be derived from an enhanced public outreach program. They have produced a handbook advising NEPA practitioners on procedures for gaining the most out of collaboration with stakeholders and list the following benefits (Council on Environmental Quality 2007):

- Better information
- Fairer process
- Better integration
- Conflict prevention
- Improved fact-finding
- Increased social capital
- Easier implementation
- Enhanced environmental stewardship

Another benefit of positive and productive stakeholder participation is their support for implementation of the proposed action and they can serve as advocates throughout the process. In the case of the Boston Harbor Cleanup EIS (see Chapter 10 for a summary discussion of the EIS), the U.S. EPA and the EIS teams worked closely and frequently with the CAC and TAC. The EIS team met at least bimonthly with both groups during critical phases of the EIS preparation and they played an active role in alternative development, identification of issues of concern, and methods for impact analysis. A key issue was the location of the treated effluent outfall (i.e., a pipe or tunnel structure that conveys the effluent to the point of discharge in the receiving waters) and outfall locations were considered within Boston Harbor, at varying distances from the northern and southern entrances to the harbor. Committee members with different concerns and points of view were represented on the committees, including representatives from communities in proximity to each of the alternative outfall locations.

The final decision was the northerly location for the outfall, approximately 14.5 kilometers from the mouth of the outer harbor. Although the representative from the committee in closest proximity to the outfall (it was still several kilometers from her community) was not thrilled with the selected location, she supported the decision. She recognized the objective and unbiased scientific and engineering studies that had gone into evaluation and prediction of impacts. Also through the course of more than a year of working cooperatively with the EIS team, a sense of trust, mutual respect, and recognition of issues had been established. Incorporation of the CAC concerns in the outfall selection process contributed to the acceptance by the CAC in general and the representative from the closest town in particular.

After the ROD was issued and even after review by the courts, an issue surfaced with the residents of Cape Cod. The Cape Cod communities were more than 40 kilometers from the selected outfall location but they were concerned that the discharge would affect their beaches and marine mammals in the area. Reports and presentations by the EIS team and supporting scientists, gleaned from the studies supporting the EIS, did little to convince the more skeptical and less rational elements in the Cape Cod communities. They would not initially accept that adequate analysis had been done to understand the impacts and confirm the prediction of no measureable adverse impacts to beaches and marine resources at the closest locations much less 40 kilometers away. However, when a former activist and outfall opponent representing the closest community stood up at meetings and issued statements, even the greatest skeptics took notice. She told the Cape Cod residents that she had been involved throughout the process and was convinced by the studies and analyses that her town, only about 8 kilometers from the discharge, would not be impacted. The audience was then receptive to facts documenting that their communities and resources lying five times that distance from the closest and unimpacted community would not be at substantial risk. As described in Chapter 10 (Section 10.1), subsequent monitoring studies of the outfall have confirmed the EIS predictions, and no impact has been detected on Cape Cod resources in 20 years of operation and monitoring.

4.4.3 Public Outreach Tools

There are many tools available to the environmental impact analysis practitioner to maximize the benefits of a public outreach program. Many of the primary tools are described under scoping (Section 4.3.3) and include meetings, workshop/poster sessions, CAC, TAC, newsletters etc. However there are additional important tools available that are most appropriate during other steps in the environmental analysis following scoping, and these are discussed below.

For many environmental impact analyses, there is more than one public, and tools must be developed for each group. There are both active and inactive publics with the active group dominated by stakeholders in the immediate geographic area with the potential to be affected on a day-to-day basis by one or more of the alternatives. Direct contact, such as meetings and local media postings, are appropriate tools to use for such groups that can be focused on "Not In My Back Yard" (NIMBY). They are generally involved early in the process and remain involved until their issue is resolved or the area of impact stimulating their initial involvement is dropped from detailed consideration.

Unless appropriate provisions are made and steps taken, the inactive public does not surface until the end of the process when they suddenly realize they could be affected by the proposed action. More aggressive and proactive tools are required to engage this group, such as seminars and other educational events. Also seeking out parties potentially affected and actively involving them in the process is a proven technique to engage the inactive public and minimize surprises at the end of the process.

Closely related to the distinction between active and inactive publics are the local and national interest groups. These two groups similarly require different public outreach tools and techniques to solicit their input and realize the benefits of cooperative public participation. Meetings generally reach the local groups but more broad-based and creative techniques are necessary to engage the national stakeholders, including web-based tools, articles in popular or even peer-reviewed publications and in national advocacy group newsletters.

There is no one set of right tools to use for public outreach. Successful outreach is a program with the tools adapted to the specific public(s) with vested interests and those potentially affected by the proposed action or any of the alternatives. The need to adapt the tools to the specific audience is illustrated by technology, specifically electronic communication tools. There has been a major effort in Chile to involve the public in environmental impact analysis through Internet-based and other electronic tools. However, research reveals (Lostranau 2011) that national communication techniques are only available to the public as follows:

- Television 80%
- Radio 80%
- Telephone 70%
- Internet 10%
- Newspapers 55%

Thus using the Internet as the primary public outreach tool would miss the vast majority of the population. Lostranau (2011) also found that a public program highly based on technology excluded the most affected and vulnerable segments of the public.

There was a similar situation, experienced by the author, in neighboring Peru involving construction of a natural gas pipeline. The pipeline was from the headwaters of the Amazon River, over the Andes to Lima, where the gas was to be processed and distributed, and the route (selected with little or no public input) was through sensitive and vulnerable indigenous lands. The pipeline development consortium received World Bank and other international funding for the project, but the funding came with the obligation to involve the indigenous populations and engage the Peruvian government to implement the outreach programs to the affected stakeholders.

With the best of intentions, the government established several initiatives to engage the public and mitigate the impacts. One of the more creative and potentially beneficial programs was to fund the indigenous communities to establish plant nurseries along the pipeline route so there would be stock readily and immediately available to revegetate the pipeline following construction. This had the multiple benefits of including the local people as active and cooperative partners, training, economic support, and mitigating erosion in the rainforest ecosystem following clearing for the pipeline.

In spite of these efforts, there were observed cases of significant erosion impacts resulting from no revegetation following pipeline construction. In one case, the pipeline route was observed as a 10-meter eroded channel through the rainforest because the route was not revegetated after construction. The eroded soils were deposited in the downgradient stream, where they silted in the substrate and destroyed the aquatic invertebrate habitat. These invertebrates were the primary food source for the fish populations in the stream, which in turn were an important source of protein to the inhabitants in the neighboring village. Without an acceptable aquatic invertebrate habitat, the fish stock collapsed and along with it an important food source for the indigenous people. When this was reported to the government, they were genuinely surprised that the villagers' abutting the route had not taken full advantage of the nursery and replanting initiative. They even went outside their planned web-based public outreach program and visited several villages to investigate the problem.

The villagers were excited to be visited by the government and anxious to hear what they had to say. They even fired up their generator, which normally was only operated a few hours a week because of the scarcity of fuel, so the government and project proponents could make their computer presentation. After the presentation, in response to one of the villager's questions as to how they were supposed to know of the opportunities for nurseries and revegetation, the government presenter showed a slide of the project website and went into great detail describing all the information on the site. The more astute members of the government/proponent public outreach team looked at the old extension cord stretched from the barely operating generator to the traditional open air pavilion where the meeting was held and realized Internet-based communication alone was not sufficient as public outreach for the project.

Even in highly technologically advanced settings, such as the United States, social media and web based public outreach tools alone may not be sufficient. If there are affected publics who are not technologically oriented because of age, economic status, or interest there must be other tools and communication avenues made available. Some other tools that have proven successful to inform the public and generate constructive/cooperative interaction include:

- Organized site visits led by the project proponent and environmental analysis team. These visits can be to alternative sites for a proposed action or a site where a facility or activity similar to that proposed is already in operation.
- Informational workshops. An expert or knowledgeable person not associated with the project, plan, or policy can be engaged to present

a broader perspective without the perception of bias. Also as described in the Boston Harbor example above, a CAC member from the same or similar project can discuss her/his perspective without the stigma of a project proponent's bias.

- Information repositories. Local libraries, proponent's offices, or government agencies offices are useful and accessible locations to make relevant documents available if Internet-based repositories alone are not sufficient.
- Formation of ad hoc committees to address critical issues. These groups typically are composed of a combination of proponent/environmental analysis team members, citizens or other stakeholders, and outside experts. Frequently there is an effort to populate the committee with stakeholders of different perspectives on the benefits of the proposed action (i.e., advocates and those in opposition). They can even include a facilitator to keep order and formulate a consensus position on the contentious issues.
- Independent consultants to the CAC or other stakeholder groups. The proponent or an involved government agency can fund an independent consultant, reporting to the CAC, to assist with reviewing investigation results and other documents. This approach has proven very successful as part of the Superfund program to clean up hazardous waste sites in the United States, where the investigations, science, and remedial alternatives can be very complicated and technologically complex.

Tools alone cannot produce a successful public outreach program that benefits the environmental analysis and proposed action implementation. Adapting the tools to the situation and the potentially affected public is a key to success. Equally important is that all parties fully understand their role, responsibility, and authority. More important is for each party to fulfill its commitment and function in an open and honest manner.

4.5 Development and Preliminary Evaluation of Alternatives

Without the concept of alternatives integrated into the process, the environmental impact analysis would only be documentation of the impacts, much like documentation of damage in a postaccident report filed for an automobile accident with no provisions for minimizing the damage. Presenting alternatives and understanding the environmental implication of each alternative are necessary components for incorporating environmental considerations into decision making and establishing mechanisms for environmental impact avoidance and mitigation. Without full commitment to alternative development and analysis, the environmental impact assessment methodologies presented in this and similar books as well as environmental regulations would be nonproductive at best, and in most cases impossible. In fact, the CEQ Regulations implementing NEPA state: "This section [alternatives] is the heart of the environmental impact statement" (CEQ Regulations Sec. 1502.14). As discussed above and highlighted in the NEPA CEQ Regulations, a primary purpose of any environmental impact analysis process or document is to support decisions, and the key decision supported by the process is alternative selection. Thus the identification, description, evaluation, and comparison of alternatives are central and guiding threads of any sound environmental analysis.

The role of alternatives in environmental analysis can be viewed as analogous to a high school junior or senior selecting a college. The college search begins with the prospective students first considering what they want out of an advanced education experience, which is analogous to the environmental analysis definition of purpose and need. In environmental analysis, identifying the purpose and need is problem identification and developing alternatives is the first step in problem solving. College selection is a similar process. Once the purpose and need are identified, then the student identifies a number of alternative colleges and universities which might meet his goals for higher education, similar to developing alternatives which may meet the purpose and need for a proposed action in an environmental analysis. A next step for the student might be to narrow the list of schools by determining which ones would potentially accept him, which is similar to determining which alternatives efficiently satisfy the purpose and need in an environmental analysis. Then the school search comparison can begin in earnest by determining where each school falls with respect to the student's goals such as: urban versus rural; academic ranking; extracurricular opportunities; cost; living arrangements; student mix; etc. Similarly the environmental analysis determines the impact of each alternative on the resources of concern, such as transportation, endangered species, water quality, and air quality. Once the impact of alternatives or the attributes of each potential school in the search are known, an informed decision focused on the critical factors can be implemented.

The approach to addressing alternatives in the environmental impact analysis is discussed in the following sections. Various approaches to the treatment of alternatives and the proposed action are presented first, followed by a discussion of methods to develop and describe alternatives. The section then presents screening of alternatives so that only feasible alternatives that fully satisfy the purpose and need are carried forward for a full and detailed evaluation in the environmental impact analysis. The comparison of alternatives and selection for implementation are highly dependent on the impacts resulting from each alternative, thus a full determination and analysis of impacts (as discussed in Chapter 5) must be completed before there is a final comparison and selection of alternatives. However, the final comparison is an extension of the screening process and thus is presented as the final discussion in this section.

4.5.1 Proposed Action and Alternatives Treatment

There are two primary schools of thought on the treatment of alternatives as part of environmental analysis. One is to identify a proposed action that is the preferred alternative the proponent intends on implementing to meet the purpose and need. After this alternative is developed and designated as preferred, development of alternatives to the proposed action proceeds, but generally only to comply with the consideration of alternatives mandated by regulations. However, specific alternatives are sometimes developed in response to issues raised or impacts identified late in the environmental analysis. The other approach is to begin with a blank slate and characterize the full-range of practical options that meet the purpose and need as equal alternatives. Both approaches are consistent with the letter and intent of NEPA and most other environmental analysis regulatory programs. Each approach has advantages and constraints, as discussed below and the choice is dependent on the specific project, proponent, and stakeholders.

If the NEPA or other environmental analysis process is initiated late in project planning and there is not an anticipated active, informed, committed, and cooperative stakeholder population, the approach of identifying the proposed action during the scoping process and in the draft document is a common approach. This acknowledges to the stakeholders, and potentially to the courts if there is a challenge, that the project proponent does in fact have a preference and encourages reaction to the preference. It also allows and perhaps encourages the proponent to develop and refine the proposed action so that sufficient detail is available to fully describe environmental impacts and even develop environmental impact mitigation measures. The downside of the approach is that there can be the perception that the proponent has already made a selection and is simply using the environmental analysis to justify their decision, thus an "after the fact" environmental analysis.

Another advantage to the initial designation of a proposed action that is only rarely used maximizes public input to identify alternatives. In such cases, the proponent develops the preferred alternative in detail, including at least a qualitative determination of associated impacts. It is then presented to the public, and if unforeseen issues and impacts are identified, additional alternatives are developed to avoid the impacts and address the issues. These alternatives developed based on public input are then compared in detail with the original proposed action in the environmental impact analysis, and a final selection is made based on the results of the analysis.

The other approach to alternatives analysis (equal treatment without a designated proposal) is supportive of environmental analysis as an integral part of the project planning process. Under this approach, transparency is essential and the project proponent should not have internally identified a preferred course of action. A full range of alternatives is identified through technical evaluation and stakeholder input, and each is treated equally with respect to description and analysis. The one that emerges from an objective comparison and decision process then becomes the proposed action and the preferred alternative.

If the project and stakeholder conditions are ripe for the equal identification and treatment approach, but the regulations or project proponent's organizational constraints (e.g., agency guidance or precedence) dictate designation of a "Proposed Action," the object of transparency and equal treatment can be achieved in other ways. One possibility is to characterize the "Proposed Action" as the purpose and need or the intended goal of the action and then develop alternative approaches to achieve the goal. For example, if the issue is congestion in a small Maine town, because the regional highway goes through the center of the town, the proposed action could be reduction of in-town congestion. Equal alternatives could be developed including:

- Widening the street through the town
- Varying traffic patterns so there are more lanes going toward the ski resort outside of town on Friday evening and more lanes going the other direction on Sunday evenings
- Restricting logging truck traffic during peak recreational traffic periods
- Construction of a bypass around the town center via a number of alignment options

Another widely accepted example of defining the proposed action generally is given in CEQ's "Memorandum: Forty Most Asked Questions Concerning CEQ's NEPA Regulations (46 Fed. Reg. 18026 March 23, 1981, as amended 51 Fed. Reg. 15618 April 25, 1986, and summarized in Table 2.3 of this book) as Question 5. In this example the proposed action is to grant a National Pollutant Discharge Elimination System (NPDES) permit to a non-federal entity, and the equally treated alternatives could include:

- Various wastewater treatment technologies (e.g., conventional aeration treatment, biological nutrient removal, or activated carbon filter)
- Effluent reuse
- Method of discharge (e.g., land application or aqueous discharge)
- Various discharge locations

Another possibility of applying the transparent and equal treatment approach is to develop alternatives in an appendix to the EIS or other environmental analysis document. This appendix could include: identification of a full range of alternatives; alternatives screening, including environmental criteria, initial comparison of alternatives, identification of alternatives for detailed evaluation, and then designation of one of the alternatives as the proposed action. If this approach is followed to select a single preferred alternative, it is critical that environmental considerations be a component of the comparison of alternatives.

The alternative development and comparison appendix could also be constructed based on nonenvironmental factors and identify the alternatives that are practical and fully meet the purpose and need. The environmental implications of each alternative and the identification of the proposed action could then be the focus of the environmental analysis. If this approach is followed and all the alternatives considered preferably from a nonenvironmental perspective (technical, economic, implementation, etc.) result in a significant environmental impact to one or more environmental resources, it may be necessary to revisit the initial alternative screening and carry at least one alternative forward that mitigates environmental impacts.

There is another approach to identifying the proposed action that incorporates active public involvement in the alternative selection process. The alternatives can be developed and a full technical, economic, implementability, and environmental analysis included in the draft document. The alternatives can be compared on each of these bases and the public encouraged to weigh in with comments on the draft, or even workshops with active stakeholder participation conducted on the draft environmental analysis. The public input during the workshops and comments on the draft environmental analysis can then be integrated into the alternative comparison and the proposed action identified in the final environmental analysis.

Any alternative development, analysis, comparison, and selection process as part of an environmental impact analysis must at a minimum have two alternatives: proposed action and no action. In cases where history has shown there is a successful approach with little or no environmental impact to satisfy a similar proposed action, incorporating the minimal number of alternatives can be adequate and the environmental analysis can focus on site-specific conditions and mitigation of minor impacts. Similarly, if the purpose and need are of relatively small magnitude and preliminary environmental analysis and project planning have clearly demonstrated that there are no significant adverse impacts associated with the proposed action, the analysis can be limited to just the proposed action and no action as alternatives. This condition could apply to a simple EA as opposed to a full EIS for a NEPA environmental analysis. In such cases, the courts have ruled in NEPA compliance cases (Smith 2007) that if a clear explanation incorporating environmental concerns is provided as to why other alternatives were not considered in the environmental impact analysis, evaluation of just the proposed and no actions is adequate treatment of alternatives. Such explanations can be part of the analysis document, an appendix to the document, or earlier investigations such as a feasibility study referenced in the environmental impact analysis.

The no-action alternative is necessary from a technical environmental impact analysis perspective and as discussed in Chapter 4 also as a requirement for actions subject to NEPA. The no-action alternative is similar to the description of the affected environment or existing conditions, but they differ in ways critical to environmental impact analysis (Smith 2007). As discussed in Chapter 5, the description of the affected environment is a snapshot of each critical environmental resource as it exists prior to implementing the proposed action or any of the alternatives. It is the base condition to measure change or impact resulting from the proposed action or alternatives. In contrast, the no-action alternative represents a prediction of the condition and characteristics of each environmental resource at some point in the future if the purpose and need are not satisfied and no alternative is implemented. For example, in an area of traffic congestion and anticipated significant future growth, the existing transportation condition might be described as mildly congested with wait times at stop lights averaging 0.5 to 1.5 minutes. However, when the projected future growth is added to the traffic projections, wait times might be 2 to 4 minutes and congestion described as approaching gridlock. If the projected traffic impact from the proposed action was a wait time of 1 minute at stop lights, and the comparison was only to existing conditions, the projected impact to other environmental resources might suggest little benefit compared with adverse impacts. However when compared with the no-action alternative, the future benefits of cutting traffic-light wait times by 50%–75% (i.e., from 2 to 4 minutes with no action for 1 minute for the proposed action) would portray the impact trade-off in a different light. Appropriate presentation and treatment of the no-action alternative can be a very useful and productive approach in clearly illustrating the benefits of a proposed action to stakeholders and gaining support for implementation (Eccleston 2008).

The classic case of utilizing the Hudson River in New York for power generation illustrates another example of the difference between no action and the affected environment. The Hudson River estuary is a breeding ground for the commercially and recreationally important striped bass or rockfish (*Morone saxatilis*) and it was also identified as the primary source of water for nuclear power plants and other electrical power generating operations for the New York metropolitan area. These two attributes were in conflict because withdrawing the water for cooling or other power generating purposes also entrained or entrapped the immature forms of the bass and resulted in significant mortality.

The impacts on striped bass were evaluated initially for each individual power generating facility in comparison with existing conditions. The existing striped bass conditions were studied in detail and the conclusion was that the Hudson River estuary supported millions of immature striped bass and more than 90% of them never made it to maturity due to predation, natural death, lack of adequate food to support such populations, etc. The environmental analysis for each individual power generating facility environmental analysis predicted that only a small percentage (generally less than 5%) of the immature striped bass would be entrained or entrapped as part of their operation and this small percentage was included in the 90% that would not survive anyway. Thus the conclusion was that there was no reduction in the breeding population, the population was sustainable, and there would be little or no adverse impact on the Hudson River estuary striped bass breeding success. The academic and regulatory community initially agreed with this analysis because it was based on a wealth of high-quality data and state-of-the-art ecological population modeling. It even included adequate allowance for uncertainty and conservative probability analysis.

However, a coalition of Hudson River environmental advocacy groups pointed out a flaw in the analysis. Only the existing conditions were considered and there was no evaluation of the impact of each power plant based on future conditions when other power generating facilities were in operation. When the striped bass models were rerun using a diminished number of immature bass resulting from the future operation of other facilities, an impact on overall populations was detected. After this impact analysis comparing the proposed action with the no action was completed, the power generating community agreed to a reevaluation, and the result was a negotiated agreement among the power generation community, the environmental advocacy groups, and the regulators as to the total amount of water, and associated immature striped bass that could be withdrawn, and then this total number was proportioned among the proposed projects (this could also have been resolved with a comprehensive cumulative impact analysis, see Section 5.5). It was then incumbent on the environmental analysis of each individual facility to demonstrate that their facility design, operation, and mitigation would not result in immature striped bass mortality greater than their allotment based on future conditions.

4.5.2 Development of Alternatives

In any situation after the problem has been defined, identification of alternative solutions is the first step in problem solving. This is certainly true of environmental analysis and a primary, and usually the first alternative identification approach is to make use of the environmental analysis and project teams' technical expertise to identify tried and true methods to satisfying the purpose and need. This is frequently accomplished in advance or during the earliest stages of environmental analysis in the form of a technical feasibility study for a project or similar analysis for plans or policies. Once the members of the team with technical or policy expertise have identified the classic and innovative project-specific alternative solutions to the purpose and need, the environmental practitioners can begin identifying the environmental resources potentially affected.

The next chronological step in alternative identification is through scoping. As discussed in Section 4.3, both social and technical scoping can be very useful in the identification of alternatives that the environmental analysis team and project proponent have overlooked or dismissed prematurely. Also as discussed above, the development and description of classic and other alternatives by the environmental analysis team can provide food for thought and stimulate ideas by the stakeholders, producing new alternatives or modification of previously identified approaches that can enrich the environmental analysis, demonstrate trade-off, and produce a superior proposed action.

As the environmental analysis progresses, it is important to have alternatives that represent a clear choice regarding area, magnitude, and intensity of impact. This sometimes necessitates developing new alternatives or modifying existing ones to ensure there are true choices. Keeping in mind that a primary purpose of environmental analysis is to provide environmental input to decision making, the purpose is not achieved if the decision involves only alternatives that have the same or similar impacts on the same environmental resources. Thus there should be alternatives that have markedly different impacts on the primary environmental resources of concern identified during scoping. Typically there is at least one alternative with minimal cost and one with minimal overall impact. It is also common to include in the environmental analysis an alternative with minimal impact on each of the critical environmental resources of concern (e.g., wetlands, traffic, cultural resources, and water quality). Similarly, identifying alternatives that achieve the purpose and need to various degrees is a common approach to providing input to meaningful decisions. The objective of this approach is to support informed trade-offs among adverse impacts and benefits on the various resources of concern.

Similar to providing a clear choice among alternatives, the vulnerability of a resource can also be an important consideration in the identification of alternatives. If scoping or the environmental analysis has identified a particular vulnerable attribute (e.g., air quality or endangered species) potentially at risk, alternatives can be included that minimize impact on these resources. Typically, such an alternative will have higher costs or more severe impacts on another environmental resource, thus illustrating the trade-offs and providing meaningful and substantive choices for the decision-making process.

In many situations, segmentation and simplification of the alternative development and evaluation process can be important to a successful environmental analysis. For complex projects, plans, and policies with many components, complex and multifaceted alternatives that fully address every aspect of the purpose and need can be unmanageable, thus defeating the objective of incorporating stakeholder input and providing meaningful choices for environmental trade-offs. In such situations, it is frequently necessary to segment the project, plan, or policy into independent components or at least aspects of the action with minimal dependence on other components. This approach permits presentation of a simple, clear, and focused analysis to decision makers and also allows stakeholders to be actively and cooperatively involved with aspects within their area of concern and knowledge without being overwhelmed with issues and information beyond their interest or experience. The importance of taking this approach of segmenting complex and critical environmental problems into manageable, bite-sized pieces is illustrated by the Boston Harbor Cleanup EIS case study, where alternatives were developed separately for the semi-independent components of wastewater treatment location, wastewater treatment processes, wastewater residuals (i.e., sludge) processing, residuals disposal, and wastewater effluent discharge (see Section 6.3.1).

Another example of segmenting the action for alternative development is a project in Haiti to upgrade an agricultural irrigation system (USAID, in progress). This project has three basic components: upgrade the agricultural road servicing the area, repair and enhance for low maintenance the irrigation canal system, and replace and upgrade the river diversion to the canal. The river diversion is necessary to get the water into the canal system and the improved canal is necessary to get the water to the existing and new agricultural lands. The road improvement is necessary to develop and maintain the diversion and canal, and also to transport agricultural materials to the area and produce to market. Thus, all of these components are integrated, and successful implementation of each is essential to the overall purpose and need to stabilize agricultural land use and expansion of food production in the region. However, exactly how and where the road is upgraded does not affect the diversion or canal. Similarly, the type and method of river diversion do not affect the location or type of road. Therefore, it was possible to identify, and ultimately analyze, independent alternatives for each component.

There are potential pitfalls associated with segmenting a project for alternative identification, and steps must be taken at various stages to avoid compromising the environmental analysis. The first step is to call on the expertise, particularly technical expertise, of the environmental analysis team and appropriate stakeholders to determine the interrelatedness of the components. This is particularly important in alternative screening (see Section 4.5.3) to ensure screening out an alternative for one component does not unintentionally eliminate or handicap an alternative for another component of the project. Another potential concern, relevant to impact prediction is that the combining of selected alternatives for all components creates additional impacts not inherent in any individual component. This is not a fatal flaw, but must be addressed by impact prediction of comprehensive alternatives, usually conducted as a final stage of impact prediction in the environmental analysis.

Once the alternatives have been identified, they must be described and the description of alternatives in an environmental impact analysis is a balancing act. The description must ensure there is enough detail provided to predict impact but not so much as to indicate the project, plan, or policy is nearing implementation without full environmental evaluation or stakeholder input. The "credit card" diagram (Figure 4.1) is a simple visualization of the goal of alternative description. The features of the alternative that can interact or overlap with environmental resources should be described in detail, while the characteristics with little potential to affect environmental resources require little or no description.

The process must be relatively equal and balanced among all alternatives evaluated in detail. For example, providing close to a full engineering design for one alternative and a general qualitative description of others can indicate an alternative preference or even de facto selection, thus compromising the goals of environmental analysis. Similarly, the level of detail does not have to be the same for all aspects of the alternatives. As discussed above, where there is overlap with environmental resources, the description should be more detailed. It is also important for the description to be a balance of technical detail and explanation understandable and useful to all stakeholders.

4.5.3 Screening of Alternatives

Alternative screening can take several forms and levels. The first step in any alternative screening process as part of an environmental impact analysis is to determine if an alternative under preliminary consideration meets the purpose and need. If the purpose and need cannot be accomplished, then the alternative either must be modified so that it does accomplish the intended purpose or eliminated from detailed consideration in the environmental impact analysis. Screening against the purpose and need can efficiently be accomplished by establishing a list of criteria with specified thresholds needed to meet the purpose and need. This list guides the initial description of alternatives because there must be sufficient definition and detail of the alternative developed to compare with the purpose and need screening criteria. The screening of alternatives for the Washington Aqueduct Water Treatment Residuals Management EIS (see Section 10.4) and the USCG DCR EIS (see Section 10.2) are examples of establishing criteria and thresholds to compare alternatives with the purpose and need and eliminate those which do not measure up. As with these examples, if an alternative is eliminated from detailed evaluation, it is critical that the reasons for elimination are well documented and available to stakeholders.

After the alternatives that meet the purpose and need have been identified, the list can sometimes be further screened by a "fatal flaw" analysis. This form of screening maximizes environmental analysis efficiency and focus by establishing criteria that would singly eliminate an alternative from further consideration. Thus if an alternative did not satisfy any one of the specified criteria, it would be considered "fatally flawed" and not carried forward. Evaluation of the alternative with respect to detailed description, costs, and impacts in other environmental resources would not be necessary, thus focusing the environmental analysis on viable alternatives and conserving the resources available for the analysis. Common examples of fatal flaws used in alternative screening are:

- Exceedance of an environmental resource standard or criteria (e.g., water-quality standards for one or more contaminants)
- Nonconformance with environmental permitting requirement (e.g., involving the fill of more than the threshold area for a necessary wetlands permit)
- Economically infeasible (e.g., more than double the estimated cost of any other alternative or far in excess of available funds)
- Not implementable (e.g., not able to acquire the land or implement within a mandatory schedule)
- Unacceptable level of anticipated impact to a critical environmental resource, particularly as defined during scoping (e.g., destruction of critical endangered species habitat)
- Technology not proven or accepted standard practice (e.g., if technology failure results in catastrophic unacceptable consequences such as for bridge construction methods or water treatment)

Side-by-side comparison is another accepted method of alternative screening for complex projects, plans, and programs. This approach requires an early, generally qualitative, and relative evaluation of each alternative for each critical factor, such as economic viability, implementability, impact to environmental resources potentially at risk, etc. If an alternative is predicted to have relatively severe adverse impacts or significant economic/engineering limitations and there is another alternative that has less impact in these areas and equal or greater attributes in all other areas, the inferior alternative can be eliminated at the screening stage. For example, in the USCG DCR EIS alternative screening process, there were numerous initial alternatives for collecting the spilled DCR on the vessel decks. One was an industrial grade mini street sweeper stored on the ship's deck and another was a broom and shovel combination. Preliminary analysis concluded that the mini street sweeper would be very expensive; prone to maintenance issues; difficult to empty while the ship was underway; require operator training; and relatively ineffective in collecting DCR around the hatches. In comparison, the broom and shovel combination was cheap, required little training, could easily be emptied, and was maintenance free. Thus the mini street sweeper was eliminated at the screening stage and never subjected to detailed analysis.

4.5.4 Comparison and Selection of Proposed Action

The final selection of a preferred alternative is accomplished only after the full evaluation of impacts because the comparison is centered on a comparison of impacts among alternatives. Thus the final comparison cannot commence until the impacts are fully assessed and evaluated in combination with the other critical aspects of the project, plan, or program such as economic viability, engineering or technical feasibility, implementability, and stakeholder input. Once all of these considerations have been developed, the alternatives can be compared and an informed decision rendered.

The comparison of alternatives can take many forms, but in many cases it is a simple extension of the alternative screening process. Generally the same considerations, such as criteria measuring achievement of purpose and need, are used in the comparison. The final comparison of alternatives does differ from the screening in several important ways. First there is much more detailed information available on all aspects of the alternatives including impacts, costs, permit requirements, engineering details for projects, and similar detail for plans and policies. Thus the final analysis can be much more detailed than screening where the differences in impact intensity among alternatives and environmental resources can be quantified as opposed to the typical binomial yes or no or retain or discard conclusions in screening.

The final analysis can also incorporate impact significant criteria (see Section 5.3.3) into the comparison, and these are typically not available at the alternative screening level. These criteria are developed for each environmental resource to determine whether the effect on the resource from an alternative is significant, and frequently there are multiple criteria for each resource so a consistent level of impact (e.g., no impact, minor impact, moderate impact, or significant impact) among resources can be expressed. This facilitates comparing alternatives with impacts on different environmental resources because from an environmental perspective, an alternative with significant impacts on multiple resources would be inferior to one with only minor impacts on any resource.

Alternative comparison employing an impact intensity scale can be further enhanced by applying the relative importance of each resource. This can be included in the final comparison with the benefit of extensive stakeholder input identifying the relative importance of various factors to the people directly affected. Thus those aspects of greater concern to stakeholders can be given greater weight in the comparison of alternatives (see Section 4.3.1 and Table 4.1) where the importance to stakeholders was quantified and input into the comparison using a decision science approach.

Decision science is a tool first developed for business-oriented decisions that incorporates many factors to assist in an objective approach to selecting from among many options (Tillman and Cassone 2012). The approach has been adapted to environmental decisions (Gregory et al. 2012) and can be very successful for a large, complex alternative selection in an environmental impact analysis. One of the basic methods applied to alternative selection combines the relative importance of each decision criterion with a measure of the intensity of impact to produce a total score for the alternative, relative to environmental concerns. Complex algorithms can be incorporated to include uncertainty, relative bias, and balance of unequal quantification, but the basic approach is

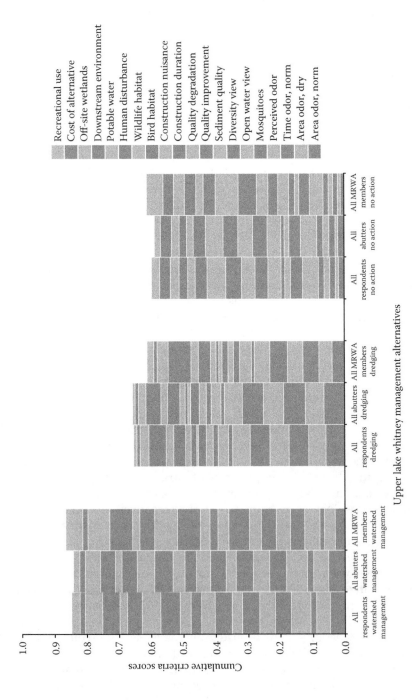

FIGURE 4.4
Comparison of upper Lake Whitney management alternatives: all survey respondents and by subgroup. *Abbreviation:* MRWA, Mill River Watershed Association.

degree of impact × importance of criteria, summed over all criteria. In the RWA lake management example discussed above (Section 4.3.1) the elements for a quantity comparison of alternatives were in place. The selection criteria had been identified and their relative importance determined (Table 4.1). The environmental analysis was then focused on the most important resources and criteria to produce a quantitative impact prediction. These were then combined for each alternative and the watershed management alternative was a clear winner among all stakeholders and each stakeholder group (abutters and Mill River Watershed Association Members) separately (Figure 4.4).

References

Bergman, J.I. 1999. *Environmental Impact Statements*. Boca Raton: Lewis Publishers.

Carnes, S.A. 1993. Social impact assessment and public involvement. In: *Environmental Analysis: The NEPA Experience*, ed. Hildebrand, S.G, and J.B. Cannon. 221–332. Boca Raton: Lewis Publishers.

Council on Environmental Quality. 2007. *Collaboration in NEPA: A Handbook for NEPA Practitioners*. Available from: http://www.ceq.eh.doe.gov/ntf.

Eccleston, C. 2008. *NEPA and Environmental Planning: Tools, Techniques, and Approaches for Practitioners*. Boca Raton: Taylor & Francis.

Fox, C. and P. Murphy. 2012. Sometimes less is better: Ethics of public participation. *Environmental Practice* 14(3): 212–219.

Gregory, R., L. Failing, M. Hartsone, G. Long, T. McDaniels, and Dan Ohlson. 2012. *Structured Decision Making: A Practical Guide to Environmental Management Choices*. West Sussex, UK: Wiley–Blackwell.

Kvaerner, J., G. Swensen, and L. Erikstad. 2006. Assessing environmental vulnerability in EIA; The content and context of the vulnerability concept in an alternative approach to standard EIA procedure. *Environmental Impact Assessment Review* 26: 511–527.

Lostranau, C., J. Oyarzun, H. Maturana, et al. 2011. Stakeholder participation within the public environmental system in Chile: Major gaps between theory and practice. *Journal of Environmental Management* 92: 2470–2478.

Maughan, J.T. 1993. *Ecological Assessment of Hazardous Waste Sites*. New York: Van Nostrand Reinhold.

Menzie, C., M.H. Henning, J.T. Maughan, et al. 1996. Special report of the Massachusetts weight-of-evidence workgroup: A weight-of evidence approach for evaluating ecological risk. *Human and Ecological Risk Assessment* 2(2): 277–304.

Morgan, R.K. 2001. *Environmental Impact Assessment: A Methodological Perspective*. Norwell, MA: Kluwer Academic Publishers.

Mulvihill, P.R. 2003. Expanding the scoping community. *Environmental Impact Assessment Review* 23: 39–49.

Noble, B.F. 2006. *Introduction to Environmental Impact Assessment*. Ontario: Oxford University Press.

O'Faircheallaigh, C. 2010. Public participation and environmental impact assessment: Purposes, implications, and lessons for public policy making. *Environmental Impact Assessment Review* 30: 19–27.

Schoepfle, G.M., E.J. Szarleta, and S. Schexnayder. 1993. How severe is severe: Public involvement and systematic understanding of wilderness as a resource. In: *Environmental Analysis: The NEPA Experience*, ed. Hildebrand, S.G., and J.B. Cannon. 289–312. Boca Raton: Lewis Publishers.

Smith, M.D. 2007. A review of recent NEPA alternatives analysis case law. *Environmental Impact Assessment Review* 27: 126–144.

Snell, T. and R. Cowell. 2006. Scoping in environmental impact assessment: Balancing precaution and efficiency? *Environmental Impact Assessment Review* 26: 359–376.

Stern, M.J., S.A. Predmore, M.J. Mortimer, and D.N. Seesholtz. 2010. The meaning of the National Environmental Policy Act within the U.S. forest service. *Journal of Environmental Management* 91: 1371–1379.

Tillman, F.A. and D.T. Cassone. 2012. *A Professional's Guide to Decision Science and Problem Solving*. Upper Saddle River, NJ: Pearson Education: Inc.

USAID. 2005. Environmental Compliance Procedures: United States Agency for International Development Title 22, Code of Federal Regulation, Part 216 Agency Environmental Procedures.

5
Conducting the Environmental Impact Analysis and Assessment

5.1 Environmental Impact Analysis Components

The planning, public input, and compliance with regulations, as discussed in the preceding four chapters, are all critical to the acceptance, implementation, and overall success of an environmental impact analysis. However, all of these activities will bear no fruit without actually conducting an objective, technically sound, and well documented analysis of the impacts. The analysis is the detailed and hard work component of the process and is analogous to conducting the experiments to test the hypothesis as part of the scientific method. The analysis applies physical, biological, and social sciences as well as engineering, and land use planning. These tools are used in combination with the experience of the environmental assessment team to understand the implications to each environmental resource of concern for each of the alternatives, thus helping to predict the impacts. This understanding and predictions of impacts are the critical environmental input to decision makers as they select and implement the proposed action.

There are three primary components of the environmental analysis: description of the affected environment, prediction of impacts, and impact mitigation. The first of these, that is, the description of the affected environment, characterizes the baseline or existing conditions upon which the proposed action and each alternative will be imposed. The prediction of impacts is a determination of how the existing condition of each environmental resource will be changed or impacted by the implementation of the proposed action and each alternative. The final step takes advantage of everything learned during the process in order to mitigate the identified impacts. Each of these analysis components are discussed in this chapter.

5.2 Affected Environment

The description of the affected environment documents the conditions of each critical resource as it exists prior to any activity implementing the proposed action. The terms existing conditions or baseline conditions are also used to identify the affected environment. However, the term affected environment is used here because it emphasizes that an environmental analysis should focus on the environmental resources that are potentially affected and not all existing conditions. As illustrated in Figure 4.1, in the "credit card" diagram, the affected environment is confined to the environmental resources, which overlap with the various components of the alternatives and do not include existing conditions unrelated or not potentially altered by any alternative action under consideration. The following sections describe the steps and approaches for addressing the affected environment in a comprehensive environmental analysis, with the first step determining which environmental resources comprise the affected environment.

5.2.1 Environmental Resources Comprising the Affected Environment

One of the criticisms of the early environmental analyses was that they were encyclopedic with little analysis or interpretation of the data provided to help decision makers understand the environmental consequences of the proposed action and alternatives. The criticism arose because the early documents were dominated by lists and detailed descriptions of conditions such as biological species, water quality, geology, and climate. The discussion of these resources dominated the description of the affected environment, not necessarily because they might be affected by any of the alternatives, but because there was data available and such descriptions were within the comfort zone of the first environmental analysis practitioners. The early environmental analysis teams had no experience, guidance, or examples of how to conduct environmental analysis, so they reverted to what they knew how to do as scientists: a detailed description of the most obvious features.

As environmental analysis has matured with more guidance documents and successful examples, it has become more focused. The focusing of the analysis begins with the identification of the environmental resources potentially at risk for inclusion in the description of the affected environment. Both technical and social scoping (see Sections 4.3.1 and 4.3.2) are primary tools for identifying the resources at risk and inclusion in the environmental impact analysis. The impact prediction conceptual model (see Section 5.3) that hypothesizes where an impact could potentially occur is critical for identifying the resources potentially at risk. There are other tools that can be used, but the critical process is for the environmental analysis interdisciplinary team to consider input from scoping, past experiences, and preliminary analysis to identify the areas where impacts could occur, and these areas constitute the affected environment.

A screening approach was successfully employed in the U.S. Coast Guard (USCG) dry cargo residue (DCR) environmental impact statement (EIS; USCG 2008) to identify environmental resources potentially at risk, thus warranting inclusion in the affected environment section (see Chapter 10 for description of the DCR EIS). All of the environmental resource areas typically covered in USCG and similar EISs were identified, supplemented with input from internal scoping, and then a preliminary and qualitative evaluation was done on each by the environmental analysis team to determine whether they were potentially affected by any of the alternatives under consideration. The results of the evaluation that eliminated many environmental resources from detailed consideration in the EIS were reported at the beginning of the affected environment chapter of the EIS (Table 5.1 summarizes results of the resource screening process). The evaluation identified the environmental resources that would be addressed in detail as part of the affected environment chapter and analyzed in the EIS: Great Lakes sediments, water quality, biological and related resources; and socioeconomic environment. The DCR EIS was particularly amenable to this approach because it was confined to a narrow range of activity (i.e., management of DCR) and the waters of the Great Lakes. A critical element in the success of the approach is that it was presented during both social and technical scoping with a solicitation of comments on resources potentially at risk, which had been missed or prematurely screened from detailed consideration. This approach yielded buy-in from the major stakeholders, even those with serious concerns over any DCR discharge to the Great Lakes. For the DCR EIS, as in preparation of any environmental analysis, once the resources of concern have been determined and stakeholder buy-in has been received, the detailed description of each resource as it relates to the alternatives under consideration can begin.

5.2.2 Mandatory First Step in Describing the Affected Environment

In almost every case there is a wealth of readily available information on the affected environment. In the United States the federal, state, and local governments have expended considerable resources monitoring and describing environmental conditions. They have also required private entities to perform extensive environmental monitoring under various regulated activities. In addition, academic research frequently includes description and analysis of existing conditions. In most cases, the combination of data and information on existing conditions from all these sources is broader, more detailed, at a higher density, and of longer duration than information that can be originally developed within the scope, budget, and schedule of an individual environmental impact analysis.

TABLE 5.1

Environmental Resources Eliminated from Detailed Evaluation in the U.S. Coast Guard DCR EIS

Environmental Resource	Preliminary Evaluation Conclusion
Geology, topography, soils, hydrology, and floodplains	The purpose and need are limited to ship board activities and thus there are no potential ground disturbing activities
Air quality	Potential air quality implications of the loading/unloading process of all dry cargo, so overwhelm any implications of just DCR alone that any effects on air quality from DCR management alternatives would be *de minimis*
Noise	Noise impacts occur when sound levels exceed thresholds of noise-sensitive receptors, and all DCR alternatives involve shipboard activities where there are no noise-sensitive receptors and there is high ambient noise. Thus no noise impacts will occur.
Potential hazardous material	By definition, hazardous materials are excluded from the DCR management rulemaking thus none are involved with any of the alternatives.
Land use and housing	Any proposed activities are shipboard focused, so no land-use alterations or implications are envisioned.
Cultural resources	All activities to address the purpose and need would occur in the open water shipping lanes of the Great Lakes, and a survey of the lanes demonstrated there were no submerged historic resources.
Visual and aesthetic resources	Alternatives under consideration involve only shipboard activities where there are no visual or aesthetic resources.
Land-based traffic	Alternatives under consideration involve only shipboard activities, and no land based transportation resources would be involved.
Water-dependent recreation	Any activities will occur at marine terminals or in active shipping lanes, which are devoid of water-dependent recreation.
Population and services	Only vessels in transit or at port are involved, and these will not affect populations in the study area.

Source: U.S. Coast Guard. Final Environmental Impact Statement: U.S. Coast Guard Rulemaking for Dry Cargo Residue Discharges in the Great Lakes. 2008.

It is imperative that the environmental impact analysis team conduct a thorough and comprehensive search of available literature and other information on the affected environment before embarking on or even designing an original data collection effort. Compliance with this axiom not only makes common sense, improves the quality of the analysis, and has been proven effective overtime, but it embodies several benefits for the project proponent and many other stakeholders. Typically, the data collected or required of private entities for environmental purposes by government agencies and for academic research is of a longer duration than one-time data collection for a specific environmental analysis. Thus it can indicate trends and long-term averages rather than just a snapshot of conditions at the time of the environmental analysis. The data from such sources are also generally of high quality because they have been generated by standard and accepted methods and are frequently subject to peer review. Thus they are a sound representation of the current and historic conditions of the environmental resources potentially affected by the proposed action or any of the alternatives. Existing data, particularly through an established monitoring program, are also more likely to be accepted by stakeholders, particularly members of the technical advisory committee (TAC) and regulatory agencies, and it is less likely to be the subject of bias allegations by skeptical stakeholders or advocacy groups. Maximizing the use of previous collected information and data is much more efficient than expending precious environmental analysis resources for expensive one-time data collection programs for only a snapshot of existing conditions. Given all these advantages, it is surprising that it is not uncommon for an environmental analysis to be conducted in a vacuum. Frequently the environmental analysis team does not acknowledge or take advantage of all the work that has been done previously. This can result in embarrassment and loss of credibility if one-time studies conducted solely for the analysis are contradicted by the results of long-term and accepted data collection programs.

The search and review of available baseline information are also mandatory inputs to designing any site-specific environmental analysis-specific data collection program. A comprehensive understanding of what information is already available is critical to designing a data collection program so that it can "fill the gaps" or confirm previously established trends rather than duplicating information that is already available. Collection of specific information should also be consistent with the available information by following the same data collection procedures, analysis methods, and where appropriate, matching previously used sampling locations so that valid comparisons can be made over time and geographic locations.

There are many available sources of environmental baseline information and data in most areas of the United States and increasingly in other countries throughout the world. The data from these sources can be utilized to describe existing conditions and if additional data are necessary they can be used to design the data collection program. The environmental analysis team must be creative and resourceful to identify all the potential sources of information but should not ignore readily available sources such as:

- Previously conducted environmental analyses. As is discussed in Chapter 6, these can often be incorporated by reference.
- Monitoring programs required as part of wastewater discharge permits. These are conducted by the dischargers but submitted to state or federal regulatory agencies for review and they are subject to split-sampling and oversight.

- Ongoing regional monitoring required by the Clean Water Act and Clean Air Act at established long-term sampling sites.
- Hazardous waste cleanup efforts. As part of remedial site investigations, the conditions of critical environmental resources associated with the site, and sometimes the adjacent areas, are described in detail (e.g., biological conditions as part of the ecological risk assessment required for most site cleanups, Section 7.2).
- Academic research. Investigations done for some graduate theses frequently include description of existing conditions as part of testing a hypothesis.
- Government-sponsored environmental monitoring programs. For example, the U.S. EPA sponsored the Environmental Monitoring and Assessment Program (EMAP) that focused on aquatic environmental resources. The goal of EMAP was assessment of current ecological condition and forecasts of future risks to natural resources. EMAP collected field data from 1990 to 2006. Currently, the National Aquatic Resource Survey, conducted by EPA, has incorporated much of the former EMAP's monitoring efforts of the nation's aquatic resources. There are similar programs in place by other resource and regulatory agencies at the federal, state, and local levels.
- Master plans developed by regional planning agencies, states, and local municipalities.

There is another source of available information that should be incorporated into the affected environment description. This information comes from stakeholders and can be valuable, worthless, or anything in between, but no matter its value, stakeholder input must be considered. As discussed in Section 4.3, scoping input by stakeholders can be an important source of information useful in describing the affected environment. Non-environmental professionals can have valuable information regarding the local area and conditions that can be helpful and not readily available from other sources. This can include personal observations, anecdotal reports, gray literature (e.g., high school or nonprofit monitoring data), news reports, etc. Caution must be exercised in incorporating such information because frequently the information reflects bias, posturing, or misinterpretation.

Interviews with local fishers are an example of stakeholder input to the description of the affected environment. People often spend decades fishing and just observing natural environmental conditions in local aquatic systems and can provide a wealth of knowledge such as:

- Habitat conditions
- Observable water quality conditions (e.g., turbidity, aquatic plant growth, odors)

- Presence and apparent density of fish and other aquatic species
- Mammals and birds dependent on aquatic resources
- Changes in water quality and the aquatic community over time

Information gained from fishers and similar sources warrant particular scrutiny because the providers of information have a vested stake in the resource. They can exaggerate the positive attributes in an attempt to influence decision makers to avoid the resource, or they can minimize the attributes to prevent exploitation of the resource and diminish recreational fishing success. It is also important when using information on existing conditions from fishers and similar resources to keep in mind the old adage learned from decades of experience: "fishermen know a lot about fish, and some of it is true."

Data and other reports of existing conditions identified by stakeholders must be addressed in the description of the affected environment. If the information is ignored, the environmental analysis team will spend more time and effort responding to comments on why it was not used than if they had considered it in the description. Also not addressing the information can damage credibility and spoil the stakeholder relationship and support. But, the source of the information, inherent limitations, and potential uncertainty must be recognized, and if the information is critical to prediction of significant impact or a decision regarding the proposed action, it should be verified.

5.2.3 Original Investigations to Describe the Affected Environment

After the available information has been gathered and reviewed for each environmental resource potentially affected, it is frequently necessary to conduct at least confirmatory original investigations of existing conditions to fill data gaps. For any such investigation, "proceed with caution" is good advice both when making the decision to collect original data and designing the data collection program. There are many pitfalls awaiting the overeager environmental analysis team embarking on data collection programs:

- A little knowledge is dangerous. A quick peek or snapshot of the affected environment can often present a distorted picture, and interpretations made on such information are often unacceptable to adversarial stakeholders (and all too frequently not correct). The risks of this pitfall can be minimized by using original data collection to relate a specific situation to a larger data set even if the existing data only represents a similar adjacent location.
- Proving the negative is difficult, particularly within the scope and time frame of an environmental analysis. The classic example is a survey conducted at a proposed site to demonstrate that rare or endangered species or even their habitat is not present. Such one-time

efforts or even surveys conducted over a year can rarely convince the doubter and will almost always generate comments, like the survey was not comprehensive enough, conducted at the wrong time, and observers were biased. Relying on the long-term record, general habitat description, or just making the conservative assumption that the species could be present in the long run is often the most efficient and productive approach.

- Use of inappropriate methods. This is a common shortcut taken as part of affected environment investigations and can be easily avoided by identifying which methods have historically been used and accepted. The risk of such pitfalls can also be minimized by review of the proposed methods with TAC as part of the investigation planning and design. If available resources and schedule do not allow use of accepted methods, it is frequently better in the long run to rely on existing data and conservative assumptions rather than waste resources on a data collection program that will not be useful or accepted to support decisions.

- Collection of too much or inappropriate data. It is not uncommon to see pages and pages of data in the affected environment section of an environmental impact analysis and then never see it again. Frequently, it is not even referenced in the decision-making efforts or other critical aspects of the analysis such as: impact prediction, comparison of alternatives, impact mitigation, or documentation of decisions. A good way to minimize the risk or repercussions of this pitfall is to always ask the question, "how will I use the results of the data collected?" before collecting the data. Unless there is a definitive answer, the data collection effort needs to be redesigned or dropped.

- Failure to collect information for subsequent environmental approvals. As discussed in Section 9.2, environmental analysis is just the first step in gaining environmental acceptance and approval. Following the environmental impact analysis, there are environmental permits required and frequently, public support is essential to implement the proposed action. Identifying the necessary permits and public concerns and their associated data input requirements should be incorporated into any investigation of existing conditions. Most environmental permits have very specific baseline information requirements and methods to be used to gather the information. Significant time delays and wasted resources can result if it is necessary to collect separate permit-specific baseline information when the necessary information could have been part of the affected environment data collection with proper planning. In the worst case, data collected for permits and that collected as part of comprehensive environmental

analysis could be contradictory, thus calling the whole process into question.
- Expending all resources and time on affected environment investigations and not leaving adequate time or resources for the critical impact prediction and alternative comparison steps in the environmental analysis process.

Once all necessary precautions have been exercised and the appropriate stakeholders consulted, the designing of the affected environment data collection program can commence. In most cases, the investigations address environmental resources commonly monitored and there are standard methods and protocols that should be followed. The members of the interdisciplinary environmental impact analysis team with expertise in the environmental resource under investigation should be familiar with the standard techniques and apply them to the design of the affected environment investigation.

A primary, affected environment data-collection challenge presented to the team is to collect the information that is needed at the required level of detail. The combination of issues, potential impacts, and specific environmental resources involved is infinite, so there is no standard affected environment investigation template. Every investigation is unique and must be designed by the environmental analysis team to efficiently meet only the specific needs of the overall environmental analysis. Summarized below are affected environment investigations from different environmental impact analyses that illustrate different approaches based on different issues and site-specific conditions. A common thread to these case studies is a phased approach, beginning broadly and then narrowing in on areas of uncertainty or greatest potential impact. In each case, this approach was successful in maximizing efficiency and collecting only useful information.

5.2.4 Estuarine Wastewater Discharge-Affected Environment Investigation Case Study

The town of Scituate, Massachusetts, is a community south of Boston that derives its identity from coastal resources. It is bordered on the east by Massachusetts Bay beaches and the southern border of the town is the North River, a major tidal river with numerous tributaries and adjacent broad salt marshes (Figure 5.1). These resources support extensive recreation and commercial activity including fishing, lobstering, clamming, boating, and swimming, and these activities have increased the town's popularity and stimulated construction of new homes and conversion of summer "camps" to full-time residences. During the 1970s and 1980s, the popularity and conversions had a downside as the on-lot wastewater systems (i.e., cesspools and septic tanks) adequate for the camps and low density development were not capable of handling the domestic sanitary waste generated from the higher

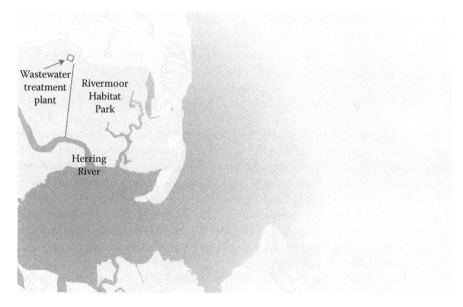

FIGURE 5.1
North River and nearshore estuarine system.

density year-round inhabitation. As a result, faulty and failing on-lot wastewater treatment systems were threatening the environmental health of the very marine resources that characterized the town.

The solution to protecting the coastal and estuarine resources was to expand the town's limited existing public sewer system service area and upgrade the treatment facilities. Two primary alternatives to the treatment and disposal of the wastewater from the expanded system were identified during scoping. One was to provide an advanced level of treatment and maintain the existing discharge to a small tidal tributary through the *Spartina* spp. salt marsh adjacent to the North River. The added flows from the expanded system would increase the current nutrient and organic load to the system, which in turn would exacerbate the eutrophication and associated degradation of water quality and estuarine habitat in the marsh, river, and nearshore environment (issues identified during scoping). An advanced level of treatment was necessary to address this issue if the discharge of treated wastewater were to remain at the existing location within the North River system.

The second wastewater treatment and disposal alternative identified during scoping was to relocate the discharge via an outfall pipe and effluent diffuser beyond the North River and the adjacent nearshore environment. This would reduce the nutrient and organic load to the system, thus arresting the accelerated eutrophication process and addressing the issues of water quality and estuarine habitat degradation. The question was how far offshore must the discharge be in order to avoid enriching the river and nearshore system?

This situation reflects a meaningful comparison of alternatives and ultimately a decision centered on the major issues, as discussed in Section 4.5. Both alternatives (advanced treatment with nearshore discharge or extended effluent discharge) met the purpose and need of eliminating failing septic systems and the associated degradation of the estuarine environment while not exacerbating and in fact, potentially improving conditions resulting from nutrient and organic loading. So the comparison of alternatives was based on cost, reliability of treatment, construction impacts, and other benefits associated with each alternative.

The challenge of describing the affected environment was to fully understand the nutrient and organic loading to the system and how it affected the eutrophic status of the estuary and nearshore environment (Cibik and Maughan 1991). This information would then be the primary input in determining the exact level of advanced wastewater treatment for the North River discharge alternative and exactly how far offshore the outfall must be for the other alternative. The challenge was approached consistently with the described process (Section 5.2.2) of thoroughly reviewing available information and supplementing with original data collection only where necessary. This mandatory first step in describing the affected environment was all that was necessary for the advanced treatment/marsh discharge alternative. The existing information available from such sources as the state's monitoring program and studies commissioned by the North River watershed association revealed that the North River system was healthy, but very near its assimilative capacity. Any substantial increase in nutrient or organic loading could push the system beyond its capacity to process the load and result in excessive plant growth and oxygen demand. Thus the literature review fully defined nutrient and organic loading to the system and how it affected the eutrophic status of the estuary. It also provided the necessary input to fully develop the details of the advanced treatment alternative: in order to avoid degradation of water quality and estuarine habitat, the nutrient and organic load associated with the increased wastewater flow had to be offset by a wastewater treatment process with a higher removal rate. In fact, the affected environment evaluation revealed that the system was close enough to the assimilative capacity that a margin of safety was appropriate and the total nutrient and organic load to the system must actually be less than the current load. Using this information, the wastewater treatment process needed to accomplish the reduced load could be defined and the cost of implementation calculated.

The existing information was not adequate to understand offshore conditions and an original investigation was necessary. The challenge to the environmental analysis team was that it is a "big ocean" (actually Massachusetts Bay) out there, and available financial resources, personnel, and schedule limited the scope of any original investigation and required that data collection be focused on a narrow set of questions. Thus concise questions were posed:

- How far offshore does the nutrient and organic loading from the North Shore system enrich the nearshore environment?
- What level of increased loading, if any, can the nearshore environment accommodate without exceeding the assimilative capacity?
- How far offshore must a discharge be to prevent the nutrients and organic matter associated with the discharge from being entrained with the incoming tide and entering the North River and nearshore systems?

Even with these concise questions, an efficient investigation design was necessary to meet the schedule and resource limitations for the investigation. The question of entrainment with the incoming tide could be relatively easily addressed by a series of current measurements over a full monthly tidal cycle. When the measured tidal currents were input to a standard estuarine mathematical hydraulic model, the output adequately determined the offshore "break point" where introduction of nutrients and organic matter into the North River system was minimal.

The questions associated with the enrichment issues of the offshore affected environment were not so easily addressed. For a classic research investigation, the nutrient concentrations, plankton dynamics, and chemical reaction rates would have been measured over at least a full calendar year in both the North River and offshore areas. These measurements might even have been supplemented with laboratory or *in situ* algal growth experiments to determine the nutrient concentration's limiting growth. All of this information would then form the input to a complex ecological model to determine the existing relationship of nutrient concentrations, organic load, algal growth, dissolved oxygen levels, deposition of carbon in the sediments, etc. Such an investigation would make a great master's thesis, but available resources and schedule for the Scituate wastewater environmental impact analysis prevented such a comprehensive investigation, and creativity was necessary to conduct an affected environment investigation design to address the questions.

Two creative approaches were identified to design an affected environment investigation to adequately address the critical enrichment questions and complete the investigation within the resources and schedule available. The first approach was to shorten the investigation to a "snapshot" rather than an investigation of the dynamic seasonal characteristics of the system. Review of the available literature (see Mandatory First Step, Section 5.2.2) and extensive work on estuarine wastewater discharges and similar projects by members of the environmental evaluation team with marine science expertise revealed that nutrient and organic loading frequently left a "footprint" in the sediment as the dense plankton growth resulting from enrichment died and settled to the bottom. Thus delineation of the footprint in the sediment (if there was one) would define the area affected by the nutrient enrichment and the intensity of the footprint would reveal the degree of enrichment. The advantage of this method was that the sediments represented an integration of processes that occurred over time as opposed to the one-time snapshot of very dynamic water column processes and nutrients available from a limited collection of water samples.

Once the sediments were identified as the focus of a creative investigation to understand the enriched dynamics of the affected environment, the second approach to maximize efficiency came into play. A phased approach was selected to investigate the sediments at various levels of detail at different special scales. The first phase was to tow a video camera on a sled positioned just above the bottom over a large area (Figure 5.2) to visually characterize the bottom with respect to deposition of organic matter. The video camera could be towed at a speed of 1 to 3 knots so a very large area of the bottom could be surveyed in just a few days. The resolution of the video imaging was adequate to characterize the bottom type as:

- Definitively not an enriched area with sediment consisting of coarse sand, gravel, boulders, or exposed bedrock
- Definitely an enriched area with sediment consisting of black, highly organic, fine particles with bottom contours relatively even and flat
- Indeterminate areas with both enrichment and non-enrichment characteristics.

The video camera provided a real-time feed to monitors on the research vessel towing the sled. This allowed modification of the originally planned coverage to fully define the limits of organically enriched sediment deposition. It also allowed adjustments to the preplanned survey lines to best define the boundaries between depositional and nondepositional areas.

The first phase video imaging was adequate to grossly define the various offshore areas. Some, with coarse sediments were not subject to enrichment effects and thus suitable for a wastewater discharge. Other areas were already obviously enriched by the nutrients and organic matter discharged from the North River and locating a wastewater discharge outfall in these areas could exacerbate the problem. There were also areas where the exact boundaries and intensity of enrichment were not apparent from the video imaging alone and a second phase of investigation was necessary.

The second phase of the investigation was to deploy a sediment camera (Figures 5.2 and 5.3) over a smaller area where the sediment characteristics could not be fully determined by the towed video camera. The sediment camera was capable of characterizing the sediment more precisely and in much more detail than that achievable with the towed video camera. Deployment of the sediment camera preempted the need and avoided the expenditure of resources and time necessary to actually collect and process samples over the

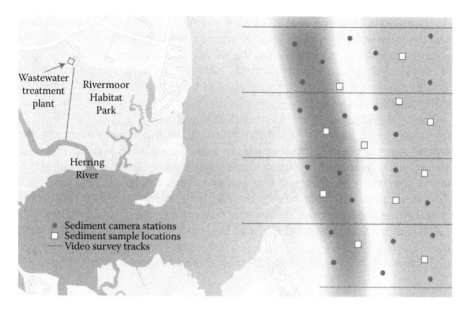

FIGURE 5.2
Video survey, sediment camera imaging, and sediment sampling locations.

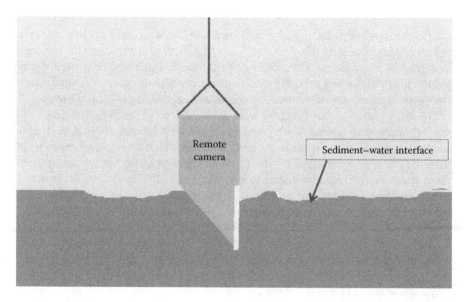

FIGURE 5.3
Sediment imaging camera.

entire study area. This camera penetrated the sediment–water interface, thus allowing a detailed inspection of the sediment surface and also giving a wealth of information of sediment characteristics at depth. Thus, a distinction could be made regarding areas with just a thin surface of enriched sediment, which was most likely transitory and organically rich sediments over depth indicative of long-term and persistent enrichment.

Another advantage of the investigation method was that it was not necessary to photograph every area that could not be characterized from the video imagery. Where there were multiple areas with very similar appearance in the video image but undetermined enrichment characteristics, only selected areas had to be photographed with the sediment camera. A subset of areas could be photographed and the interpretation of the results applied to areas appearing similar in the video image.

The combined results of the video imaging and sediment camera photographs gave a precise demarcation of the offshore area of enrichment (Figure 5.4). The results also gave a very good indication of the intensity of enrichment based on the visual observation of *in situ* sediment pictures and the depth of enriched sediments. From this information, the outfall location that avoided enriching sediments beyond the assimilative capacity and producing adverse impacts to the estuarine system could have been determined. However, since the treatment/discharge decision was nonreversible and required a significant commitment of engineering, construction, and operation resources, the stakeholders supported a final phase to confirm the

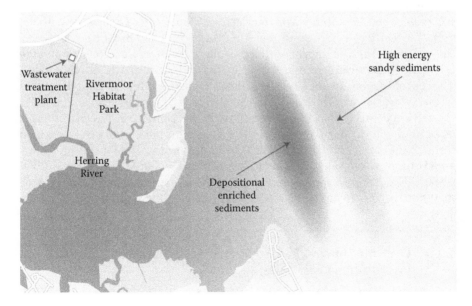

FIGURE 5.4
Scituate offshore affected environment.

description of the affected environment and minimize uncertainty. The final phase consisted of collection and processing sediment samples from a range of areas along an enrichment gradient defined by the first two phases of the investigation for both chemical and biological characteristics (Figure 5.1). Thus there were verifiable laboratory data to confirm conclusions from the first two phases of the affected environment investigation. The laboratory analysis was able to quantify the degree of enrichment as the organic carbon content of sediments in the enriched area approached 4% as opposed to less than 1% in other areas. The analysis also documented a different biological assemblage in the area, indicating that although not currently stressed, substantial additional organic and nutrient inputs to the area could result in degraded conditions.

The approach used to describe the affected environment for the Scituate wastewater environmental impact analysis was creative, conclusive, and efficient. The major area of concern (the sediment), which was also indicative of related enrichment conditions, was identified and the investigation was focused on this area. The approach was phased with the early efforts providing general observations over a large area and later phases focusing on the specific areas of concern and providing more detailed information. The result was a detailed description of a small well-defined subset of affected environmental conditions (area of impact in the "credit card" Figure 4.1) that were subject to impact. The description of the affected environmental also provided the necessary input to compare alternatives and make an informed decision. In the final analysis, based on the description of existing conditions, there was a clear choice between advanced waste treatment and an offshore discharge with the outfall extending well beyond the enriched area at the mouth of the North River (Figure 5.4). The impacts of the two alternatives on water quality and the estuarine system were similar, but they differed significantly in other areas. The cost of the offshore outfall and diffuser were much higher than the cost of advanced wastewater treatment because the outfall pipe had to extend beyond the enriched area. Also there were potentially significant adverse impacts associated with the construction of the outfall pipe, particularly in the section through the salt marsh, barrier beach, and sand dunes. Thus the decision was an advanced wastewater treatment plant with a discharge to the North River system. The plant is now in place and the reduction of nutrients and organic matter to the system has been effective in protecting the health of the estuarine ecosystem.

5.2.5 Over-the-Horizon Radar-Affected Environment Investigation Case Study

As part of the Cold War defense system, the U.S. Air Force committed to installing an over-the-horizon (OTH) radar system providing detection capability up to 5000 km beyond the shores of the continental United States. Four transmitting and four receiving facilities located throughout the country were called for, and the installation of these large facilities (up to 1200 hectares each) necessary for such an extensive range of coverage was considered a federal action with potentially significant impacts requiring NEPA compliance and an EIS. The investigation of the affected environment for the transmit OTH radar in Modoc National Forest (northeastern California) was focused on the critical issues and demonstrates an imaginative and creative approach that can provide insight to structuring affected environment investigations for other situations.

The preferred location for the Modoc National Forest OTH transmitting radars was in potential conflict with two important and environmentally sensitive resources identified during scoping: large mammal (pronghorn antelope and mule deer) migration and Native American cultural resources. Thus the description of the affected environment had to address the existing migration patterns of the large mammals and the cultural resource characteristics of the area. The mandatory first step of the investigation (literature review) identified important information on the two critical resources such as the type of cultural sites to expect and a general east–west migration pattern for both the antelope and mule deer. The animals traversed the study area in the spring on the way to their feeding or breeding grounds and in the fall on the way to their wintering areas. They were present in the study area only for these very short migration periods, but their continued survival was dependent on successfully migrating through the area twice each year. However there was no site-specific information available for their exact migratory routes, thus an investigation of the existing conditions of these resources had to be designed and implemented.

The standard methods to describe large mammal migration were not compatible with the schedule of the environmental impact analysis; thus creative measures had to be derived. The common methods for mammal migration investigation were real-time observations (via on-ground spotters and aerial surveillance) and radio tagging and tracking of the animals. Both these methods required multiseason investigations, but the environmental analysis had to be completed before fall and spring observations could be conducted. After consultation with stakeholders, including the Forest Service and state fish and game wildlife specialists, an effective and efficient method that did not require real-time observation during both migration seasons was developed. The entire study area (over 2000 hectares) was surveyed by biologists (primary graduate students) by walking regularly spaced transects. Although the survey was conducted during the summer, when there were no large mammals in the area, the migration routes were discernible by fecal pellets left by the antelope and deer as they migrated. The pellets left by the two species were very similar, and it was not practical to determine in the field which were left by which species. But given the objective and the narrow focus of the investigation, it did not matter because the objective was to ultimately understand the impact of the radar placement on large mammal migration, independent of species. Although the method was labor intensive, it could be conducted within the schedule of the environmental impact analysis, and it proved effective in identifying the migration routes.

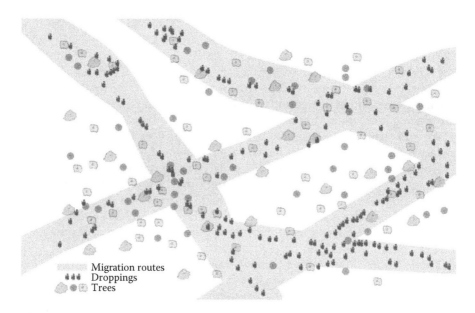

FIGURE 5.5
Over-the-horizon radar study area antelope/deer migration routes.

Four well-defined routes were defined by the high density of fecal pellets in discernible patterns (Figure 5.5). It was not clear which of the four routes were traveled by mule deer or pronghorn antelope in spring or winter but it did not matter because the focused question was not which migration route, by which large mammal would be impacted, but whether any large mammal migration route would be obstructed by the OTH radars.

The affected environment related to cultural resources is not transient as is the mammal migration, so conventional investigation methods could be applied. Unfortunately, as determined by review of available information, the cultural sites most likely to be present in the study area were small and could not be identified by aerial survey or photographs. The most common cultural sites in the area were temporary Native American hunting locations where a day camp was established and the hunters shaped obsidian rock (naturally occurring volcanic glass formed as an extrusive igneous rock) into points (i.e., arrow and spear heads). The actual sites consisted of small areas with high concentrations of obsidian flakes, which were the by-products left by the hunters when they shaped the obsidian into points.

The identification of cultural resource sites was accomplished by an "on the ground survey" in a similar manner to the large mammal fecal pellet method, except anthropology graduate students instead of biologists conducted the survey. The survey identified distinct sites characterized by the obsidian flakes and potentially left by two different Native American groups (Figure 5.6). But as with the mammal migration investigation, for the purposes of affected environment investigation and impact prediction, it did not matter if there were

Conducting the Environmental Impact Analysis and Assessment

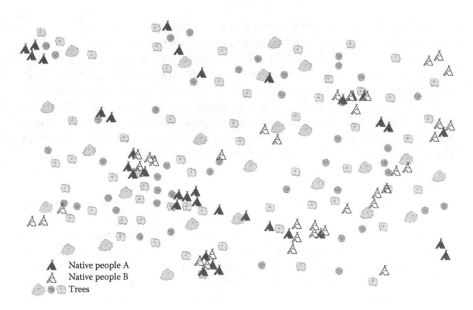

FIGURE 5.6
Over-the-horizon radar study area cultural resource (obsidian flake) sites.

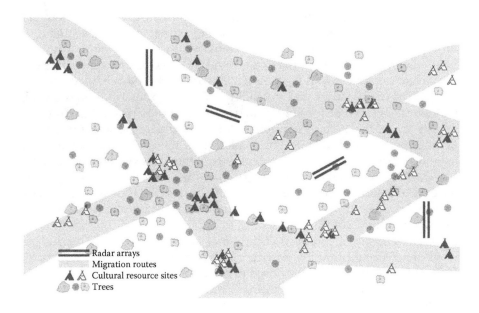

FIGURE 5.7
Over-the-horizon configuration to avoid cultural resource and mammal migration impacts.

two groups, and the location of the cultural sites was the critical input to the description of the affected environment, prediction of impacts, and mitigation.

The cultural resources/wildlife-affected environment investigations made for a busy summer with biologists bumping into anthropologists, and the soles of many pairs of boots were ground to dust from walking over many kilometers of the high-desert lava surface. But the two surveys proved to be confirmatory because the identified cultural resource sites coincided with the migration paths identified from the fecal pellet surveys (Figures 5.5 and 5.6). After the information was assimilated, this made sense because for centuries the mammals followed the exact same migration routes and the Native Americans established their hunting locations along the routes. This confirmed that the results of the pellet survey identified present, historic, and persistent migration routes and added a level of confidence to the findings. The results of the affected environment investigations allowed a proposed radar configuration that both avoided impacts to cultural sites and did not interfere with large mammal migration (Figure 5.7). Using the information from the affected environment investigation to determine the radar layout made the prediction of impacts very simple: the critical resources were avoided and the impact was minimal.

5.2.6 Dry Cargo Discharge to the Great Lakes Affected Environment Investigation Case Study

The investigation of the affected environment for the DCR management EIS (U.S. Coast Guard 2008) accomplished not only a description of existing conditions but also an effective approach, similar to the OTH example, to impact prediction. As described in the background for the DCR management environmental impact analysis (Section 10.2), since the late 19th century the residue spilled from loading and unloading DCR (i.e., primarily iron ore, coal, and limestone) has been "swept" into the waters of the Great Lakes. Thus the description of the affected environment reflects the impacts resulting from decades of discharging DCR. The challenge for the environmental analysis team was to identify the specific areas within the vast expanses of the Great Lakes where DCR had been discharged and then describe the conditions of the critical environmental resources within the discharge areas. This characterization would suffice as description of the affected environment (because future sweeping would occur in the same locations), and comparison of these areas to similar portions of the Great Lakes with no DCR discharges would lend insight into the impacts caused by decades of DCR sweeping. This comparison could then be used to predict future impacts.

A phased approach similar to the one described for the Scituate wastewater-affected environment investigation (Section 5.2.4) was followed for the DCR EIS description of the affected environment. As emphasized throughout this chapter, the first step was to review existing literature, and two types of useful available information were identified. The first source of information was ecological monitoring conducted by state and federal agencies in the Great Lakes, which described general conditions in various Great Lakes areas and also provided information for some specific locations associated with DCR discharge.

The second source of existing information was the logs of Great Lakes carriers where the latitude and longitude of the starting and ending points of DCR discharge were recorded. Review and plotting of the information from the ship's logs revealed that generally the same locations were used year after year by the vast majority of Great Lakes carriers for deck washing and discharge of DCR. Thus there were concentrated areas where the discharges had occurred.

The ship's logs identified these concentrated areas but the vastness of the Great Lakes made pinpointing the exact location of a DCR discharge difficult at best and a further refinement was necessary to determine where, within the concentrated areas, DCR and associated impacts could be expected. This refinement was accomplished by the next phase of the affected environment investigation: a towed side-scan sonar system was used (Figure 5.8) to identify discharged DCR within the concentrated areas of deck sweeping. This device used sound waves to distinguish the hard residue of coal, iron ore, and limestone from the relatively uniform and soft lake bottom.

Once these likely "targets" of discharged DCR were identified with the towed sonar array, actual sediment sampling similar to the Scituate investigation was conducted to both confirm the presence of DCR and describe the environmental conditions in the vicinity of the past and potentially future DCR discharges. These samples confirmed that the locations identified by

FIGURE 5.8
Sonar L-3Klein System 3000 towfish used to map Great Lakes affected environment (Courtesy L-3Klein, Inc.)

the side-scan sonar system were in fact the result of DCR discharges, as actual cargo such as taconite pellets (which are the preprocessed iron ore transported on the Great Lakes) were apparent in the sediment samples (Figure 5.9). The samples were also processed to determine the biological, chemical, and physical properties of the sediments in the areas of DCR discharge. These areas of historical DCR discharge would be the same areas of future discharge because the Great Lakes carriers would follow the same courses they have for decades; thus the investigation fully described the affected environment with respect to sediments for all DCR management alternatives. Comparison of conditions in discharge areas with similar areas receiving no DCR discharge was also a critical input to predicting impacts associated with the alternatives.

5.2.7 Affected Environment Investigation Summary

The description of the affected environment lays the foundation for the entire environmental impact analysis. It is typically the component of the overall environmental impact analysis that requires the most time and resources, and generates the maximum data. It should also be the simplest or at least most straightforward because it most often involves employing standard research methods that are well established and familiar to the members of the environmental analysis team with expertise in the environmental resource under investigation. However, it is also the component of the analysis most vulnerable to numerous pitfalls, including expending excessive time

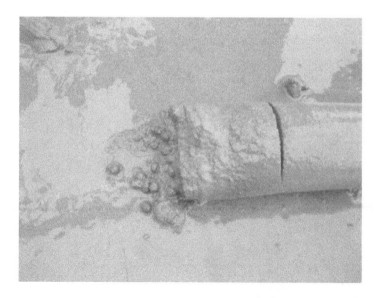

FIGURE 5.9
Great Lakes sediment sample with discharged iron ore (taconite) dry cargo residue.

and resources early in the analysis before the critical issues are identified. Also it is the first technical environmental effort in a series of environmental analysis steps, and misdirection at the beginning can set the entire process off course.

The investigation of the affected environment must establish the existing conditions of the environmental resources potentially subject to change by the proposed action or any of the alternatives, but the investigation can serve multiple purposes as pointed out in the above case studies. In the Scituate example, the investigation of existing conditions was critical in establishing the primary alternatives such that meaningful comparisons focused on the critical resources and issues. Similarly, in the OTH radar example, the investigation enabled structuring a proposed action that avoided impacting the resources of greatest concern. A critical component of impact prediction was accomplished by the affected environment investigation in the DCR example because the evaluation of existing conditions reflected decades of DCR discharge, which was embodied in one form or another in each of the DCR management alternatives.

The most important considerations in the planning and implementation of an affected environment investigation are:

- Completely identify and thoroughly review all available information on each critical environmental resource vulnerable to impact before proceeding with an original data collection program.
- Don't collect too much data on existing conditions and make sure the intended use of data is clear before it is collected. Additional data can always be collected, particularly in a phased approach with more detail collected in critical areas or locations, but once the resources and time have been expended to collect unnecessary information, they cannot be retrieved.
- Where possible, structure the investigation of the affected environment to achieve multiple goals, including impact avoidance, alternative development, impact prediction, and use of data for future environmental purposes such as permit applications.

5.3 Impact Prediction

Predicting impacts of the proposed action and alternatives is the hardest work but generally is the most rewarding aspect of an environmental impact analysis. Many of the other environmental analysis components (e.g., scoping, alternative development, and impact mitigation) are equally, or perhaps, more important because they address tasks critical to the success and focus the analysis. But impact prediction is typically the greatest technical challenge of the analysis and provides the requisite answers and input to decisions.

Forecasting the condition of numerous natural and built environment resources following construction and operation of facilities or infrastructure, or implementing of a policy or regulation, is no simple task. If one could accurately and simply know the future outcome of one's actions, not only environmental conditions but also human history would be quite different. But the challenge of predicting impacts to the extent practicable is the essence of environmental analysis and why it is so critical. Thus the challenge is to make the best prediction possible as to the condition of each critical environmental resource following implementation of an alternative given the funding, time, technical expertise and methods available. Added to this not so simple challenge is the necessity to focus on the critical resources where impact is most likely to occur and expending resources in proportion to the magnitude and likelihood of impact. This section presents available approaches to meet the challenge and case studies where the challenge has been successfully met, and others where the effort has fallen short.

5.3.1 Impact Prediction Conceptual Model

5.3.1.1 Model Development

The first step in predicting impacts for any alternative, for any purpose and need, is the development of an impact prediction conceptual model, qualitatively evaluating and illustrating where impacts could potentially occur. Developing this model involves predicting the consequences of an action through a number of steps beginning with a proposed activity and terminating with an impact receptor (e.g., fish spawning, traffic patterns, or National Park visitors). One of the final goals of environmental impact assessment and analysis is to quantify, to the extent practical, each sequential cause/effect step in the conceptual model to determine probability, magnitude and extent of the changed conditions for the receptor. Before significant amounts of energy, time, and funds are spent for the detailed investigations and models needed to quantify impacts, it is important to confirm that the impact could possibly occur. The impact prediction conceptual model is essential in making the determination of likely occurrence of an impact. The conceptual model also forms critical input to the planning of the affected environment investigation because it identifies the resources that must be addressed and helps identify the intensity and level of detail required for each environmental resource at risk.

The impact prediction conceptual model can be considered comparable to a computer program logic flow or a wiring diagram. The model can take a number of forms from the simple to the complex and are sometimes considered "casual networks" (Perdicoulis and Glasson 2006). In the conceptual model or casual network each potential impact is initiated with some aspect of the proposed action or alternative, just as a wiring diagram begins at the source of the electricity. The potential impact then proceeds through nodes again similar to the nodes of switches, fuses, and lights in a wiring diagram. The nodes in an impact prediction conceptual model might be environmental attributes such as a specific water body, assimilative capacity of the water body, or the aquatic organisms in the receiving waters, which could be the impact receptor.

An impact prediction conceptual model for a proposed multiuse land development (e.g., condominiums, elderly housing, retail stores, and recreational facilities) on existing agricultural land could be quite complex. The potential impact would be initiated at the site by factors such as development layout and area of each specific land use. A section of the impact prediction model would be developed for each area of impact such as traffic, noise, visual aesthetics, utilities, community services, wildlife habitat, and storm water. The conceptual model in each section would then be constructed showing the nodes between the source of impact and the receptor. For example, the storm water section of the conceptual model would show the source of impact as the roof and paved areas generating runoff. Landscaped areas might also be a potential impact source due to fertilizers and pesticides and the potential impact receptors would be aquatic biota, recreational users, and raw drinking water supply. The pathways and nodes in the model would include water quality and quantity of runoff, storm water conveyance system (including any proposed treatment or infiltration systems), quality and quantity of storm-water discharge, drinking water intakes in receiving waters, recreational areas in receiving water, and water quality/aquatic biota conditions in receiving waters. First, the investigation of the affected environment and then the impact prediction and quantification would focus on each node, and the impact conceptual model would be reviewed and revised as warranted as each affected environment and impact evaluation was completed. If the evaluation of drinking water sources revealed no sources downstream of the storm water runoff or no discharge points for the development, the path of impact to potable water would not be completed and no additional investigation of impact to drinking water would be necessary. For areas of impact that illustrate completed pathways, analyses must be designed and implemented to quantify the transfer among nodes to determine impacts.

The impact prediction conceptual model is an enhancement of the frequently used environmental checklists (Eccleston 2008) as a critical step in the impact prediction process. At its most basic level, the environmental checklist is a generic compilation of all possible environmental resources, both natural (e.g., water quality, terrestrial wildlife, endangered species) and built (e.g., traffic, community services, and infrastructure), that could conceivably be affected by any type of project in any location. The list is then reviewed in light of the alternatives and the resources potentially at risk are checked on the list. At a more detailed level, the single-column check list can be expanded to a matrix, with the resources on one axis and the type of activity (e.g., construction, operation, wastewater discharge) on the other axis. The checklists are designed as a tool for preparing environmental analyses to facilitate identification of potential impacts. If used appropriately, the checklists can minimize overlooking an area of impact by prompting consideration of each resource and mentally project the proposed action overlaying the resources. The resources confirmed as potentially affected are the environmental receptors at the bottom, or end of the impact prediction conceptual model (e.g., the impact receptors air quality, traffic delays, economic viability). The checklists provide an opportunity for discussion and consideration by team members of the proposed action and alternatives in relation to a laundry list of resources. However, if the checklist is completed in a vacuum by someone with expertise in only one or a limited number of environmental resources, the checklist completion task can become nothing more than a required exercise with little or no benefit. Also, generic checklists developed to accommodate all projects in all types of environments tend to be either so general that they are of little value (air quality, as opposed to mercury vapor concentrations) or so long that they are overwhelming, and critical resources at risk are not given adequate consideration.

The most advantageous use of the checklist is as a component in the development of a comprehensive and dynamic impact prediction conceptual model. A project- and location-specific checklist can be developed, perhaps referencing and editing a more generic list. The project-specific list can provide a starting place in the identification of environmental receptors and perhaps it can serve as input for an open discussion and workshop for the entire environmental analysis team and other stakeholders as they jointly develop the conceptual model.

5.3.1.2 Impact Prediction Conceptual Model Example: USCG DCR Environmental Impact Analysis

As described in detail in Chapter 10 (Section 10.1) the shipping of dry cargo (e.g., iron ore, coal, and limestone) from the locations where they are mined to manufacturing centers has occurred on the Great Lakes for more than a century. The loading, unloading, and transport of the cargo result in spillage on the deck and within tunnels of the ships, and this DCR was historically swept overboard. Under pressure from Great Lakes' environmental stakeholders the U.S. Congress recognized that this practice may have environmental impacts, violate the Clean Water Act, and be inconsistent with treaties and international agreements addressing Great Lakes environmental stewardship. As a result they charged the USCG with developing a suitable management plan and governing regulations, taking into account the impacts of past DCR discharge practices. Since the regulations were a major federal action with potential environmental impacts, NEPA applied and an EIS was required.

As discussed earlier, the first step in predicting environmental impacts is the development of an impact prediction conceptual model, and for the DCR EIS this was done initially for the continuation of existing DCR management practices, which was one of the alternatives considered in the EIS. In the impact prediction conceptual model (Figure 5.10) a potential impact is initiated during loading or unloading of the dry cargo onto a Great Lakes carrier. If during

Conducting the Environmental Impact Analysis and Assessment

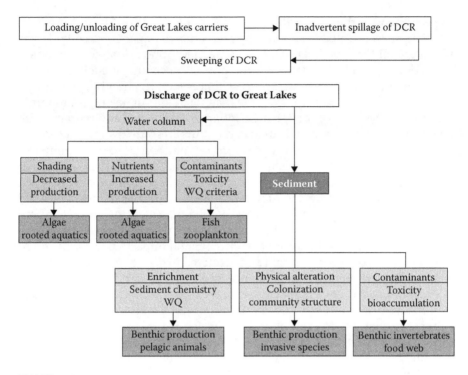

FIGURE 5.10
A DCR discharge impact prediction conceptual model for natural resources. *Abbreviation:* WQ, water quality.

loading or unloading, residual cargo is spilled on the deck or in the hold of the ship as indicated by the arrows in the figure, a series of events can be initiated that could potentially reach an ecological receptor and produce an impact. The sequence of events potentially resulting in an impact includes sweeping of the DCR into the waters of the Great Lakes, DCR dispersed in the water column and sediments, alteration of ecosystem attributes, and ultimately contact of DCR with receptors (e.g., pelagic animals, invasive species, or the aquatic food web). If all of the events in the series are completed, an impact will occur but if any are constrained, the impact will be mitigated or eliminated. An objective of the EIS was to determine the likelihood of completion of the series of events and the nature and extent of impact on the ecological receptors that would result. Thus as discussed in Section 5.3.4, a series of investigations were conducted to determine which of the steps in the impact prediction conceptual model were realized, and if so, to what degree.

5.3.2 Impact Prediction Process

The process of predicting impacts or consequences of the proposed action and alternatives for input to decisions is the stated object of environmental impact analysis; thus all of the other steps described in this and previous chapters are conducted to support the prediction of impacts. Successful completion of all the other steps (e.g., scoping, description of affected environment, development of alternatives, and impact conceptual model) frequently results in an impact prediction process that is simple, straightforward, and virtually automatic. Take for example the OTH radar in northern California (Section 5.2.5) where a thorough description of the affected environment with respect to large mammal migration and cultural resources supported development of a proposed action with no impact on these resources. Similarly, the understanding of the existing conditions and causes of eutrophication in the Scituate, Massachusetts, nearshore environment (Section 5.2.4) made it very clear that locating a wastewater effluent outfall close to shore would exacerbate an already marginal water quality and estuarine ecosystem condition, and locating the outfall beyond that point would improve conditions. Thus in the planning phase of the environmental impact analysis, the eventual prediction of impacts should be considered in the design of all precursor steps to support, simplify, and enhance impact prediction.

The USCG DCR environmental evaluation (see Section 10.2 for a discussion of background on the DCR EIS) presents an example similar to that of Scituate in that the description of the affected environment was the major input to impact prediction. Since the discharge of DCR had occurred for over a century, the description of the affected environment was a critical input to impact prediction for many environmental resources because the impact of past DCR practices could be directly measured. The premise that a reduction in DCR discharge would result in a proportionate reduction in impact to many of the critical environmental resources was presented to and accepted by stakeholders, including the expert advisory committee (equivalent to a TAC). Thus the impact of all alternatives, which represented various levels of DCR discharge reduction could be quantified and compared based on the measured impact from past actions for many of the environmental resources.

Another commonly used approach for impact prediction that is simple and efficient is reliance on prediction of impacts in similar environmental impact analyses. For many types of projects, plans, and programs there have been many analyses conducted representing a range of existing conditions and magnitude of actions. The power industry has conducted environmental evaluations for thousands of projects ranging from multiunit nuclear power plants to solar generation facilities and everything in between. For many components of power generation, impact predictions methods have been developed including:

- Thermal discharge of cooling water
- Non-ionizing radiation from transmission lines
- Bird strike risk from wind turbines
- Land use changes from transmission corridors

- Management of fly ash (i.e., waste from coal fired plants)
- Fish passage for hydroelectric generation

There have been standard techniques of impact prediction for other industries and actions, such as wastewater treatment (as described in Section 5.3.3), transportation, and land-use planning. When embarking on an environmental analysis of a project, plan, or program in one of these areas, review of impact prediction for similar actions can result in not only efficiency but also improved technical quality of the analysis, and use of previously successful impact predictions can support acceptance among stakeholders. In order to successfully apply previously employed impact prediction methods, the environmental analysis team, relying on appropriate expertise within the team, must determine the applicability of previous methods to the action under consideration, scale the approach to the current action, and then develop and quantify as appropriate the input for the various variables in the impact prediction procedure.

Evaluation of impacts centers on predicting what the environmental conditions would be if a particular action were taken. Similar to predicting the weather (but hopefully with more success and public acceptance), impact prediction frequently involves models. Such models can range from the most simple (conceptual impact models, Figure 5.10) to very sophisticated mathematical models such as the one used in the Boston Harbor Cleanup EIS (see Section 5.3.5). The models can also be empirical and based on measurement of past similar actions as in the USCG DCR EIS (see Section 5.3.4). In most disciplines, models are available for predicting future conditions given a set of input values, and the experts on the environmental analysis team and/or TAC should be familiar with the models available and their appropriate application to specific situations. However, having the model is the easy part of impact prediction, and the hard parts, identifying and providing the input variables, interpreting the results, and understanding the uncertainty require the hands-on participation of the environmental analysis team members with appropriate expertise.

It is beyond the scope of this book, or in fact any book, to even attempt an identification of models or other impact prediction methods applicable to all, or even many, specific combinations of environmental conditions, types of actions, and locations. However, the general approaches summarized above can be applied by environmental professionals with the appropriate technical expertise to most specific situations. Case studies are presented for the USCG DCR management EIS and the Boston Harbor Cleanup, which can serve as food for thought in designing impact prediction approaches for other actions and environmental resources. Common and critical to each of the case studies and methods presented earlier is the use of significance criteria; so once predicted the impacts and other changes in existing conditions can be judged, compared, and used as input to decision making. As discussed immediately below, a comprehensive understanding of impact significance criteria is essential to adapting expertise in various environmental disciplines to successful, meaningful, and useful impact prediction.

5.3.3 Impact Significance Criteria

Virtually any activity will impact, to some degree, the environmental resource touched by the activity, but the challenge of environmental impact analysis is to determine the level of impact and its significance. The level of impact determination is dependent on the specific action, resource, and level of intensity, and as discussed in this chapter, methods are available or can be developed to measure the impact. But the measurement is only part of the picture, and it is necessary to understand the significance of the predicted measurement to fully address the goals of environmental impact analysis.

The level of impact can range from no observable change to collapse of an environmental resource on a regional scale, and the position on this scale of the predicted impact is the useful outcome of comprehensive impact prediction. Assigning objective significance criteria identifying the threshold between insignificant and significant impact levels is an approach proven successful in positioning the degree of impact on a common scale among resources and alternatives (Figure 5.11). The thresholds or significance criteria vary for different environmental resources and types of impacts, but applying the significance criteria to every resource places impact prediction on a common impact scale. Determining the level of significance on such a scale has multiple advantages:

- It is useful to decision makers with little or no expertise in the environmental resources at issue. For example, informing a decision maker that a particular alternative will generate an additional 20 trips on the adjacent highway or decrease dissolved oxygen by 0.5 mg/l is virtually useless unless the predicted change is placed in context.

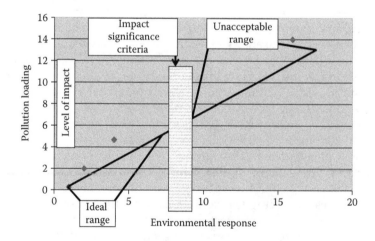

FIGURE 5.11
Impact significance criteria for defining relative level of impact.

- Impacts to vastly different environmental resources can be compared on a common basis. In the above example, it is impossible to determine which is more detrimental, 20 trips or 0.5 mg/l reduction given just the measurement. But if both are assigned a level of significance on a common scale they can be directly compared.
- The process allows another avenue for stakeholder participation by encouraging input in defining threshold levels of impact. It is an active role for the stakeholders and by supplying meaningful input to impact determination they receive a positive return on their investment of time and energy into the process. It also enhances support for the final comparison and decision because of their acceptance and participation in defining the level of impact.
- It helps the team focus on the environmental impact analysis. Whether conducted internally by the environmental analysis team or with a larger set of stakeholders, trying to determine what constitutes a significant impact forces a close look at the resource and places an emphasis on quantitative analysis. Identification of threshold levels of impact can also point out that a resource will not really be impacted and can be eliminated from detailed consideration.
- Impact prediction and affected environment investigations can be structured around the impact significance thresholds. For example, if a significance threshold for impact on a vegetated wetland from a water supply well is less than a 0.3 meter drawdown of the groundwater table, field investigations and groundwater modeling can be designed to determine the level of withdrawal that will reduce the groundwater elevation at each alternative location.

As discussed below, there are numerous approaches to developing impact significance criteria in order to maximize these advantages.

5.3.3.1 Environmental Standards as Significance Criteria

Standards and criteria are ready-made impact significance criteria. For example, water quality criteria are generally available for most contaminants and often expressed as concentrations associated with both acute and chronic effects. A predicted concentration below both levels could be designated as no impact, a concentration above both a significant impact, and a concentration between the two as a moderate impact. There are similar criteria for other environmental resources including air, public health, sediment quality, noise, risk of cancer, and radiation exposure. Thresholds established for environmental permits and regulations are another source of predetermined significance criteria. For example, the impacts to wetland resources can be easily predicted from a layout of the project and description of the affected environment if the significant criteria for wetland impacts are based on the square meters of fill identified in Section 404 of the Clean Water Act permit requirements:

- No wetlands within the facility foot print—no impact
- Allowed by permit—*de minimis* impact
- Allowed if there is two for one replication—insignificant impact
- Not allowable—significant impact

Another example is air quality permitting under the Clean Air Act, which differentiates areas as either compliant or noncompliant areas, and the allowable activities are different in each. An alternative in a compliant area could have a low level of significance with similar activities in noncompliant areas having different levels of significance. An activity which creates a new noncompliance area could have an even greater level of significance.

There are many advantages in using standards and criteria to define levels of impact significance. Environmental analysis team members should be familiar with such criteria within their area of expertise, and they can efficiently define and adapt them to specific conditions for use in the environmental analysis. Also using such criteria all but guarantees acceptance by stakeholders and provides a level of confidence in decisions made based on established standards and regulations as significance criteria.

5.3.3.2 Significance Criteria for Specific Environmental Impact Analyses

Unfortunately, standards are not available for all environmental resources and types of impacts. In these cases the expertise, experience, and creativity of the environmental analysis team and stakeholders come into play. Frequently, there are research techniques that are familiar to the discipline experts on the team that can be useful in establishing significance criteria, such as grading the level of service or wait times at intersections for traffic impacts. Also, similar or comparable criteria used in other environmental analyses involving similar actions can be considered. Stakeholder input and polling is another method that has been used successfully to determine the level of impact in specific cases. The DCR EIS (U.S. Coast Guard 2008) applied various techniques to develop significance criteria (Table 5.2) for each type of impact identified during scoping and through the impact prediction conceptual model (see Section 10.2 for a description of the EIS and background information). Use of significance criteria also facilitates presentation of predicted impacts in a simple and usable comparative form for a variety of projects, such as a hypothetical highway bypass (Figure 5.12).

Original research as part of a specific environmental impact analysis can also be used to develop significance criteria. The USCG DCR example (Table 5.2) and impact significance criteria related to invasive mussels discussed in Section 5.3.4 represent research done for the environmental analysis to determine impact significance. Similar studies can be conducted as described in Section 7.2 for ecological risk assessment to determine acceptable levels of impact. As with original investigations as part of affected environment description, such research should be approached cautiously. The resources and schedule of most environmental impact analyses are not compatible with extensive impact criteria research, and if not adequately planned, conducted, and reviewed, the results may not be acceptable to stakeholders. An alternative, as described in the next section, is to implement a comprehensive research effort to develop significance criteria applicable to a broad program or conditions applicable to multiple actions and then apply the criteria to specific environmental analyses.

5.3.3.3 Programmatic Significance Criteria: Clean Water Act Section 201 Wastewater Management

Next to NEPA, the passage of the Clean Water Act in 1972 (33 U.S.C. §1251 et seq. 1972) was the largest environmental protection leap forward in the early years of environmental awareness. The Act had the lofty goals of creating fishable swimmable waters within 10 years of passage and achieving zero discharge of pollutants five years later. The Act included the following major initiatives to achieve the goals:

- Regulation of all industrial and municipal discharges to waters of the United States
- Development of water quality standards by each state
- Promoting research into water quality and wastewater treatment
- Federal financial assistance for municipal wastewater management including treatment plant construction

The funding of municipal wastewater management under Section 201 of the Clean Water Act was the largest government-financed environmental improvement program ever undertaken. The "201 Grant Program" with the addition of state assistance provided up to 90% of the funds for every municipality in the United States to plan, design, and construct new or upgraded wastewater treatment facilities. Since at the time there were very few cities or towns that had adequate wastewater treatment systems in place, the program required a huge commitment of funds. For example, most of New York City had only primary treatment and the entire west side of Manhattan had no sewage treatment at all and discharged raw sewage to the Hudson River. An estimated $202.5 billion in federal funds and an additional $50 billion in state and local funds were necessary to fulfill the municipal wastewater management mandate of the Clean Water Act (Copeland 2010).

EPA promulgated regulations to implement the Clean Water Act 201 Grant program, which included a mandate for all municipal wastewater systems

TABLE 5.2
Example Impact Significance Criteria Used in the U.S. Coast Guard DCR EIS

Area of Impact	No Impact	Insignificant Impact	Significant Impact
Sediment chemistry	Chemical composition of DCR below threshold effect levels or reference area levels	Chemical composition of DCR above threshold effect levels but less than probable effect levels	Chemical composition of DCR above probable effect levels
Sediment physical structure	Grain size distribution of sediment samples from DCR discharge area similar to reference areas	Grain size distribution of sediment samples from DCR discharge area different from reference areas but benthic communities similar	Grain size distribution of sediment samples from DCR discharge area different from reference areas and benthic communities
DCR deposition rate	DCR rate within the range of natural deposition rate	Combined natural and DCR rate no more than 10% greater than maximum natural deposition rate	Combined natural and DCR rate greater than 10% maximum natural deposition rate
Water chemistry	No predicted concentrations greater than Great Lakes Initiative (GLI) chronic values	Mixing zone (MZ) predicted concentrations greater than GLI chronic values but less than acute values	MZ predicted concentrations greater than GLI acute values or GLI chronic values exceeded outside MZ
Nutrient enrichment	No stimulation of algal growth with exposure to 100% DCR slurry	No stimulation of algal growth with exposure to predicted DCR concentrations	Greater than 10% stimulation with exposure to predicted DCR concentrations

Special status species	No special status species are present	Species present but no effects on individuals, populations, or habitat	DCR could potentially jeopardize continued existence of species
Protected and sensitive areas	No DCR discharges within protected or sensitive areas	DCR discharge within area but no alteration of resource	DCR discharge within area and alteration of resource
Benthic community	Both of the following: community structure in discharge area similar to reference; no adverse effects in DCR or discharge area sediment bioassay tests	Any of the following: community structure in discharge area different from reference but not impaired; difference in growth but not survival in DCR bioassay tests; DCR present in tissue but below harmful levels	Any of the following: community structure in discharge impaired; difference in survival in DCR bioassay tests; DCR present in tissue at harmful levels
Invasive mussels	Any of the following: mussels did not preferentially attach at anticipated DCR densities; mussel distribution limited by nonsubstrate factors; mussels currently at maximum density	Mussels preferentially attach at anticipated DCR densities but less than 10% greater than native soft sediment	Mussels preferentially attach at anticipated DCR densities at a rate greater than 10% for native soft sediment

Source: U.S. Coast Guard. Final Environmental Impact Statement: U.S. Coast Guard Rulemaking for Dry Cargo Residue Discharges in the Great Lakes. 2008.

Resource category	Proposed action	No action	Altern. A	Altern. B
Built environment				
Traffic	○	●	◔	○
Housing	○	○	○	○
Infrastructure	◔	◐	○	○
Water resources				
Water chemistry	◔	○	◐	●
Seasonal flow	○	○	◐	◔
Biological resources				
Special-status species	○	○	●	○
Wildlife	○	○	●	○
Aquatic biota	◔	○	◐	●
Cultural resources				
Historic structures	○	○	○	◐
Archeological sites	●	○	○	●
Recreation	○	●	◐	○

Legend: ○ No impact; ◔ No to insignificant impact; ◐ Insignificant impact; ● Significant impact

FIGURE 5.12
Comparison of impacts using significance criteria among alternatives for hypothetical highway bypass project.

to achieve at least secondary treatment. This requirement was to reduce the organic and solid waste load through biological treatment, as opposed to primary treatment that achieves only minimal reduction of solids (Metcalf & Eddy 1991). Initially, this requirement was accepted as an absolute impact significance criterion and if secondary treatment was achieved, there would be no significant adverse impact. But as the secondary treatment initiative was implemented, it became apparent that secondary treatment alone was not always sufficient to maintain adequate water quality and healthy aquatic ecosystems. Also the location of the discharge was in many cases as critical as the level of treatment, and the mandate for secondary treatment did not address the discharge location issue.

With a program of this financial magnitude, participation by potentially every municipality in the United States, and potential effects on the nation's aquatic resources, EPA saw a need for additional science-based criteria in conjunction with a secondary treatment and water quality standards to guide the spending of over a quarter of a trillion dollars of public funds. Specifically the agency identified the need for guidelines, or impact significance criteria, to establish the need for treatment beyond secondary and identifying discharge locations compatible with the goals of fishable swimmable waters without unnecessary sewage treatment and effluent outfall construction costs. Part of EPA's program to establish guidance in impact significance criteria was to address the multiple related questions:

- How do estuarine systems respond to nutrient loading?
- What nutrient loading rates are beneficial, acceptable, and detrimental to the estuarine environment?
- How do estuarine systems respond to sewage solids loading?
- What loading rates of sewage solids are acceptable and detrimental to the estuarine environment?

EPA funded research to address these questions. This was done by developing impact significant criteria to assist in implementing municipal wastewater treatment requirements mandated by the Clean Water Act and maximizing the water quality and other environmental benefits achieved by the $200+ billion 201 Grant program.

One of the research programs EPA funded was at the Marine Ecosystem Research Laboratory (MERL) at the Graduate School of Oceanography (GSO), University of Rhode Island. MERL is a research facility envisioned and funded jointly by GSO, the EPA, and other government agencies located on the combined National Oceanographic and Atmospheric Administration, GSO, and EPA Estuarine Laboratory complex on Narragansett Bay. The charter of MERL was to research the effects of human perturbations with a goal of providing scientific input to support sound, efficient, and effective environmental regulations, enforcement, and investments. In terms of environmental impact analysis terminology, MERL's mission was to develop environmental impact significance criteria.

MERL is a series of 14 identical large-scale tanks that simulate conditions in a mid to northern North American Atlantic estuary, using the adjacent Narragansett Bay as a template (Pilson et al. 1979). Each tank is a mesocosm with all the components of an estuarine ecosystem (Figure 5.13), the forcing functions (e.g., mixing, temperature, exchange rate, depth, controlled, and

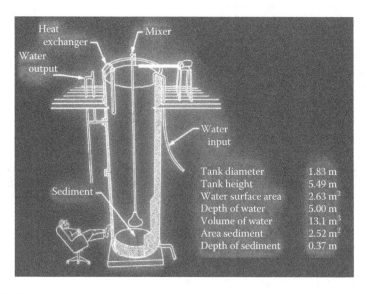

FIGURE 5.13
Marine Ecosystem Research Laboratory (MERL) experimental tank. (Adapted from Sampou, P. and Oviatt, C.A., *J. Mar. Sci.*, 44, 1991.)

the ecosystem characteristics (e.g., water chemistry, productivity, biological populations) monitored periodically, and in some cases daily. The research facility was designed and is ideally suited to measure the response of estuarine ecosystems to various degrees of human perturbations because conditions are virtually identical among mesocosms except for the variable under consideration.

During the late 1970s and 1980s, a series of MERL experiments were conducted to measure ecosystem response to loading gradients of the primary pollutants associated with treated municipal wastewater (i.e., nutrients and organically rich solids). One set of experiments focused on nutrients only (Oviatt et al. 1986) and another on the organic load and nutrients associated with wastewater solids (Oviatt et al. 1987, Sampou and Oviatt 1991, and Maughan and Oviatt 1993). The purpose of these experiments was to determine whether there was a direct and quantifiable relationship between pollutant loading and the response of individual components of the estuarine ecosystem (e.g., chemistry, water column biota, production, and benthic community). The measured responses of the individual components could then be evaluated to determine the integrated ecosystem response. The nutrients and solids were added daily to the mesocosms at rates from double up to sixteen times the estimated input to a relatively unstressed estuary (Figure 5.14) using lower Narragansett Bay, with low residential development density and little agricultural input, as a standard for a low stressed condition.

The results of all the experiments demonstrated a strong relationship between pollutant loading and ecosystem response and clearly indicated

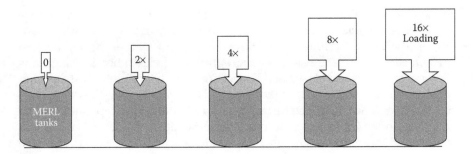

FIGURE 5.14
MERL nutrient and solid enrichment experiment and pollutant loading.

what loading rates were associated with extreme ecosystem degradation (Oviatt et al. 1987, Oviatt et al. 1986 and Maughan and Oviatt 1993). Thus it was possible to identify the impact significant criteria for the loading of many pollutants with respect to the bright line between acceptable and unacceptable impact on estuarine systems. For some pairings of pollutant loading rates and responses, there was a predictable gradient and it was possible to identify a series of quantifiable significant impact criteria (in terms of loading rates of a pollutant) associated with various levels of impact.

The relationship of the benthic community to organic carbon loading from the combination of carbon in the solids and from the increased primary production stimulated by nutrient additions is an example of a quantifiable gradient response (Maughan and Oviatt 1993). At low organic carbon loading rates, the benthic community characteristics were similar to those in the control mesocosms and lower Narragansett Bay, which received no nutrient additions. However at the highest loading rates, the community was significantly different from the controls with a total density of benthic organisms up to four times the density measured in controls, domination by pollution-tolerant species with such species occurring at over 300 times the density seen in controls, and at the end of the experiment the assimilative capacity of the ecosystem was exceeded and mass mortality of the benthic community occurred. Some of the pollution-tolerant species showed a strong linear correlation with the combined organic carbon input to the mesocosm (correlation coefficient greater than 0.9), which permitted prediction of benthic community characteristics at given organic loading rates.

The MERL experiments gave solid scientific backing for the identification of pollution loading rates associated with various levels of environmental impact from municipal wastewater treatment and discharge approaches (i.e., impact significant criteria). However, there was a degree of uncertainty with the results because of inherent limitation with mesocosms and concerns with applicability to ecosystems different from Narragansett Bay. These concerns were addressed by a comparison of MERL experimental results to field investigations reported in the literature, and in most cases, the results were found to represent a commonly observed relationship. For the relationship between organic carbon loading and the benthic community quantified in the MERL experiment (Maughan and Oviatt 1993) there was a strong agreement with other investigations (Bascom et al. 1978 and Person and Rosenberg 1978) that permitted establishing impact significant criteria strongly supported by scientific investigations (Table 5.3). These criteria could then be used in the environmental impact analysis of the EPA-funded 201 program to upgrade and expand the nation's municipal wastewater treatment system. As discussed in Section 5.3.5, these criteria were used directly in the Boston Harbor Wastewater Cleanup EIS, which addressed one of the largest wastewater treatment facilities in the country with one of the longest wastewater effluent outfalls and diffusers.

5.3.4 Impact Prediction of DCR Management in the Great Lakes

An interesting and informative approach to impact prediction was used in the USCG environmental analysis of DCR discharged to the Great Lakes (see Section 10.2 for a summary of the background on the DCR EIS). The analysis was conducted to meet a Congressional directive, achieve NEPA compliance, and promulgate rules and regulations governing the management of DCR. The approach is described in the next section, followed by a comparison with another and quite different NEPA approach, addressing impacts to similar environmental resources.

TABLE 5.3

Impact Significance Criteria for Estuarine Benthic Communities Derived from Environmental Impact Research

	MERL Control	Normal[a]	MERL Low Organic Loading	Changed	MERL Medium Organic Loading	Degraded	MERL High Organic Loading
Excess carbon loading above background m^2/day	0	<0.1	0.3	0.1–0.5	1.3	1.5–3.1	2.7
Pollution-tolerant species, % of total fauna	34	88	51	36	71	11	76
Percentage of biomass in normal area	100	100	136	270	209	470	244

Source: Maughan, J.T. and C.A. Oviatt. *Water Environment Research*. 1993.

[a] Bascom, W., A.J. Mearns, and J.Q. Ward. Establishing boundaries between normal, changed and degraded areas. W. Bascom, ed. *Coastal Water Research Project Annual Report* 81. 1978.

5.3.4.1 Impact Prediction Approach

Impact prediction for the DCR EIS had an advantage that most EIS scientific investigations do not share. The practice of dry cargo sweeping has occurred in the Great Lakes for over 100 years and any impact on ecological receptors from DCR discharge has occurred and continues to occur with the continuing DCR sweeping practice. Thus in this case measurement of existing conditions is an excellent predictor of future impacts of various DCR management alternatives.

An impact hypothesis was proposed: the degree of impact was generally proportional to the mass of DCR discharged. This hypothesis was tested through literature review and discussions within the project team and other experts. It was also tested through the design and implementation of numerous scientific investigations and found to be supportable. Thus if an EIS alternative for DCR management reduced discharge by 25% compared with existing practices, any ongoing or historic impact measured would similarly be reduced by approximately 25%.

Even with this impact prediction advantage, not shared by most scientific studies for EISs, there was uncertainty in the impact prediction. In order to address this uncertainty, a series of scientific investigations was conducted (Table 5.4), which identified at least two types of investigations to address potential impacts on each ecological receptor (as identified in the impact prediction conceptual model, Figure 5.10) and more in areas where the impact was considered to be more likely or more severe.

For example, because the discharged DCR was much heavier than water, the iron, coal, and limestone sweepings would rapidly descend to the lake floor and accumulate over time. Thus the organisms that live on the bottom (the benthos) were at the greatest risk of impact, and multiple studies were designed to evaluate impacts on this segment of the aquatic ecosystem. These investigations and the predicted mechanism of impact on the benthos tested were as follows (Table 5.4):

- *Sweepings Characterization*: The physical and chemical attributes of iron ore (as partially processed taconite), coal, and limestone (the three types combined represent over 95% of the dry cargo shipped on the Great Lakes and also the dry cargo with the greatest impact potential) transported and potentially discharged by Great Lakes carriers were determined. These characteristics were then compared with the requirements and tolerances of the critical benthic species inhabiting the floor of the Great Lakes. The degree to which DCR attributes were incompatible with the species' chemical tolerance (i.e., sensitivity to toxicity) or habitat requirements (physical attributes) represented the degree of impact.
- *Physical characterization of deposition area*: As discussed above, the physical characterization of the actual DCR is important to predict impacts to benthic organisms, but the characteristics of the *in situ*

TABLE 5.4

Investigations to Predict Impacts from DCR Management Alternatives

Scientific Investigations	Water Chemistry	Enrichment and Nutrients	Sediment Chemistry	Sediment Alteration and Deposition	Benthos	Pelagic Organisms	Water Fowl
Sweepings characterization	X	X	X	X	X	X	X
Sweepings discharge analysis	X	X	X	X		X	
Historic deposition analysis				X			
Physical characterization of deposition area				X	X		
Chemical characterization of deposition area			X		X	X	X
Toxicity tests			X		X	X	
Benthic community structure evaluation					X		
Nutrient enrichment	X	X					

Source: U.S. Coast Guard. Final Environmental Impact Statement: U.S. Coast Guard Rulemaking for Dry Cargo Residue Discharges in the Great Lakes. 2008.

Note: "X" indicates impacts to indicated resource addressed by scientific investigation.

benthic habitat once the DCR has been introduced are perhaps a more important predictor of impact to sediment-dwelling organisms.
- *Chemical characterization of deposition area:* The chemical composition of the DCR is important in determining impacts, but similar to physical characteristics, the chemistry of the sediments where the DCR has been deposited is an important factor that could impact the benthos. Chemically characterizing the sediments in the DCR deposition areas allows determination of factors such as bioavailability and thus the actual chemical exposure and impact to the benthos.
- *Toxicity tests:* Actually exposing organisms to the DCR and measuring their survival and other responses (growth, reproduction, etc.) allowed assessment of the toxicological impacts to the benthos.
- *Benthic community structure evaluation:* Collecting sediment samples for the DCR deposition area and examining which animals are living in the sediment, compared with similar areas without DCR is a classic method of determining impacts, and was very useful for the DCR environmental analysis. However, it is not always definitive because it is only a snapshot of a community that is naturally highly variable in time and space, thus it was necessary to supplement the sampling results with information reported in the literature.

An advantage of multiple studies to evaluate impacts on a single receptor is that it affords the opportunity to apply a weight-of-evidence approach. Since there is variability and uncertainty in the characteristics and requirements of natural systems, detection of an effect in one incidence is not always assurance that it will frequently occur or have ecosystem-wide implications. Detection of effects through multiple lines of evidence is a much more certain and accurate predictor of impacts.

Based on the multiple investigations into the various areas of potential impact (Table 5.4) only limited impacts were predicted as summarized below:

- *Water chemistry:* The concentration of potentially toxic chemicals in the DCR was found to be low and when mixing models were applied to the DCR discharge, the concentrations in the waters of the Great Lakes were predicted to be well below levels harmful to aquatic organisms.
- *Enrichment and nutrients:* The nutrient concentrations of the DCR were low and addition of DCR slurry to phytoplankton cultures did not stimulate growth to a level of concern. Thus no impact to enrichment was predicted.
- *Sediment chemistry:* The model used to predict sediment concentrations based on the chemical characteristics of the DCR and the measurement of sediment chemistry in the areas of heavy DCR

discharge both revealed sediment concentration below levels of concern. Also the exposure of sediment-dwelling organisms to DCR in the laboratory did not produce an adverse response. Thus no sediment chemistry impact was predicted.

- *Sediment alteration and deposition:* The prediction of deposition and resulting physical characterization of the sediment did not identify a change in the sediment to a level of concern. However, measurement of sediment in the areas of heavy DCR discharge and conditions reported in the literature (Mudrock et al. 1988) did indicate a potentially small degree of sediment alteration. As discussed below, this led to follow-up investigations to determine the effect this might have on invasive mussel species.
- *Benthos, pelagic organisms, and water fowl:* With the exception of potential sediment alteration, no change in the chemical or physical characteristics of Great Lakes habitat attributes was predicted. Thus no impact to the biological resources was expected with the possible exception to changes in invasive mussel population that warranted additional investigation as described in the next section.

5.3.4.2 Prediction of Invasive Mussel Impacts

The scientific investigations (Table 5.4) conducted to evaluate the potential impacts identified in the conceptual model (Figure 5.10) revealed an interesting and potentially significant ecological impact. The initial scientific results were reviewed in light of the conceptual model by the project team and a TAC formed for the project. The review identified a potential expansion of invasive mussel populations in the Great Lakes resulting from the predicted alteration of the sediment physical characteristics. This concern was based on the fact that mussels prefer the hard substrate offered by the DCR. It then became necessary to amend the original scope of the scientific investigations to take a hard look at the potential for DCR management practices to cause an expansion and density increase in the Great Lakes' invasive mussel population.

The nonindigenous zebra (*Dreissena polymorpha*) and quagga mussels (*D. bugensis*) have invaded freshwaters throughout the United States from New England to California. Once introduced, they breed rapidly with a fully mature female mussel capable of producing up to one million eggs per season (Claxton and Mackie, 1998) and in many habitats very dense populations can become established. One of the critical mussel habitat characteristics is a hard substrate (Czarnoeski et al. 2004) and introduction of the hard DCR into the otherwise soft muddy bottom of the Great Lakes could create a habitat conducive to high mussel population densities. The high density of these invasive mussels can preclude establishment of native species, affect water quality, and also become so dense that they clog water intakes for power plants and municipal water supplies.

Thus, the expansion and stimulation of invasive mussel populations in the Great Lakes from DCR management practices was determined to be a potentially serious impact and thus under the requirements of NEPA warranted a "hard look" by the USCG as they developed regulations for the management of DCR.

In order to investigate and predict the potential for invasive mussel impacts, a series of scientific investigations were designed and conducted (U.S. Coast Guard 2008). The investigations included literature review to document whether mussels were already established in the Great Lakes, factors limiting their presence and density, their life history, and environmental requirements at each life-history stage. The investigation also included an extensive series of laboratory tests.

The first laboratory test conducted was Phase I Adult Mussel Attachment. This experiment consisted of introducing both zebra and quagga mussels to aquaria with each type of DCR and native sediments. The percentage of mussels attaching to each sediment type was determined and attachment was found to be much more prevalent in aquaria with DCR compared to those with just native sediment.

After reviewing the Phase I results that showed a very strong mussel attachment preference for DCR compared with native sediments, the project team suspected that the study design may have been too conservative (i.e., potentially over predicting impacts). To address this concern, a Phase II adult mussel attachment study was designed to more closely simulate *in situ* conditions. On the lake floor, the DCR would generally be covered by a thin layer of natural sediment due to the heavy nature of DCR sinking into the sediment surface. Also, bioturbation (i.e., burrowing and feeding activity of animals living in the sediment or at the sediment–water interface) working and resuspending the sediment tends to cover the DCR with a thin layer of native sediment. The attachment experiment was repeated to address this issue by covering the DCR with a 0.5 to 1.0 mm layer of Lake Superior sediment. The Phase II results (Figure 5.15) strongly indicated a preference for adult mussel attachment to DCR, even when covered with a thin layer of sediment, compared with the natural soft sediment substrate. Thus, the adult mussel attachment studies indicated that discharge of DCR could result in an impact by increasing mussel attachment and thus a potential increase in mussel population density and range.

Attachment to DCR by adult mussels was identified as a concern; however, the potential for impact from increased colonization by larval mussels, or veligers, could be an even greater concern because the veligers are more mobile than the adult mussels. As one would imagine, the heavy shells of the adults limit their mobility and ability to spread and colonize new areas. Their evolutionary development has accommodated this limitation and both zebra and quagga mussels have mobile, free-floating larval forms which swim and drift with the current and can travel dozens of kilometers. This enables a rapid spread of the population, if there is suitable substrate for veliger attachment once they reach maturity.

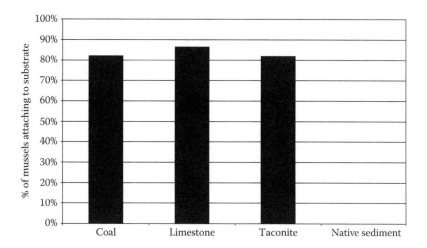

FIGURE 5.15
Attachment of invasive mussels (zebra and quagga) to DCR and Native Sediment Phase II Investigation: U.S. Coast Guard DCR EIS (see text for reference).

A series of veliger studies was designed and conducted to evaluate the potential impact of spreading and increasing the density of the invasive mussels in the Great Lakes via larval dispersal. Veligers (both zebra and quagga) were introduced to aquaria with native sediment and the three DCR types: some uncovered, some covered with 1 mm of sediment, and others covered with 3 mm of sediment. The results showed that up to 6% of the veligers present in the experiment attach to DCR, compared with virtually zero percent attachment to natural sediments (Figure 5.16). Even a 1 mm cover of natural sediment over the DCR, which is less than the cover observed on the lake bottom greatly inhibits veliger attachment to DCR to less than 0.2% of the veligers present and a covering of 3 mm is an even greater attachment inhibitor.

Full evaluation of the literature, laboratory study, input from the expert committee, and field evaluation provided a hard look at the impacts associated with invasive mussels from past and future DCR management practices and alternatives. This hard look reached the following conclusions, which were presented in the DCR EIS and used to evaluate alternatives, and ultimately for the USCG to issue a Record of Decision (ROD) and promulgate DCR management regulations:

- The preferable adult mussel attachment to DCR compared with native sediment was a concern. However, the adults have limited mobility and if they are already present in an area, new adults can attach to old shells or even other live mussels. Thus in areas where mussels are already present in dense concentrations, the addition of DCR is unlikely to increase mussel distribution or density via adult attachment.

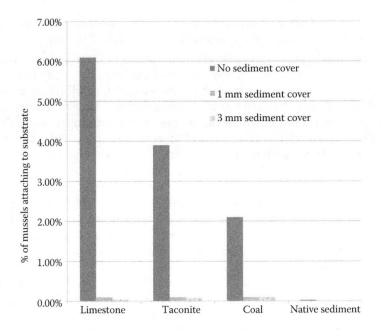

FIGURE 5.16
Percent attachment by immature (veliger) mussels to DCR with 1 mm and 3 mm sediment cover compared with native sediments.

- Veligers prefer to attach to DCR and thus the presence of DCR could potentially result in the spreading and increased density of invasive mussels in the Great Lakes. However, the probability of this occurring is greatly reduced by two factors: the *in situ* covering of DCR which naturally occurs greatly decreases the likelihood of veliger attachment and in areas which already support a high density of mussels, the old shells and live mussels provide ample opportunity of veliger attachment and addition of DCR will make very little difference.
- Lakes Erie and Ontario currently support wide zebra and quagga mussel distribution and dense populations. Thus there is sufficient suitable attachment substrate present and the addition of DCR in these lakes will not produce any adverse impact from invasive mussels.
- The open waters of Lake Superior do not currently support invasive mussel populations and the literature documents that their absence is due not to a scarcity of suitable attachment substrate, but rather to water softness (i.e., low concentrations of minerals including calcium) and the addition of DCR will not alter the softness of the water as the mussel population's limiting factor. Thus there would be no effect on

the population from any change in DCR management practices and no adverse impact from invasive mussels in Lake Superior.
- Zebra and quagga mussels are present in Lakes Michigan and Huron currently, but at least in some areas of the lakes they may not be at maximum density or distribution. No specific factor limiting mussel distribution or density has been identified and suitable attachment substrate cannot be ruled out as one of the limiting factors. Thus DCR discharge has the potential to increase the quality and extent of suitable mussel habitat in these lakes. However, the natural covering of DCR with native sediment, limited mobility of adults, and the presence of DCR on the lake bottom from over 100 years of dry cargo transport indicate that any increase of mussel density or distribution resulting from the continued discharge of DCR would produce only a minor, or insignificant impact from invasive mussels on the aquatic ecosystem of Lakes Michigan and Huron.

Using the information developed from the invasive mussel investigations, the USCG concluded that the insignificant impact prediction of mussel population expansion from ongoing DCR management practices in Lakes Michigan and Huron was acceptable. The studies demonstrated that, at least in the short term, the continuation of the current practice would not result in the increase in invasive mussels to the extent that there would be any ecosystem-wide adverse impacts. This decision did incorporate a degree of judgment but the conclusion is based on hard scientific evidence and more than fulfilled the "hard look" mandated by NEPA, as discussed in Chapter 3, Section 3.7. The hard look at invasive mussel impact provided in the DCR EIS differs from other approaches as summarized for the Hughes River environmental impact analysis in the following section.

5.3.4.3 Comparison of DCR Impact Prediction to North Fork of the Hughes River EIS

One might take issue with the interpretation of the invasive mussel investigations or the DCR management rule-making decision reached by the USCG. But it is unarguable that the USCG took a "hard look" at the impacts related to the invasive mussel situation before they issued the final EIS or rendered a decision on the DCR management rule making. Thus they met the impact prediction of NEPA mandate as interpreted by the courts. This was not the case in another proposed federal action with potential invasive mussel impacts.

In the early 1990s, after years of consideration, the Natural Resources Conservation Service (NRCS) and the U.S. Army Corps of Engineers proposed a dam on the North Fork of the Hughes River in a mountainous area of northwestern West Virginia (project history and court decision presented in *Hughes River Watershed Conservancy v Glickman*, 81 F.3d 437, 445 [4th Cir. 1996]).

It was a project with multiple beneficial goals, including economic stimulation. It also involved infringement on significant environmental resources including extraordinary scenic value, habitat of an extensive variety of fish and wildlife, and a population of 22 freshwater mussel species, including two species under consideration for listing under the Endangered Species Act. The project involved multiple federal actions including joint funding by the NRCS and Appalachian Regional Commission (a federal agency established to assist in the economic development of the Appalachian region), and importantly a wetlands permit issued by the U.S. Army Corps under Section 404 of the Clean Water Act. Thus with the inherent potential impacts associated with damming a river an EIS under NEPA was required.

Even with serious objections raised by the U.S. EPA and Department of the Interior, Fish and Wildlife Service (FWS) on both draft and final EISs, the NRCS and U.S. Army Corp issued a ROD to build the dam on the North Fork of the Hughes River. The EPA even informed the Corps that the final EIS was inadequate and a supplemental EIS should be prepared before going forward with the project.

Concurrent with the final EIS and ROD, the U.S. Army Corps announced the intention to issue a Draft 404 wetlands permit for the project and noted that the NRCS had already prepared an EIS for the project. In their review of the wetland permit, the EPA repeated comments on the EIS, and took exception to the project including warning the Corps that the dam and reservoir would probably cause infestation of the North Fork by zebra mussels. EPA comments also requested a supplemental EIS to address zebra mussel infestation, and thus take a hard look at impacts associated with invasive mussels. In response to these comments, the U.S. Army Corps stated that the District Office biologist discussed the zebra mussel issue with an employee of the Corps's water quality section who stated mussel infestation was not a concern.

The District Office of the U.S. Corps informed the EPA and FWS that they intended to issue a 404 permit and did not resolve or even address all issued raised by the other agencies. Both agencies elevated the intended district office action to the Assistant Secretary of the Army for Civil Works and requested a review of the district office's decision to issue the 404 permit. In response to EPA and FWS comments, the Assistant Secretary directed the District Office to undertake a comprehensive reevaluation of alternatives to the project.

The district office conformed to the Assistant Secretary's direction and prepared a memorandum for the record that rejected all the alternatives to the proposed action to build the dam. The EPA and FWS objected to the findings of the memorandum. Corps headquarters concurred with the District Office's memorandum for record and issued a Draft 404 permit for the project. In issuing the permit the Corps noted that the NRCS had already prepared an EIS and concluded: "There has been no new evidence or information that would require that the [EIS] be supplemented."

In response to the U.S. Army Corps' issuance of the draft permit, EPA again raised concern over invasive mussel infestation. In their comments they included supporting information from Dr. Richard Neves, a professor of fisheries at Virginia Polytechnic Institute and State University, who concluded that the dam would have a devastating effect on the downstream ecosystem. Dr. Neves discussed the issue with five other experts, and after consultation with these experts he concluded that the reservoir resulting from the dam would provide a critical habitat for mussels and result in downstream infestation. He urged the Corps to verify his conclusions by contacting the other experts he had consulted and provided their contact information.

The U.S. Army Corps reacted to the EPA's Draft 404 permit comments on mussel infestations in the same manner they addressed their comments on the intent to issue a permit. They again called a Corps employee without documenting the employee's expertise, if any, who stated mussel infestation, was not a concern. They did not conduct any investigation or analyses or even contact the experts recommended by Dr. Neves.

The Hughes River Watershed Conservancy, the Sierra Club, and the West Virginia Rivers Coalition sought the judicial review of the NRCS and U.S. Army Corps' decisions regarding the project and issuance of the 404 permit. The suit alleged that the NRCS and the Corps violated NEPA in a number of regards, including failure to adequately consider all the information and prepare a supplemental EIS in light of new information. The district court determined that NEPA and 404 regulations were properly followed and ruled in favor of the NRCS and Corps of Engineers. The environmental advocacy groups took exception and appealed the decision.

The essence of the appeal centered on the potential for invasive mussel infestation. The Watershed Conservancy, Sierra Club, and River Coalition pointed out that EPA, FWS, and others had raised the potential for the spread of invasive mussels as a result of the dam and reservoir. As a comment on the application for a 404 permit, the EPA explained to the U.S. Army Corps that zebra mussel infestation would have the following adverse effects downstream of the dam:

- Destruction of indigenous mussels (some of which are rare and endangered)
- Clogging of water intake structures by the mussels
- Ecosystem level negative impacts on the aquatic community

The district court's decision was overruled and the finding on the appeal was that the U.S. Army Corps failed to take a "hard look" and at mussel infestation resulting from the project and thus violated NEPA. The court decision confirmed an agency's primacy in evaluating technical data and conclusions even when there are conflicting opinions. However, there was nothing in the record as to the basis for the Corps' employee's opinion on potential mussel infestation or the qualifications of the employee who made the statement. Without documentation for the opinion, lack of stated qualifications of the employee, and the absence of pursuing opposing expert opinions by Dr. Neves and others, the court found the decision was not supportable under NEPA and section 404 regulations and case history. The Corps had failed the "hard look" test mandated by practice and NEPA litigation.

The USCG DCR and the U.S. Army Corps/NRCS North Fork EISs' treatment of invasive mussels illustrate two ends of the impact prediction spectrum. The Corps' telephone conversation with an employee of unknown, or at least undocumented, qualifications to address serious concerns raised by environmental advocacy groups and two other federal agencies with expertise in the field (and supported by outside experts) fell well short of the NEPA-required "hard look" at impacts. In contrast, the USCG's approach included:

- A thorough literature review of the issues
- Proactive presentation to an independent board of expertise in the field and their input
- Three phases of laboratory experiments to analyze the potential impacts of the alternatives

The impact prediction should be more intense, proactive, complete, and well documented than that used to address invasive mussels in the North Fork EIS, but it does not always have to be as thorough as the approach used in the DCR EIS. The level of investigation and the magnitude of the "hard look" should be based on the intensity of the potential impacts, the probability the impact will occur, the importance and vulnerability of the environmental resource, and implication if the impact does occur. The greater the potential, probability, and implication of impact the greater should be the level and intensity of investigation and accuracy of impact prediction.

5.3.5 Impact Prediction of Boston Harbor Cleanup Alternatives

The cleanup of Boston Harbor was a long running, controversial, politically, charged, and heavily litigated endeavor (see Section 10.1 for a summary of the background), and consequently both the draft and final EISs (U.S. EPA 1988a and 1988b) addressed a wide spectrum of impacts to a long list of environmental resources. However in the final analysis of wastewater treatment and disposal options there were two resources that proved to be:

- Extremely vulnerable
- Exhibited substantially different responses to different wastewater treatment and disposal alternatives

- Were indicative of potential broad-scale and significant ecosystem-wide impacts
- Were suitable for quantitative and comparative impact prediction, analysis, and comparison among alternatives
- Could be compared with accepted impact significance criteria

These two environmental resources were the enrichment or eutrophication of the Boston Harbor water column and the enrichment and degradation of the benthic environment. Not coincidentally, these were also two of the critical impact areas in the entire nationwide effort to address gross pollution of the waters of the United States by municipal wastewater treatment and disposal through the Clean Water Act 201 funding to municipalities. As a result they were also the topics of significant research to develop impact significance criteria (see Section 5.3.3.3). So the approach to impact prediction for these two critical environmental resources for the Boston Harbor Cleanup was to determine the extent of nutrient and sediment enrichment for each alternative and compare it with the significance criteria.

There were standard methods of hydraulic modeling available for predicting conditions in both the water column and the sediments resulting from wastewater discharges. Extending these methods to a system as large and complex as Boston Harbor (e.g., numerous rivers entering the system, extreme vertical stratification, and a tidal range up to almost 4 meters) was a challenge, but with the advice and expertise available on the environmental analysis team and the TAC with stellar academic credentials (see Section 4.3) the challenge was successfully met. After an extensive affected environment data collection program to determine currents, tidal exchange, boundary conditions, and thermal stratification as input parameters, a mathematical hydraulic model was developed, calibrated, and verified. The model was then used to predict the nutrient concentrations, sediment deposition rates, and deposition area from each of the alternative wastewater effluent discharge locations and wastewater treatment processes.

The results of the predictions were then compared with the impact significance criteria developed by Clean Water Act initiated research to evaluate municipal wastewater manage issues (see Section 5.3.3.3). From this comparison, the area of Boston Harbor exceeding the criteria for various levels of nutrient and sediment enrichment impact could be determined for a wide variety of alternative discharge locations and levels of wastewater treatment (Figures 5.17 and 5.18 are example combinations of discharge location and level of treatment). This information then provided objective and quantitative expression of impact in these two critical areas so that the decision makers could make informed decisions. The decision makers and the other stakeholders were able to quantitatively consider the tradeoffs of costs and environmental benefits of higher levels of treatment and superior dispersion and dilution provided by discharge locations in deeper water farther from shore. In the final analysis, there was a consensus decision among:

- The Massachusetts Water Resource Authority, the owner, and operator of the wastewater system
- The Massachusetts Department of Environmental Protection, the state regulatory and permit issuing authority
- The U.S. EPA, the federal regulatory and permit issuing authority
- The CAC and TAC
- The litigants in the lawsuit that forced the court-supervised cleanup of Boston Harbor

All of these stakeholders agreed that degraded benthic habitat of 1.2 km^2 or greater and changed habitat greater than 12.2 km^2 were unacceptable (Figure 5.17). Similarly, the consensus decision was that any area of degraded water column with respect to nutrient enrichment and a changed area greater than 4 km^2 were an unacceptable level of impact and did not meet the purpose and need (Figure 5.18). Thus secondary treatment with a discharge at Site 5 resulted in impacts below the significance criteria and was selected as the proposed action.

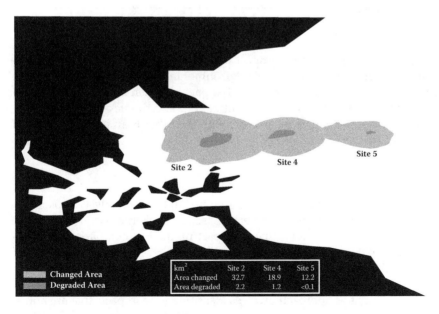

FIGURE 5.17
Areas of Boston Harbor exceeding impact significance criteria for benthic habitat.

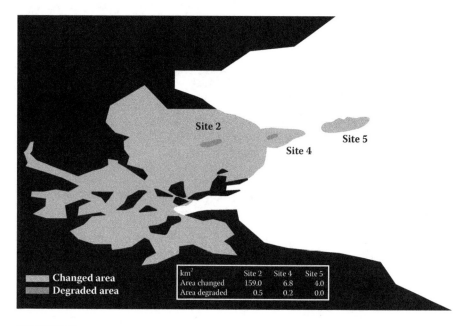

FIGURE 5.18
Areas of Boston Harbor exceeding impact significance criteria for nutrient enrichment.

5.4 Impact Mitigation

The process of mitigating the impact from a proposed project, plan, or program is an inherent component of environmental impact analysis. However, it can only be initiated once a potential impact has been identified, and there are two basic approaches to impact mitigation. The classic approach to mitigation occurs at the end of the process, and the other is an integrated analysis. Each of these approaches is discussed in detail.

5.4.1 Classic Approach to Mitigation

The classic process through all the environmental impact analysis steps is discussed in Chapter 4 and this chapter (e.g., scoping, alternatives development, description of affected environment, impact prediction, comparison of alternatives, and designation of proposed action), which results in prediction of impacts and then considers mitigation. This approach was introduced as part of NEPA (see Section 3.1.3) and was incorporated into early environmental impact analyses. The classic approach involves identifying the impacts that are anticipated from implementation of the proposed action and then evaluating them individually to determine what, if anything, can be done to lessen the impact.

The classic mitigation evaluation typically follows prescribed steps which are some combination or disaggregation of the following (adapted from Erickson 1994):

- *Avoidance:* Elimination or large-scale modification of the component of the proposed action causing the impact. Avoidance can be accomplished by physical relocation of the offending component of the proposed action to avoid intersection with a sensitive environmental resource.
- *Minimization:* Reducing the size, intensity, duration, or magnitude of the proposed action component producing the impact.
- *Rehabilitation, restoration, or replacement:* Returning a damaged environmental resource to the same or improved condition of the resource prior to the implementation of the proposed action. The classic example is construction of a new or improvement of an existing wetland to compensate for impacts suffered during construction of the proposed action.

Following a full determination of the potential avoidance, minimization, and restoration options, an evaluation of costs compared with benefits is conducted for each impact anticipated from the proposed action. The evaluation includes an assessment of how inclusion of the mitigation measures might compromise achievement of the purpose and need or the effectiveness of the proposed action. A decision is then made as to what, if any, mitigation measures are justified and if they can be justified, the proposed action is modified to include a commitment to implementing the measures. NEPA mandates this classic mitigation process of identifying potential mitigation measures, determining which to implement, and explaining why others will not be included in the proposed action (see Chapter 3, Section 3.1.3).

This classic approach to impact mitigation served an important function and resulted in reduced impacts in the early days of environmental impact analysis. Before environmental analysis was an integral part of infrastructure, utility, military, and land use planning and development, there was a strongly preferred course of action at the earliest stages of planning, and comparison of alternatives rarely resulted in a change in the originally conceived proposed action. However, in such cases, the legitimate consideration and incorporation of mitigation measures was a concession to environmental concerns, and some aspects of an action were frequently modified to mitigate the most egregious and avoidable environmental impacts. These included measures such as:

- Minor realignment of highways to avoid wetlands or other environmentally sensitive areas

- Reducing the capacity of a water, wastewater, or power generation facility to more realistically represent future demands and realistic conservation measures
- Marginal support for public transportation as part of a traffic management plan centered on extensive highway expansion
- Public education on recycling and waste reduction as part of a new or expanded solid waste landfill
- Water conservation education as part of a new or expanded water supply source
- Limiting takeoff and landing times as part of airport expansion to mitigate noise impacts

However, the classic approach to impact mitigation had several substantial limitations. One of the most glaring deficiencies was that the after-the-fact add-on measures often created unforeseen impacts. Because these measures were incorporated into the proposed action after any environmental analysis or stakeholder scrutiny, the implications of the actions beyond the narrow focus of the impact they were designed to mitigate were not examined. In hindsight such add-on mitigation measures seem shortsighted and defy common sense (summarized from Erickson 1994):

- Impact to tortoise migration was predicted to result from construction of a military installation in the desert and a fence was constructed to guide the migration safely through the facility. However the fencing was unsuitable and tortoises were found entangled in the mesh with resultant fatal cuts and suffocation.
- In order to mitigate wildlife habitat impacts resulting from the preferred alignment of a new highway, a section of the highway with a wide median strip was planted with long-duration flowering shrubs that produce berries with high food value for local bird species. Initially the mitigation was quite successful; however during periods of drought, the berries of the chosen shrub species ferment and the birds feeding on the berries became intoxicated. So when these birds fly within the travel portion of the highway, many traffic accidents occured. Thus, the entire median strip had to be stripped of vegetation.
- The selected alternative for a new highway alignment bisected an area supporting a productive deer population and prevented migration between two critical habitat types. A large underpass was constructed to facilitate safe deer migration with fencing on either side to guide deer to the underpass and critical habitat on the other side of the new highway. Local wildlife enthusiasts confirmed that the underpass was effective in reducing roadkills but the deer population declined

concurrent with the highway development. Investigation revealed that the decline was due to significantly increased hunting success, by taking advantage of the funneling of deer through the underpass.

The deficiencies of only considering mitigation following comprehensive environmental analysis, and only for the proposed action, are the primary causes of the environmental protection missteps illustrated by these examples. Another shortcoming of after-the-fact mitigation is that it frequently is not a factor in the alternative selection and comparison. If a particular alternative has numerous cost, performance, and environmental attributes but an environmental impact fatal flaw (such as generating noise above a significance criterion), it might be dismissed and a less advantageous alternative in every way but noise selected. If a measure to mitigate noise to an acceptable level had been developed and incorporated earlier in the process, the designated proposed action may have been different, more environmentally beneficial, and preferable to all stakeholders. Given these downsides of the classic mitigation approach, it is not surprising that a more proactive and integrated approach has been developed.

5.4.2 Integrated and Proactive Approach to Mitigation

As environmental impact analysis matured, the shortcomings of after-the-fact mitigation outlined above became apparent, and a more progressive approach evolved. The enhanced approach centers on identifying impacts early in the process, even at the initial impact prediction conceptual model (see Section 5.3.1) and the scoping stages (see Section 4.3). Once the potential for substantial impact has been identified, existing alternatives can be modified or new alternatives developed that mitigate the anticipated impacts. This approach addresses most of the disadvantages of the classic approach and most importantly it supports a full evaluation and comparison of both the positive and negative aspects of the mitigation measures.

The integrated approach produces a full range of alternatives, some with and some without mitigation. The prediction of impacts for some alternatives reflects consequences without mitigation, and some reflect the mitigation, both on the target impact and collateral damage to other environmental resources. This supports a full, concurrent, and relatively equal evaluation of all impacts of all alternatives and associated mitigation, taking advantage of impact prediction tools and approaches discussed above including:

- Application of impact significance criteria
- Review of prediction methods and results by the CAC and TAC
- Modeling that includes all aspects of an alternative, including mitigation
- Design of affected environment investigations to support impact prediction with and without mitigation

A major advantage of integrating mitigation into the full environmental impact analysis process is in the alternative comparison process. The impacts of an alternative can be very different with and without mitigation, as can the costs. Take for example, an alternative that has many benefits and much lower costs but results in significant traffic that causes serious safety concerns for an adjacent school with a large pedestrian population. This alternative might not be selected or even screened out early in the process because a critical public safety significance criterion was exceeded. But formulation of an additional alternative incorporating public safety mitigation such as a relatively inexpensive pedestrian overpass, electronically controlled crosswalk, or other measures to meet the public safety significance criteria could be the alternative with the least cost, most benefits, and fewest impacts. There are numerous similar examples where "fatal flaw" wetland impacts can be avoided by incorporating well-designed, constructed, and monitored created wetlands as part of the alternative.

Another advantage of the integrated approach to impact mitigation is that it takes into account the complexity of environmental factors contributing to an impact or existing conditions. If for example there is a predicted impact on a fisheries population resulting from an increased silt load to the river, mitigation of the siltation impact alone may have no actual benefit (Erickson 1994). There may be other existing factors such as habitat limitations, food availability, and breeding locations, which are already limiting the fish population density or distribution. Therefore, changes in silt load or the resulting suspended sediment concentration would not have any effect on the fish population. Similarly, if there are predicted impacts from an alternative resulting from both reduction in dissolved oxygen concentration and silt load, mitigating just one of these factors would not produce any benefit. Including mitigation in the full environmental impact analysis, rather than adding it after the fact would expose such false benefits and they could be integrated into the alternative comparison and other decisions.

The downside of the integrated approach to mitigation is that it can result in an unwieldy and repetitive evaluation of alternatives. This issue can sometimes be addressed at the alternative screening stage (see Section 4.5.3) by eliminating alternatives in a side-by-side comparison. If the comparison reveals that the two alternatives have generally the same benefits and adverse impacts but substantially higher costs, the one with the higher cost can be eliminated at the screening stage. Another approach to addressing the complexity of alternatives created by integrating mitigation into the full analysis is to identify and evaluate alternatives separately for independent segments of a project, plan, or program (see Section 6.3.1). With this approach, the superior alternative(s) for each component of the action can be combined into a manageable set of comprehensive alternatives.

5.4.3 DCR Management Mitigation Case Study

In the USCG DCR EIS example (background discussed in Section 10.2), the impact prediction conceptual model (Figure 5.10) can also provide a guide for impact mitigation evaluation. Where impacts are predicted (e.g., potential increase in invasive zebra mussels in the DCR example) the progression of actions which resulted in the impact can be traced in reverse to the initial action, (i.e., loading and unloading of dry cargo in the DCR example). Each of the steps and the connection between steps in the progression causing the impact can be an opportunity for mitigation. Alteration of any of the actions or an interruption in any of the connections between steps can reduce the impact. For example, the impacts originated from the spillage of DCR, which occurred primarily in the ship's tunnel during unloading. Thus alternatives were developed and mitigation measures evaluated that would reduce the amount of DCR accumulating in the tunnels from unloading events. The measures considered included:

- Pumping the sweepings from the ship's hold back onto the unloading conveyor belt
- Installing and periodically cleaning screens over the sumps which collected the wash water from sweeping activities
- Installing and maintaining scrapers to remove accumulated DCR from the conveyor belt at locations where it could be collected rather than spilled in the ship's hold
- Manual collection of large deposits of DCR before they become mixed with wash water

Consistent with the integrated and proactive mitigation approach described in Section 5.4.2, each of these measures were incorporated into alternatives. As part of the impact prediction step in the DCR environmental impact analysis, the quantity of DCR generated and discharged was estimated from each of these measures and the resulting impacts on the environmental resources for each measure fully evaluated.

5.4.4 RWA Lake Whitney Mitigation Case Study

The environmental impact analysis for returning Lake Whitney to a water supply reservoir by the South Central Connecticut Regional Water Authority (RWA) included a successful application of the classic approach to impact mitigation. The environmental impact analysis identified the most serious impact of the proposed action as a lowering of dissolved oxygen (DO) if the flow in the Mill River, controlled by releases for the Lake Whitney Dam, was reduced. This impact was identified as critical in the impact prediction conceptual model, so an investigation was conducted to quantify the DO concentration reduction as a function of river flow.

The investigation took advantage of a planned lowering of the lake level for dam maintenance. Rather than gradually refilling the reservoir after the maintenance and maintaining a downstream release, the release was severely curtailed and the reservoir captured all the flow. The DO concentrations in the river were monitored at increasing distances downstream from the dam. The investigation was very successful, in that it demonstrated the dam release rate associated with unacceptably low DO to be below approximately 20,000 cubic meters a day. However, the enthusiasm and scientific curiosity of the environmental analysis team did create some issues as the dam release was throttled below the volume considered adequate to maintain acceptable DO concentrations as a validation of the prediction. The prediction was indeed validated, but the resulting minor fish kill was not appreciated by the downstream stakeholders that frequented the park adjacent to the Mill River.

With a validated impact prediction quantifying lowered DO with river flow, mitigation measures could be investigated and evaluated. An integrated approach was not applicable to this impact because years of water-needs studies, alternative source investigations, regulatory intervention, and litigation had clearly demonstrated that the only feasible and environmentally sustainable water supply alternative to supply south central Connecticut (centered on New Haven) was the reactivation of Lake Whitney as a water supply reservoir. Thus, after-the-fact mitigation of the predicted DO impact was the only option available.

The mitigation measure selected was a lake management plan. The plan was a negotiated compromise among stakeholders and included:

- Residents and the watershed association around the lake, who were strongly opposed to lowering the lake level because of aesthetic issues
- Users of Mill River downstream of the dam who used the adjacent park and correctly feared that extremely reduced dam releases would impact DO and result in aquatic biota and wildlife impacts
- State regulatory officials charged with ensuring an adequate water supply
- State regulatory officials charged with ensuring water quality, including DO, in the Mill River downstream of the dam
- State regulatory officials charged with ensuring fish migration
- RWA customer representatives concerned with the price of water, particularly to economically depressed neighborhoods in central New Haven.

A lake management plan was developed and accepted by stakeholders to mitigate DO impacts and consisted of the following major elements:

- Pumping water from the reservoir over the dam if the lake level dropped below the dam and river flow was below a specified level

- Creating an artificial water flow to simulate "water over the dam" to maintain aesthetics if lake levels dropped below the dam spillway
- Exceptions to the minimum release provisions in the lake management plan to accommodate specified drought and available water supply volume conditions
- Continued monitoring of Mill River conditions, particularly salinity and DO, and periodic reevaluation of the lake management plan in consideration of new data
- A lake management study to evaluate the relationship between lake level, aesthetics, and alternatives to improve the lake quality (including exposed mudflats perceived by some stakeholders as aesthetically adverse)

The Lake Management Plan was developed in the late 1990s, and continuous monitoring of DO and the aquatic community has proven that the plan is successful. There have been minor modifications to the plan based on the monitoring data, including a slight modification of the DO concentration provisions. The lake management study included mitigation of lake aesthetic impacts and an ecologically friendly storm water management plan. The plan also included an expert environmental advisory team, which has met at least annually with stakeholders to review results and over the almost 15 years of the plan, the team, RWA, and stakeholders have forged a productive and even friendly relationship, to the advantage of the environmental attributes of Lake Whitney and the Mill River.

5.4.5 Programmatic Mitigation

Mitigation that goes beyond the progressive and integration approach described above can be implemented as part of a programmatic environmental impact analysis (see Chapter 6 for a description of programmatic and similar analyses). This approach identifies impacts that are likely for a set of similar actions as part of a program. A classic example is a program for a National Forest where forest road maintenance and associated drainage is a major component with the potential to create substantial environmental impacts on vegetation, aquatic resources, soil, and land use. Although there can be a high risk of impact on these resources and the consequences could be far reaching, they often can be easily avoided or otherwise mitigated. Most of the mitigation measures are not unique to a specific road or drainage system and common measures applicable to all roads and drainage systems in the National Forest can be indentified in the programmatic environmental evaluation. If the mitigation measures are included in an individual project, the impacts are addressed and a separate environmental analysis for the project is unnecessary, or at least reduced in scope.

The programmatic analysis conducted for U.S. Army Fort Campbell Kentucky/Tennessee (U.S. Army 2004) and the WINNER project in Haiti (USAID 2011) included programmatic mitigation to address adverse impacts and thus limit the scope of subsequent environmental analyses. As discussed in detail in Chapter 6 (Section 6.3.2), the Fort Campbell programmatic environmental analysis for infrastructure maintenance and minor development projects included required measures that would ensure impacts were mitigated (e.g., qualified personnel will conduct a detailed surface runoff impact study, including calculation and evaluation of both current and proposed associated land drainage contours and impervious surfaces). The analysis also included 30 required best management practices such as:

- Vegetative ground cover shall not be destroyed, removed, or disturbed more than 15 calendar days prior to grading.
- A floating sediment boom shall be placed downstream of the construction area to collect the unsettled silt or debris. The device shall be cleaned and maintained on a daily basis.
- Construction must be sequenced to minimize the exposure time of cleared surface areas.
- Grading activities must be avoided during periods of highly erosive rainfall.
- All surface water flowing toward the construction area shall be diverted around the construction area to reduce its erosion potential, using dikes, berms, channels, or sediment traps, as necessary.
- Temporary diversion channels must be lined to expected high water level and protected by non-erodible material to minimize erosion.
- Rock, log, sandbag, or straw bale check dams shall be properly constructed to detain runoff and trap sediment.

The USAID Watershed Initiative for National Natural Environmental Resources (WINNER) program in Haiti to protect water resources while implementing projects to support economic sustainability, particularly agricultural irrigation, also included programmatic mitigation measures applicable to individual projects comprising the program (Table 5.5).

The advantages of program-wide mitigation measures, as illustrated by the hypothetical Nation Forest, Fort Campbell, and Haiti examples are as follows:

- Effort is not diluted by having to address the same or very similar actions, environmental resources, impacts, and mitigation measures numerous times.
- Effective measures can be universally required and ineffective measures readily dismissed.
- Measures proven to mitigate impacts and which can be readily implemented can be repeated with a high degree of confidence.

TABLE 5.5

Haiti USAID Watershed Initiative for National Natural Environmental Resources (WINNER) Programmatic Example of Mitigation Measures

Activity	Impact	Mitigation
Sourcing of construction materials	Mining of material is not environmentally sound and results in ecological, hydrological, aesthetic, and land use impacts. Other material may not be sourced sustainably producing adverse impacts.	Independent environmental review for procurement of construction material.
Ancillary temporary facilities such as: maintenance areas, staff camps, access roads, and storage areas	Loss or degradation of wildlife, vegetation, habitat and species, changes in hydrology, impacts to cultural resources	Phase construction to minimize footprint of temporary facilities.
Decommissioning	If construction camps and other ancillary features are not dismantled in a timely manner, they will degrade over time, and can impact soil and create a public hazard.	Temporary facilities are removed immediately following construction. Removal will include disposal of solid waste, buildings, etc. A provision will be included in the engineering contract for complete decommissioning prior to finalizing payment.
Water supply for construction camps	Deplete ground and/or surface water resources and damage local ecosystems. Pose risks to human health if contaminants are present in the water and cause groundwater contamination.	Determine safe and sustainable yield and establish system for regulating use. Test seasonal water quality and examine historical water quality and quantity data before building facility. Incorporate sitting, design, and operation and maintenance practices that minimize environmental impacts.
Development and construction of standpipes	Create pools of stagnant water that could support malaria transmitting mosquitoes and other vectors.	Construct soakway or equivalent structure to divert spilled water and rainwater. Monitor and repair leaks from cracked containment structures, broken pipes, faulty valves, and similar structures.

Source: USAID. Haiti WINNER Project Programmatic Environmental Assessment. 2011.

- Since known and proven methods are part of the proposed action, the uncertainty regarding anticipated consequences is low.
- Impacts to critical environmental resources are avoided.
- As discussed in great detail in Chapter 6, unnecessary and repetitive environmental analysis and documentation inefficiently expending scarce environmental protection resources is avoided.

5.5 Long-Term Productivity, Irreversible Commitment of Resources and Cumulative Impacts

Early environmental analysis regulations attempted to provide some detailed guidance, because at that time there was no collective experience in conducting impact assessments. Most of the guidance provided has proved very successful and established a base procedure, which subsequently has been enhanced reflecting research and experience. However, some of the guidance provided in NEPA (Sec. 102 [42 U.S.C. §4332]) has proven to be cumbersome at best:

- The relationship between local short-term uses of man's environment and the maintenance and enhancement of long-term productivity
- Any irreversible and irretrievable commitment of resources that would be involved in the proposed action should it be implemented

After years of attempting to understand and then comply with these mandates, many environmental practitioners finally realized that these requirements represented an effort by the drafters of NEPA to ensure a comprehensive and interdisciplinary environmental analysis. It was the drafter's attempt to expand, rather than limit the analysis so that it was not short sighted chronologically or limited in terms of the environmental resources considered. Most environmental analysis practitioners now agree that a thorough analysis properly planned and conducted addresses the intent specified by the two ambiguous requirements listed above, and a separate and distinct analysis of "relationship between local short-term uses of man's environment and the maintenance and enhancement of long-term productivity" and "irreversible and irretrievable commitments of resource" does not serve any purpose.

Cumulative impacts are another concern addressed in the early environmental regulations (CEQ regulations §1508.7; see Chapter 2, Section 2.3 for a discussion of the regulations) which defined cumulative impacts as:

> "... impact on the environment which results from the incremental impact of the action when added to other past, present and reasonably foreseeable future actions...."

The requirement to consider cumulative impact was important in the early days of environmental analysis because there was a natural tendency to focus on the action at hand and spend less time integrating the projected effects on a broader canvas. The specific cumulative requirement forced the practitioner to take the broader perspective, and much has been written on approaches to cumulative impact analysis (e.g., Eccleston 2008). However, there is another view that cumulative impact analysis should not be distinct but an integral component of a comprehensive environmental impact analysis.

Any thorough environmental analysis should fully document and consider the environmental setting (or Affected Environment, see Section 5.2) as part of impact prediction. Certainly a complete description of existing conditions would incorporate *"other past, present...actions...."* Similarly, the description of the no-action alternative would incorporate *"reasonably foreseeable future actions,"* and these actions should be included in the prediction of impacts for all alternatives. Thus, the full impact prediction approach, procedure, and methods as described in Section 5.3 will address cumulative impacts completely, and a separate and typically less comprehensive analysis of just cumulative impacts will be redundant and nonproductive. This is not to diminish the importance of considering the impacts of the proposed action and alternatives in combination with other actions; cumulative considerations are essential to a successful prediction of impacts. They just should not be performed or considered as an add-on after-the-fact analysis of impacts.

References

Bascom, W., A.J. Mearns, and J.Q. Ward. 1978. Establishing boundaries between normal, changed and degraded areas. W. Bascom, ed. *Southern California Coastal Water Research Project Annual Report*. 81. Costa Mesa, CA.

Cibik, Steve and James T. Maughan. 1991. Environmental comparison of offshore vs. estuarine effluent disposal for the town of Scituate, Massachusetts. *Water Pollution Control Federation October 1991 meeting proceedings*. Toronto.

Claxton, W. T. and G. L. Mackie. 1998. Seasonal and depth variations in gametogenesis and spawning of *Dreissena polymorpha* and *Dreissena bugensis* in eastern Lake Erie. *Canadian Journal of Zoology* 76: 2010–2019.

Copeland, C. 2010. Clean Water Act: A summary of the law. *Congressional Research Service*. 7–5700. Available from: www.crs.gov RI 30030.

Czarnoeski, M., U. Michalczyk, and A. Pajdak-Stós. 2004. Substrate preference in settling zebra mussels. *Dreissena polymorpha. Archiv für Hydrobiologie* 159: 263–270.

Eccleston, C. 2008. *NEPA and Environmental Planning: Tools, Techniques, and Approaches for Practitioners*. Boca Raton: Taylor & Francis.

Erickson, P.A. 1994. *A Practical Guide to Environmental Impact Assessment*. New York: Academic Press.

Maughan, J.T. and C.A. Oviatt. 1993. Sediment and benthic response to wastewater solids in a marine mesocosm. *Water Environment Research* 65: 879–889.

Metcalf & Eddy. 1991. *Wastewater Engineering Treatment, Disposal, Reuse*, 3rd Edn. New York: McGraw-Hill, Inc.

Mudrock, A., L. Sarazlin, and T. Lomas. 1988. Summary of surface and background concentrations of selected elements in the great lakes sediments. *Journal Great Lakes Research* 14: 241–251.

Oviatt, C.A., A.A. Keller, P.A. Sampou, and L.L. Beatty. 1986. Patterns of productivity during eutrophication: A Mesocosm experiment. *Marine Ecological Progress Series* 28: 69–80.

Oviatt, C.A., J.G Quinn, James T. Maughan, J.T. Ellis, B.K. Sullivan, et al. 1987. Fate and effects of sewage sludge in the coastal marine environment; A mesocosm experiment. *Marine Ecological Progress Series* 41: 187–201.

Perdicoulis, A. and J. Glasson. 2006. Casual networks in EIA. *Environmental Impact Assessment Review* 26: 553–569.

Person, T.H. and R. Rosenberg. 1978. Macrobenthic succession in relaton to organic enrichment and pollution of the marine environment. *Oceanography and Marine Biology: An Annual Review* 16: 229–340.

Pilson, M.E., C.A. Oviatt, G.A. Vargo, and S.L. Vargo. 1979. Replicability of MERL microcosms; initial observations. In: Jacoff F.S., ed. *Advances in Marine Research. Proc. of a Symposium*, June 1977, EPA-600/9-79-035. U.S. EPA. Washington, D.C.: ERL-N, Narragansett, Rhode Island, 359.

Sampou, P. and Oviatt, C.A. 1991. A carbon budget for a eutrophic marine ecosystem and the role of sulfur metabolism in sediment carbon, oxygen and energy dynamics. *Journal Marine Science* 44: 825–839.

U.S. Army. 2004. *Environmental Assessment to Analyze Standard Practices for Construction Projects in the Cantonment Area*. Fort Campbell, KY.

U.S. Coast Guard (USCG). 2008. *Final Environmental Impact Statement: U.S. Coast Guard Rulemaking for Dry Cargo Residue Discharges in the Great Lakes*. DOT Document Number: USCG-2004-19621. Washington, DC: Commandant USCG Headquarters.

U.S. EPA. 1988a. *Draft Supplemental Environmental Impact Statement for Boston Harbor Wastewater Conveyance System*, Volume I and II. Boston, Massachusetts: U.S. EPA, Region I.

U.S. EPA. 1988b. *Final Supplemental Environmental Impact Statement for Boston Harbor Wastewater Conveyance System*, Volume I and II. Boston, Massachusetts: U.S. EPA, Region I.

USAID. 2011. Haiti WINNER Project Programmatic Environmental Assessment. Chemonics Contract No. EPP-I-04-00020-04. Port Au Prince Haiti.

6
Multilevel Environmental Impact Analysis

6.1 Overview

Duplication, repetitiveness, and excessive length are common problems associated with environmental impact analyses. These flaws often discourage productive stakeholder participation, and by creating confusion and boredom, compromise the purpose of the National Environmental Policy Act (NEPA) and similar initiatives. The concept of multilevel environmental impact analysis was developed to mitigate these shortcomings by producing more concise, focused, and readable evaluations.

A major objective of the approach is to streamline the process. "It takes too long and delays the project" is a criticism leveled at environmental impact analysis in general and NEPA specifically, more often than criticism that environmental impact analysis is useless or an inhibitor to progress. In recognition of this criticism, a bill to reform the federal analysis process by setting the first-ever deadlines for agencies to complete environmental reviews under NEPA was passed by the Judiciary Committee during the summer of 2012 and was forwarded to the House floor. The bill sets a 4.5-year maximum deadline to complete the NEPA review process, an 18-month maximum for environmental assessments (EAs), and a 36-month maximum for an environmental impact statement (EIS). No environmental analysis that is properly planned and executed (e.g., in line with the guidance in this book) should take even half that long. But as evidenced by the necessity for congressional intervention, there are too many NEPA analyses that take even longer. Employing multilevel environmental analysis can cut this schedule in half or more in many cases.

The underlying logic of multilevel analysis is to address an issue once, make a decision "when the time is ripe and right," and then move on to the analysis and decisions that logically follow. As the process cascades through the multiple levels, issues are put to rest and each subsequent level of analysis and decision refers to the previous analysis and becomes more and more focused on the most relevant and critical issues ripe for action. Thus, if an initial EIS or other environmental impact analysis fully addresses the implications, both beneficial and adverse, of a regional energy policy on fuel consumption and climate change, each subsequent environmental analysis on specific actions to implement the policy addresses only the impacts of the specific action (e.g., a wind turbine, regional geothermal heat, fuel efficient mass transit) and does not have to repeat the analysis of multiple fuel use policies.

In a tiered or programmatic EIS, the objective is to evaluate alternatives and address issues that are "ripe for decision," but the challenge is to determine when issues are in fact truly ripe. Having enough information to fully evaluate the environmental impacts that would result from alternative courses of action is a primary criterion in determining the ripeness of an issue for decision. A frequent conclusion is that there is enough information to compare multiple general courses of action, but there is not enough information to evaluate the implications of implementing specific actions. In such cases, implementation details can be deferred to subsequent NEPA review by a tiered EIS, supplemental EIS, EA, or equivalent NEPA process.

Another criterion in determining ripeness for decisions is potential constraints imposed by a premature decision that limits the available options for subsequent components of the action. The prematurity of the decision with respect to constraining other aspects of the purpose and need can be evaluated by first identifying a range of implementing alternatives for each component of the action and describing them in general terms. If it can be clearly demonstrated in the initial or programmatic EIS that decisions made will not constrain the range of alternatives for other aspects of the program, the decision may be ripe in this respect in the initial or programmatic EIS. However, if a number of implementing alternatives are shown to constrain follow-up decisions, then it may be necessary to defer the decision to the next tier of the NEPA process.

An example of a programmatic EIS limiting the options in subsequent tiers is illustrated by a hypothetical example for a sustainable electrical power program for a region. The purpose and need of this hypothetical programmatic EIS is to develop multiple sources of energy to provide future flexibility if conditions change, such as restricted availability of fuel for power production. One of the alternatives relies on a mix of hydropower, developed on the region's major river, and nuclear power generation. In this hypothetical example, this alternative is selected and the Record of Decision (ROD) calls for tiered EISs for the development of each separate source of electrical power.

During parallel preparation of the tiered EISs for nuclear and hydropower following the hypothetical regional energy programmatic EIS, a major conflict was uncovered. Because of substantial water needs for cooling the nuclear power plant, the only available source was the major river in the region. The other tiered EISs for hydropower was found to limit the availability of water during dry conditions, and there was no guaranteed source of cooling water for the only acceptable nuclear power plant site. In this hypothetical example, the entire program would have been required to go back to a supplemental programmatic EIS and reconsider the selected

Multilevel Environmental Impact Analysis 231

alternative. In this case, insufficient detail regarding components of the electrical energy generation alternatives was available, thus the time was not "ripe for a decision."

Actual case studies are summarized in Section 6.3 to demonstrate the efficiency and effectiveness that can result from multilevel environmental analyses. As demonstrated in the examples, a systematic and progressive approach to the development of the proposed action and environmental impact analysis is critical to success. Also important to successful implementation of multilevel analysis are positive and productive cooperation and coordination with stakeholders. The importance of the environmental analysis team developing a comprehensive understanding of the issues, a description of the action adequate for the level of decision to be made, and a sufficient technical environmental impact prediction is also demonstrated in the case studies.

6.2 Multilevel Environmental Impact Analysis Approaches

A number of multilevel environmental impact analysis approaches have been developed promoting efficiency and effectiveness and fostering productive stakeholder participation. These range from the simple to the complex, with categorical exclusion (CATEX), as discussed in Section 3.6 being perhaps the simplest, and examples of other relatively simple approaches discussed in the following sections. Strategic environmental assessment is a more complex approach and warrants a more comprehensive discussion (Section 6.4).

6.2.1 Inclusion by Reference

Reference to a previously conducted analysis that has been fully developed and accepted is another simple form of multilevel environmental impact analysis. As the requirements and methods for comprehensive environmental impact analysis have grown and matured, environmental evaluations of many diverse types of actions in many geographic locations and land use types have been conducted. This approach draws on the description of the affected environment (e.g., the type of vegetation, the current transportation system, rare and endangered species; see Section 5.2) from a recent environmental analysis for the same area which can be briefly summarized and the previous document referenced. For example, the evaluation of endangered species would not have to be repeated if a recent environmental analysis in the same area concluded there were no rare or endangered species in the area based on a comprehensive investigation. Subsequent environmental analysis for the same area could simply cite the conclusions and reference the original document after demonstrating that no land use or other changes potentially affecting the species or their habitat had occurred since publication of the first document.

Similarly, for comparable proposed actions that have been the subject of environmental impact evaluation, the non-site-specific impacts can often be referenced and not repeated. The U.S. Air Force (USAF) Over-the-Horizon (OTH) NEPA program represents an example of this approach. This program was a continent-wide radar detection effort addressing security threats from airplanes, ships, and missiles at a range of up to 5000 km from the continental borders of the United States. In order to transmit the radar beam over such a range and have the returning signal be strong enough to detect, tremendous transmission energy was required. The exposure of humans and wildlife to such high levels of energy was a significant environmental and public health concern to the USAF. To address this concern they conducted extensive research into the environmental impacts and health effects of the radar beam as a part of the Maine OTH system, which was the first of four planned systems (the others were in Minnesota/North Dakota, California/Oregon, and Alaska). The research and EIS for the Maine system established the allowable threshold exposure levels for the non-ionizing electromagnetic radiation from the radar beam and developed mitigating measures to address impacts on public health and wildlife from the radar's non-ionizing radiation. The mitigation measures included fencing to exclude wildlife within the zone of highest radiation and land use controls at greater distances within the radar beam to avoid public health impacts. The research, impact prediction, and mitigation measures developed for the radar in Maine were only summarized and referenced in the EISs for the other systems. This allowed the other EISs to focus on more site-specific impacts such as cultural resources and wildlife migration as discussed elsewhere in the book (Section 5.2.5). Addressing common impacts in the first EIS permitted streamlining the entire NEPA process for subsequent environmental impact analyses, allowing them to be completed in approximately one year for each system as opposed to the more common two- to three-year duration of the typical EIS.

6.2.2 Supplemental Environmental Impact Analysis Documents

Supplemental EISs, EAs, and other environmental impact analysis documents are a convenient and efficient way to address something that was neglected in the original document, changed conditions, or new information. It is an efficient approach because the entire process does not need to be reinitiated, and the supplemental document can focus on one or very few issues that necessitated the supplementation. The most common applications of supplemental documents are:

- There is a relatively minor change in the details of the proposed action.

- Implementation of the proposed action identified in the original document is not feasible as planned, and either significant modification of the action or reevaluation of alternatives is necessary.
- During planning or design, new information becomes available that could alter the description of the affected environment, prediction of impacts, comparison of alternatives, or required mitigation presented in the original document.
- Once the implementation of the proposed action is initiated, the existing environmental conditions as described in the original document are found to be altered to the point the impact prediction may be substantially different (e.g., a threatened or endangered species has moved into the area where none were present previously).
- Issues left opened in the original document become "ripe for decision."
- Courts rule that the preparation of the original document is inadequate (e.g., there was no "hard look" at the impacts, decisions were arbitrary or capricious, or otherwise inappropriate procedures were followed), and a supplemental EIS must address the shortcomings.

6.2.3 Programmatic and Tiered Environmental Impact Analyses

The programmatic environmental impact analysis process is designed to incorporate environmental considerations into concept-level decisions without getting bogged down in specific details. This approach is applicable to a broad-scale program, which will eventually include many individual projects or actions. A broad-scale analysis is also generally appropriate for programs, plans, and policies as opposed to individual projects. Typically programmatic analyses are applied to the purpose and need of a long-term and/or geographically extensive action. The programmatic analysis usually focuses on generally defined actions and a geographic area much larger than that actually required for specific actions and includes only limited details and evaluation of small-scale site-specific existing conditions and impacts.

A classic example of a programmatic environmental impact analysis is a National Park or National Forest land use management plan. The management plan and companion programmatic EA or EIS for the overall park or forest plan typically includes:

- The overall purpose and need, including the desired types and magnitude of recreation and other uses of the land.
- A broad-scale scoping process to identify desired uses and stakeholder concerns on a broad scale.
- A general inventory of the land use types (e.g., forest, open water, steep slopes) and environmental resources (e.g., lakes, fishing streams, and rare or endangered species) within the park or forest.

- A comprehensive literature review (mandatory first step, see Section 5.2.2) of sensitive environmental resources and receptors.
- A listing, general description, and requirements for optional land uses and activities.
- A comparison of desired activities and land uses of the available environmental resources within the available land.
- A designation of a proposed action and alternatives to assign various activities to various large-scale land tracks within the park or forest.
- A description of the types of impacts on environmental resources and mitigation measures anticipated for each alternative.

For some activities in some areas, the programmatic document can provide sufficient detail and environmental impact analysis such that a follow-up evaluation is not necessary. But this is the exception, and in other cases the programmatic environmental analysis lays a foundation for follow-up or "tiered" analyses for each individual activity within each large-scale land track or other broad-scale environmental settings. For example, if snowmobiling is designated as a beneficial land use for a certain area in the programmatic evaluation, a subsequent analysis to evaluate alternative trails, timing of activity, administrative protocols (e.g., charging for use, number of users allowed), and so on could be conducted. This would avoid getting bogged down in the details during the initial environmental analysis and put off such specifics to a follow-up tiered analysis until the decision to actually implement snowmobiling was made in the programmatic evaluation. If the detailed evaluation done in the tiered document revealed it was impossible to designate trails of sufficient length without significantly infringing on sensitive environmental resources, the programmatic environmental analysis might have to be supplemented. Other similar site-specific analyses can be prepared for other uses and areas within the context of the programmatic analysis without having to repeat the common information and analyses multiple times. Analysis at the programmatic level also facilitates the evaluation of cumulative impact evaluation on the park as a whole, because at the programmatic level, all areas of the park are taken into account, and the combined impact on such attributes as transportation, aesthetics, and other resources can be addressed once and not repeated in each tiered document.

Another frequent and productive application of the programmatic approach to environmental impact analysis is establishing environmentally friendly protocols for program components. In many cases, a programmatic analysis addresses similar actions to be executed in multiple locations, and these actions can result in similar impacts. The programmatic analysis can fully evaluate these impacts and frequently identify mitigating measures that fully address the impacts (see discussion of impact mitigation for Fort Campbell and irrigation in Haiti, Section 5.4.5). This evaluation can sometimes narrowly limit the scope of subsequent documents implementing specific aspects of the program as long as they adhere to the protocols established in the programmatic evaluation. In some cases, a commitment to following the protocols and implementing the mitigation measures can even eliminate the need for individual environmental impact analysis for certain specific actions.

A similar approach has been successfully applied integrating programmatic analysis and CATEX (see Sections 3.6 and 5.4.5). After environmental analysis, stakeholder input, and development of environmental impact mitigating measures, certain program components can be shown to have no impact. These actions can then be added to a federal agency's or equivalent organization's list of actions excluded from the need for individual environmental impact analysis. Thus, when implementation of these program components is initiated, a separate environmental impact analysis may not be necessary.

The application of a programmatic analysis can have both advantages and disadvantages that should be weighed and considered before it is chosen as an approach. On the plus side, it can be very efficient by minimizing duplication, and perhaps more importantly, it facilitates evaluating impacts from a broad perspective and a comprehensive treatment of cumulative impacts. A programmatic analysis can also be an excellent base for subsequent tiered documents, allowing them to work with specific stakeholder groups and issues.

On the downside, a programmatic approach can have schedule implications. Besides the delay in preparing, reviewing, and revising the programmatic document, subsequent site-specific evaluations can uncover details not apparent at the programmatic level. In such cases, the process might have to return to "go" for a comprehensive reevaluation of the program-level decisions and impacts. Similarly, conditions, funding, goals, etc., can change between development of the program and individual actions causing a similar return to the start. Under NEPA the programmatic analysis is frequently a full EIS, and EAs are adequate for the follow-up actions. Also, because environmental evaluations are often prepared by detail-oriented scientists and engineers and followed by stakeholders with detail concerns, there is frequently a tendency for the programmatic analysis to deconstruct to a detail rich process, following multiple "rabbit trails" and deviating from the original intent of a broad-scale analysis. This can severely delay the process and simultaneously frustrate project proponents and stakeholder groups.

The tiered environmental impact analysis is the logical follow-up to a programmatic approach. As in the above hypothetical example of the National Park or Forest land use plan, the programmatic evaluation establishes an overarching proposed action, and the subsequent tiered analyses address individual actions in detail. These individual actions are addressed by considering site-specific conditions to the point of frequently implementing data collection programs (e.g., biological sampling, water quality surveys, noise monitoring, traffic counts). It would be very rare for a programmatic analysis to include original data collection because the scope and geographic area are typically too large for representative sampling or monitoring. The tiered analysis also includes detailed development of the proposed action and alternatives frequently to the point of preliminary engineering and flagging locations in the field. It defeats the purpose of a program-level analysis to go to this level of detail.

6.2.4 Piggyback Environmental Analysis

In the early 1970s, there was no such thing as piggyback environmental analysis because NEPA was the only game in town and there was nothing to ride on the back of a NEPA EIS. However, as other environmental analysis programs, regulations, and legislation were put into place, the potential for duplicative and sometimes contradictory analyses between NEPA and other programs such as state environmental requirements developed. For example, based on comments on the draft environmental analysis document for the state program, the proposed action could be modified. The parallel federal EIS would then have to be reissued in draft form to reflect the changes and there was nothing to prevent repeating this scenario between the two separate documents, causing a do-loop and associated frustration and delays.

The need for coordination was recognized in NEPA (Section 104, 42 U.S.C. §4334), the Council on Environmental Quality (CEQ) Regulations implementing NEPA, and the several international environmental programs. Specifically, the CEQ Regulations specified in NEPA Section 1506.2 (Elimination of duplication with state and local procedures) require:

- Federal agencies to cooperate with state and local agencies to the fullest extent possible to reduce duplication.
- Cooperation be included in the planning, research, public outreach, and environmental documentation.
- Development of joint impact statements with state and federal agencies identified as joint lead agencies.
- Identifying and discussing any inconsistencies with state or local plans or laws in federal EISs and where appropriate the approach to address the inconsistencies.

However, initially there were no or only poorly developed state or local environmental programs and it was not possible for federal agencies to fully meet the intention of the CEQ Regulation with regard to coordination or "piggybacking" with other environmental regulations.

Currently, many state programs are well developed with a history of successfully meeting the goals of NEPA, and in recognition of the state programs, federal agencies have modified their approach to NEPA. If an action is strictly under the jurisdiction of federal agencies, such as significant agency action on a military base or in a national forest, the federal agency still conducts the environmental analysis and only seeks comments from the state. If the action has both substantial federal and state involvement and approval, a jointly prepared piggyback environmental analysis is conducted to simultaneously meet state and federal requirements.

However, in states with strong environmental analysis requirements and if the action requires state approval or has substantial state participation in addition to federal agency involvement, the agency often defers to the state programs. In such cases, it is not unusual to have only a single environmental analysis conducted under the state program. This involves early coordination between the state and federal agencies and a documented decision by the federal agency that they will participate in the process and that the compliance with the state program will meet NEPA requirements. This approach would typically apply if a federal agency provided a portion of the funding or a federal permit was required, but it was only a minor aspect of the overall program.

The Boston Harbor Cleanup (discussed in Section 5.3.5, Section 6.3.1, and Section 10.1) represents the maturing of the state program and state/federal coordination efforts for environmental impact analysis. For the first environmental analysis, there was only a federal EIS, which in essence directed the state on what had to be done for the cleanup. For the follow-up tiers, there were separate documents prepared under the Massachusetts Environmental Policy Act (MEPA; see Section 8.6.1) by the Massachusetts Water Resources Authority (MWRA), the state agency implementing the cleanup and under NEPA by the U.S. EPA (the federal agency permitting the wastewater discharge). Initially, there was substantial friction between the two agencies and environmental analyses, but over time as they worked together with weekly coordination meetings, there was productive and eventually even enthusiastic cooperation. During later actions in the Boston Harbor Cleanup, there were single piggyback environmental evaluations carried out jointly by the state and federal agencies.

The coordination continued to mature and climaxed for the final phase of the comprehensive Boston Harbor Cleanup, the combined sewer overflow (CSO) program. Much of the Boston metropolitan area had combined storm and sanitary sewers, which worked well in small perception events when the rain water or snow melt was conveyed to the wastewater treatment plant. However, during large rain or snow events, the combined sewers exceeded capacity, and the combined runoff and raw sewage overflowed to Boston Harbor and its tributaries. It was most complicated and difficult to address the Boston Harbor pollution problems and required a lengthy and iterative process. The U.S. EPA recognized the complexity of the problem and potential solutions and felt their joint or duplicative environmental analysis could be a counterproductive complicating factor. Thus based on the recent history of close cooperation between the federal agencies and MWRA, the proven success of MEPA, and the level of trust developed, the U.S. EPA deferred to the MWRA and MEPA, and no federal EIS or other NEPA analysis was performed for the CSO program.

6.3 Multilevel Environmental Impact Case Studies

6.3.1 Boston Harbor Cleanup

The Boston Harbor Cleanup program is an actual example of a multilevel environmental impact evaluation and EIS that developed sufficient information at each level to keep the cleanup moving forward (see Section 10.1 for an overview of the Boston Harbor Program). At issue was the wastewater generated from the Boston metropolitan area, which was almost 3.8 million cubic meters of total wastewater a day from approximately 2.5 million people, which received little or inadequate treatment prior to discharge to Boston Harbor. The wastewater was conveyed to two primary wastewater treatment plants, Nut Island for the southern portion of the system and Deer Island for the northern portion. Both plants were designed to provide primary wastewater treatment (i.e., short detention and settling of approximately 50% of suspended solids), but age, lack of maintenance, inconsistent operation, and flows in excess of capacity prevented the plants from even meeting acceptable primary treatment standards.

The problem was exacerbated by the handling of wastewater residuals (also known as sewage sludge). Following stabilization, the wastewater residuals, consisting primarily of the suspended solids that were removed during primary treatment, were also discharged to Boston Harbor. The prescribed operating procedure was to discharge the residuals only on the outgoing tide so that they would be flushed from the harbor. However, the solids washed back into the harbor when the tide reversed. Also, the operation was less than perfect and discharge of residuals often occurred on the incoming tide, leading to accumulation of solids in the harbor. This accumulation of solids had a devastating impact on Boston Harbor caused by the concentration of toxic compounds associated with the suspended solids, the depletion of dissolved oxygen as the solids decayed, and introduction of nutrients stimulating excessive algal growth.

The wastewater collection system was another contributing factor to the degraded condition of Boston Harbor. The collection system was designed early in the 20th century (and parts of the system a century before that) to accept storm water runoff from all but the largest storms and convey it to the wastewater plant for treatment prior to discharge. As the population grew, this combined collection system (i.e., conveying the combined sanitary sewage and storm water) approached capacity with just sanitary flows. Also, the lack of adequate maintenance resulted in a "leaky" collection system, and thousands of cubic meters of groundwater infiltrated into the sewers every day. With the sewers flowing near capacity, the introduction of storm water generated even during a moderate storm created an overflow and the combined raw sanitary sewage and storm water was discharged untreated directly to Boston Harbor and its tributaries.

The Boston wastewater system not only had a significant adverse environmental impact on the marine ecosystem, but it also created a serious public health threat and was in violation of numerous laws and environmental regulations. The regulations implementing the U.S. Clean Water Act necessitated a minimum of secondary treatment (i.e., physical and biological treatment that result in 90% removal of solids) and prohibited the discharge of sewage sludge into the waters of the United States. Similar regulations prohibited the discharge of storm water in sanitary sewage systems and the discharge of untreated sanitary waste. The system had been underperforming for two decades and in violation of the Clean Water Act for over five years. A major factor in the inability to resolve the wastewater management disaster and bring the system into compliance with the laws and regulations was the complexity of the situation. The problems of inadequate treatment, overcapacity collection system, and management of wastewater residuals were all linked, and the vast range and number of permutations of all possible solutions were paralyzing to the agency charged with operating the wastewater system. There were just too many major decisions to make at once, and with each potential solution alienating at least one group of stakeholders, a comprehensive single solution would have generated so much opposition that implementation would have been impossible.

Finally, stakeholders affected most directly by the pollution of Boston Harbor, including environmental groups, communities with beaches on Boston Harbor, and the U.S. EPA, could not take it any longer and filed a suit in federal court under the Clean Water Act to force action. Not surprisingly, with multiple and clear violations of the Clean Water Act, federal "Sludge Judge" Garrity ruled in favor of the stakeholders. The resolution of the suit resulted in a negotiated settlement, including a substantial fine (which was used for environmental enhancement) and two mandates critical to the ultimate resolution of the problem. The first mandate was an agreement by the Commonwealth of Massachusetts to form a separate agency to address metropolitan Boston area wastewater management. This agency, the Massachusetts Water Resource Agency (MWRA), replaced the Metropolitan District Commission (MDC), which had managed the water resources for decades but had other responsibilities including multiple parks, major highways, recreational facilities, water supply, and even their own police force. In comparison to wastewater management, all of these services had more visibility, public support, and influential stakeholder groups. These other responsibilities took precedence over wastewater because they were used in a positive manner on a day-to-day basis, and wastewater was out of sight, out of mind; thus wastewater management got little attention and less funding, leading to deterioration of the system, which accelerated over time. MWRA's mandate was to focus on managing wastewater using a dedicated revenue stream derived from sewer use charges, which could not be diverted to other purposes. Finally, because of the litigation and settlement, there was a clear mandate and resources to address the issue.

The other factor in the negotiated settlement that made a resolution of the complex wastewater management issue possible was a consent agreement on a very tight timeline with dozens of interim milestone accomplishments (e.g., complete facility engineering, finalize environmental analysis documents, and initiate prototype sludge composting). This consent agreement was executed by the MWRA and jointly monitored by the judge, the U.S. EPA, and the State Department of Environmental Protection. Several of the interim dates in the consent agreement were directed at making decisions on individual wastewater management issues and thus lessened roadblocks to a comprehensive solution. This progression of decisions to be executed when they were "ripe" was ideally suited to a multiphase environmental analysis to achieve compliance with NEPA.

The first step in the multiphase analysis was outside the scope of NEPA but had a very large impact on the subsequent NEPA process. There was an agreement by all parties to exclude the CSO issue from the initial environmental analysis and EIS. The agreement included a "set-aside" minimum volume of combined sewage to be accommodated within whatever wastewater treatment and discharge system was ultimately selected. Initially, there was concern that this approach might constrain the options available during the subsequent effort to address the CSO issue. However, it was pointed out that any environmentally acceptable resolution of CSOs that was in compliance with the regulations and laws would either eliminate CSOs to the point they were well below the set-aside volume or store the CSO volume and convey it to whatever wastewater treatment facility was selected during off-peak hours when there would be excess wastewater treatment capacity. This agreement divorced the overall wastewater management plan from the complicated, complex, and controversial issues and questions associated with CSOs including:

- What is an acceptable volume of CSO? It is impossible to capture every storm or construct separate storm and sanitary sewers throughout a city that is over 300 years old, so there had to be some provision for overflow during major storms, but was that the storm with a 10-, 25-, 50-, or 100-year return frequency?
- Where will CSO treatment facilities be located? No one wanted a combined sewage treatment facility in their neighborhood or backyard ("Not in My Backyard"), so trying to site such facilities would be a long and arduous process and a detail that would not directly influence the comprehensive wastewater management plan.

- How will the CSO plan be funded? The benefit of CSO abatement would be realized in Boston Harbor, but the majority of work and cost would be inland, associated with a very large wastewater collection system. So there was much negotiation to be done to develop an equitable funding plan.
- If the combined sewers are replaced with a separate storm water system, what will happen to the storm water from even the smallest storm? As antiquated, overburdened, and dilapidated as the collection system was, it still captured the precipitation from small storms (which are the majority of storms) and conveyed the storm water (which flushed the urban streets) to the wastewater plant for treatment before discharge away from the shore. If a separate storm water system was created, all storm water would be discharged close to shore, and planning a system to manage all that storm water would take many years and thus unnecessarily delay selection and implementation of a comprehensive wastewater management solution.

Thus, delaying the complete evaluation and decision on CSOs cleared the way to focus on overarching wastewater management issues and the provision to allow accommodations for CSO in the overall plan removed constraints on future CSO planning. With the CSO obstacle removed, there were still several critical wastewater management components to be addressed, but with the set-aside of the CSO issues, the other elements could be more easily defined and a phased process formulated to address the other issues. The remaining critical decisions for the comprehensive wastewater management plan were:

- Where should the wastewater be treated?
- How should the raw sewage be conveyed to the selected treatment location?
- What was the appropriate level of treatment?
- Where should the treated effluent be discharged?
- Where and how should the residuals be stabilized?
- What should be the ultimate treatment and fate (e.g., disposal, reuse, incineration) of the residuals?

After conceptual engineering and environmental evaluation, a consensus was reached that an initial EIS should be initiated with a focus on where the treatment facilities should be located. The other issues, as listed above, would be addressed to the extent possible. But there was the realization and acknowledgment that subsequent NEPA phases or tiers would be necessary to address all the issues.

At the time of preparation, the initial EIS was not termed a programmatic EIS, but in reality it was. It established the purpose and need for action, clearly defined the issues, and evaluated the interrelationship of the issues and need for separate actions. As part of this process, the initial EIS confirmed that establishing the location of the treatment facilities was the key issue, and all other decisions revolved around the siting of the wastewater treatment plant. All of the other issues and questions listed above were considered in light of where the plant would be sited and a determination made as to the constraints this would put on each of the other issues. No fatal flaws or unmanageable constraints to future decisions were identified to first determining the site for the wastewater treatment plant in the initial EIS and using a tiered approach for subsequent decisions.

Thus, the environmental and engineering evaluation of alternative sites was initiated. There were four primary alternatives considered:

- A number of satellite wastewater treatment plants located throughout the service area.
- A centralized facility at the existing Nut Island Wastewater Treatment Plant site that currently served the southern portion of the service area.
- A new location in Boston Harbor, such as Long Island in the center of the harbor, for a centralized facility.
- The wastewater treatment facility at the existing Deer Island Wastewater Treatment Plant site that currently served the northern portion of the service area.

A comprehensive evaluation of alternatives was conducted considering the environmental, engineering, economic, and community impacts. In addition to these factors, the initial EIS evaluated the potential constraints to the other aspects of the Boston comprehensive wastewater management plan (e.g., residuals management, level of treatment, discharge location) in the comparison and selection of alternative sites. The satellite wastewater treatment plant sites had several advantages, including retaining the treated effluent in each watershed as a water resource. This was important because many of the outlying communities derived their drinking water from local ground water sources that were in short supply. Consequently, there were frequent water use restrictions and extreme low flow stresses to the rivers in these communities. However, it also had several disadvantages including limiting the options for effluent discharge and management of residuals. The Nut Island and other Boston Harbor locations had constraints of available space and compatibility with the sewage collection system without any significant advantages. In the final analysis, the existing Deer Island Wastewater Treatment Plant site was identified as the preferable alternative because it was equal or superior to the other sites in almost every category analyzed.

Another advantage of the Deer Island selection was that it produced the fewest constraints on the range of alternatives available for the other aspects of the comprehensive wastewater management plan. But all parties acknowledged that the planning and environmental analysis process for the other aspects of the cleanup would take a long time and that inadequately treated wastewater and sewage sludge would be discharged to Boston Harbor every day while the evaluation of alternatives was completed. They also learned from the evaluations, as part of the initial EIS conducted for the site selection, that much needed to be accomplished to ready the site for the new 3.8 million cubic meters (1 billion gallons) a day wastewater treatment plant because the site had: a prison; Fort Drummond, a World War II coastal fortification; an existing wastewater treatment facility that had to maintain operation during construction; and a drumlin (geologic formation). All of these had to be removed in order to construct the new wastewater treatment facility that would be in full compliance with the Clean Water Act. In addition, the site selection process identified the traffic impacts to the adjacent town associated with transporting workers and materials as unacceptable and committed to mitigation by constructing a marine terminal on Deer Island to reduce truck traffic through the adjacent town. Thus, a decision was made to conclude the initial EIS, tier the remaining decisions, and implement aspects of the proposed action that had been fully evaluated in compliance with NEPA.

The site selection evaluation had fully considered impacts that would be associated with construction on Deer Island, as this was a major issue identified during scoping. Thus, initial site preparation could begin without additional NEPA evaluation. The initial work consisted of demolition of the prison, fort, drumlin, and portions of the existing wastewater treatment works on the site and construction of the marine terminal. However, construction of the new treatment facilities could not commence until there was additional NEPA review of other aspects of the comprehensive plan such as the level of treatment and discharge location.

The level of wastewater treatment, discharge location, and several other aspects of the cleanup required some level of NEPA review as summarized below (U.S. EPA 1988a and U.S. EPA 1988b). The decision was made to conduct follow-up supplemental EISs or EAs, but based on current practice, terminology, and interpretation, these follow-up documents would now be considered tiered EISs. They did add additional information, so they had some characteristics of a supplemental EIS; however, each one was a cascading purpose and need based on the ROD from the initial EIS. The environmental impact analyses tiered from the initial Boston Harbor Cleanup EIS were:

- *Interim residuals management:* Sufficient evaluation was conducted during the initial EIS to identify an interim solution to the discharge of sludge to the harbor and there was strong pressure to implement the interim plan to stop daily discharge to Boston Harbor. Detailed implementation of the interim plan was included in the ROD for the initial EIS with the caveat that if a detailed design indicated environmental implications not anticipated in the initial EIS, an EA or other environmental analysis document might be required.

- *Long-term residuals management:* The initial EIS identified the potential for significant issues associated with long-term management of residuals that were not fully addressed in the initial EIS, including the location of a sludge landfill or other final disposition of the residuals. The interim plan relaxed the immediate need for a long-term solution, which halted sludge discharge to the harbor. Thus, the ROD for the initial EIS identified the need for subsequent NEPA compliance for long-term residuals management. A supplemental EIS was identified as the method of compliance, but the information developed during the initial EIS was critical to scoping and the long-term residuals NEPA evaluation tiered off the decisions and analyses supporting the initial EIS.
- *Wastewater conveyance and treated effluent discharge:* Selection of the location and means of effluent discharge from the wastewater treatment plant was determined to be unconstrained and independent of the selection of the wastewater treatment plant site (after the satellite option had been screened out). Similarly, the method for conveying the raw wastewater to a central location for treatment was an independent decision. Thus, the ROD resulting from the initial EIS relegated the environmental analysis and decision on conveyance and discharge location to a subsequent NEPA process (a supplemental EIS was identified, but in current practice it was tiered from the initial EIS to address a follow-up decision).
- *Combined sewage overflows:* As summarized earlier, resolution of the CSO issue was independent of other aspects of the Boston Harbor Cleanup, once a set-aside for CSO flows was built into the specifications of the new wastewater treatment plant. Also, resolution of the complex and ubiquitous CSO problems was projected to take a very long time (evaluation was begun in the early 1980s and has continued to the publication of this book). Thus, the need for subsequent NEPA evaluation for CSOs was identified, but the time was not ripe to identify exactly what and when. Approximately five years later, the EPA determined that the state environmental regulations (see discussion of MEPA, Section 8.6.1) would satisfy NEPA for the CSO action and a separate NEPA document was not required.

Although the Boston Harbor Cleanup NEPA compliance was termed an initial EIS, two supplemental EISs, provisions for EAs, reliance on the state environmental procedure, and other environmental analysis documents and procedures have shown that it was in reality a series of tiered EISs (and other documents) cascading from a programmatic EIS. This process represents the successful use of multilevel environmental impact analysis to finally resolve a decades-old serious environmental problem efficiently and effectively. With the assistance of and push from the "sludge judge" there finally came the realization that the problem was too large, with too many controversial elements and influential stakeholders to be resolved in one bite. The creation of a new agency staffed by technical and policy professionals familiar with the situation but not rooted in the problems of the past recognized that siting the regional wastewater treatment facility was the first bite that had to be resolved to move the solution forward. Once this was resolved and gained stakeholder support, it was recognized that a permanent solution to the wastewater residuals issues was too fraught with controversy to be resolved quickly, and the resulting delays could derail the entire Boston Harbor Cleanup. To avoid these delays and keep the process moving, while addressing the immediate need to stop the sludge discharge to Boston Harbor, an interim solution with an associated environmental impact analysis was instituted while a more permanent solution was evaluated through a full NEPA supplemental EIS. As it turned out, technology and experience advanced during the preparation of the residuals EIS and implementation of the interim solution proved suitable for the long term, so a successful reuse program, rather than disposal at an unpopular landfill, was adopted. The residuals program has been successfully operated for 25 years and is currently being reviewed for minor modifications to substantially increase energy efficiency.

Following the initial (or in reality programmatic) EIS, which identified the wastewater treatment facility site at Deer Island, site development could begin while the questions of wastewater conveyance, effluent discharge options, and residuals management were evaluated through supplemental EISs. The most complex of the issues, the CSOs, were identified as a continuing effort and have been addressed at a slow but steady pace over three decades with a series of programmatic and tiered environmental impact evaluations. The multilevel environmental impact analysis, where a better understanding of the impacts was gained at each level, was no small part of this successful program, which was completed within schedule, under budget, and has produced environmental improvement exceeding the predictions made in the environmental impact analysis documents.

6.3.2 Fort Campbell Programmatic Environmental Assessment

Fort Campbell Kentucky/Tennessee is more than a U.S. Army base. It encompasses over 40,000 hectares, 5600 of which are fully developed with structures and infrastructure to support the Fort Campbell community of approximately 68,500 people. Fort Campbell has housed the 10th Airborne Division and support units since 1960 with an annual payroll of US $1 billion and an additional US $1.2 billion in construction activities (U.S. Army 2004). Fort Campbell is actually a city with all the inherent infrastructure, maintenance, and environmental issues.

One difference between Fort Campbell and a conventional city is that because it is owned, operated, and controlled by a federal agency (U.S. Army,

Department of Defense), everything that occurs within the base boundaries, from new buildings to street repair, is potentially subject to NEPA. Following decades of making frequent evaluations as to the applicability of NEPA and the appropriate approach for NEPA compliance for every activity on the base, those in charge of public works and environmental compliance took positive action to address this inefficiency, duplication, and unnecessary use of scarce environmental resources. They decided to prepare a programmatic EA to collectively analyze the environmental impacts of recurring types of base activities. This included maintenance of existing structures and infrastructure and construction of new facilities within previously disturbed and developed land.

Part of the process to address the existing inefficient and cumbersome environmental analysis process was a review of past NEPA evaluations. This included consideration of over 75 individual EAs conducted during the previous 10 years including projects such as:

- Construction of 96 army family housing units
- Upgrade of the primary power system
- Construction of an elementary school
- Addition of a boiler to the central energy facility
- Construction of a central troop medical and dental clinic
- Revitalization of Werner Park family housing
- Construction of a training and recruiting facility
- Construction of an on-post railroad
- Waterline system improvements
- Construction of a JP-8 tank farm
- Installation of energy efficient motors
- Addition of aprons to the air field

The review concluded that there were multiple EAs prepared for very similar or identical activities within areas of similar land use (e.g., nine housing projects and four school projects all in previously developed areas). The review also found that all the EAs included a standard list of required mitigation measures and resulted in Findings of No Significant Impact (FONSIs). Thus the decision was made to prepare a programmatic EA to cover routine construction and maintenance actions within the cantonment area (i.e., the developed portion) of Fort Campbell.

The purpose and need statement and the description of the proposed action in the programmatic EA were broad, allowing application of the conclusions to a wide variety of activities. Although the coverage was broad, substantial guidance regarding the types of activities covered was provided by listing the EAs for previous activities that were reviewed and referenced in the programmatic EA as well as the listing of approximately 60 anticipated future projects. Thus, when a project within the cantonment was in the planning stage, even if it was not within the list of the 60 projects anticipated at the time the programmatic EA was prepared, if it was similar in scale, function, use, surrounding land use, etc., to one of the previously reviewed or future anticipated projects, it would be covered by the programmatic EA and associated mitigation requirements.

The programmatic EA analyzed the potential for environmental impacts resulting from routine construction and maintenance activities to environmental resources (e.g., endangered species, water quality, noise, cultural resources) present within the developed portions of the base. The programmatic EA first included a very thoughtful and carefully worded description of the proposed action, such that it covered the full range of anticipated activities but was also consistent with the letter and intent of NEPA. The proposed action was not to actually perform construction but rather to develop standardized operating practices for routine construction projects. The practices were developed and evaluated by referencing the numerous previous EAs and the extensive environmental management programs that had been institutionalized at Fort Campbell over the years. The programmatic EA pointed out that through the successful and long history of the base's EA and environmental management processes and project implementation, there was a high level of assurance regarding compliance with applicable environmental laws, regulations, and policies. The proposed action also included a requirement that each cantonment area project would require the submission of a Record of Environmental Consideration to the Environmental Division for review and an appropriate determination regarding NEPA compliance.

The next steps in the development of the programmatic EA followed the CEQ NEPA guidelines as discussed in the previous chapters. The affected environment for each of the environmental resources was described by brief summaries taken from the numerous EAs previously prepared for the cantonment area (the prior EAs were included by reference). Similarly, the anticipated impacts on each resource area were briefly summarized from the environmental impact analyses done for the previously prepared EAs and standard practices developed as part of the environmental management plans. This analysis of impacts was not simply a prediction of impacts but specific reflections of what had actually occurred from similar past actions. For example, the potential impacts from fugitive fibers during asbestos removal in buildings was addressed by referencing the history of compliance and base-wide standard procedures to ensure compliance with National Emissions Standards for Hazardous Air Pollutants regulations. The programmatic EA also specified that each construction project shall abide by the current Fort Campbell Storm Water Management Plan and provided a list of commonly used best management practices for construction activities. Similarly, storm water-related impacts were addressed by including in the project description erosion and runoff control measures proven effective in past actions and included commitments to the following practices as prudent and practical impact mitigation measures:

- Qualified personnel will conduct a detailed surface runoff impact study of the proposed construction area, including calculation and evaluation of both current and proposed land drainage contours and impervious surfaces (e.g., roof and asphalt).
- The results of the study will be used to determine impact on both man-made and natural storm water conveyance systems.
- Where impacts are predicted, recommendations will be implemented to preclude or mitigate that impact.

The analysis of impacts was treated similarly in the programmatic EA for the other environmental resources and activities.

The programmatic EA concluded that based on the environmental impact analysis conducted, the findings of relevant EAs previously prepared, and the incorporation of required mitigation, the proposed action would not result in significant environmental impacts to any resource, and so a FONSI was issued. Thus, an EIS was not necessary and future individual projects and actions consistent with the statement of purpose and need and description of the proposed action in the programmatic EA could proceed without separate and project-specific NEPA documentation.

As discussed above, the conclusions of the programmatic EA were based on adherence to certain standardized procedures and compliance with environmental regulations and permits. The EA also included a list of 30 required best management practices including construction procedures such as:

- Vegetative ground cover shall not be destroyed, removed, or disturbed more than 15 calendar days prior to grading.
- A floating sediment boom shall be placed downstream of the construction area to collect the unsettled silt or debris. The device shall be cleaned and maintained on a daily basis.
- Construction must be sequenced to minimize the exposure time of cleared surface areas.
- Grading activities must be avoided during periods of highly erosive rainfall.
- All surface water flowing toward the construction area shall be diverted around the construction area to reduce its erosion potential using dikes, berms, channels, or sediment traps as necessary.
- Temporary diversion channels must be lined to the expected high water level and protected by non-erodible material to minimize erosion.
- Clean rock, log, sandbag, or straw bale check dams shall be properly constructed to detain runoff and trap sediment.

Multilevel Environmental Impact Analysis 249

- A buffer strip of vegetation at least as wide as the stream shall be left along stream banks whenever possible.
- For streams less than 15 feet wide, the buffer zone shall extend at least 15 feet back from the water's edge.
- When using fertilizer to establish areas of new vegetation for soil stabilization, use mulches to prevent the fertilizer from washing off the vegetated areas. Apply fertilizer when there is already adequate soil moisture and little likelihood of immediate heavy rain.

The programmatic EA process included a comprehensive review of base-wide environmental procedures and produced enhanced standardized construction procedures that minimized or prevented environmental impacts. It also enabled adding items to their CATEX list, including:

- Construction that does not significantly alter land use, provided the operation of the project when completed would not have a significant environmental impact.
- Acquisition, installation, and operation of utility and communication systems, data processing, cable and similar electronic equipment that use existing rights of way, easements, distribution systems, and facilities.
- Grants of easements or the use of existing rights of way: for use by vehicles; electrical, telephone, and other transmission and communication lines; transmitter and relay facilities; water, wastewater, storm water and irrigation pipelines, pumping stations, and facilities; and for similar public utility and transportation uses.

The Fort Campbell experience demonstrates the multilevel environmental impact analysis process and the benefits that can be achieved. The resources consumed by conducting an EA about every month or two for routine construction and maintenance projects for a decade were avoided. The process to make needed improvement to base operations was streamlined and proven standardized procedures that minimized environmental impacts were institutionalized and required.

6.3.3 U.S. Coast Guard Dry Cargo Residue Management Tiered EIS

As discussed earlier and summarized in Section 10.2, the U.S. Coast Guard (USCG) was charged by Congress with developing rules and regulations to govern the management of dry cargo residue (DCR) spilled during loading and unloading of Great Lake carriers. Their charge was to develop regulations that achieved a balance of environmental protection and economic viability of the Great Lakes shipping industry. Because promulgation of the regulations was a major federal action potentially affecting the environment, the rule making was accompanied by an EIS, which in general concluded that continuation of the historic DCR management practices would not result in unacceptable environmental impacts, at least in the short term (i.e., approximately five years). This conclusion was based on a comprehensive collection of field data in geographic locations and on environmental resources potentially impacted by historic practices and extensive laboratory analyses, testing, and mathematical modeling to predict impacts (see Chapter 5 for a discussion of how the affected environment [Section 5.2.6] was investigated and impacts predicted [Section 5.3.1]). The combination of all investigations and promugation of the EIS resulted in the issuance of an interim rule.

The 2008 EIS was definitive on the prediction of natural environmental impacts and identified potential mitigating measures and alternatives that would have a lesser environmental impact. However, the EIS and associated economic studies could not make similar definitive statements regarding the costs and economic impact to the shipping industry resulting from implementation of the mitigating measures and alternatives included in the original EIS. The 2008 EIS was also less than definitive on the effectiveness and implementability of some of the alternatives and mitigation measures. Thus, an interim rule was issued, and as part of the ROD resulting from the EIS, the USCG committed to the issuance of a final rule following additional study of the effectiveness, implementability, and economic impacts of the mitigation measures and certain DCR management alternatives.

Similar to the development of an interim rule, promulgation of the final rule was a major federal action with potentially significant impacts, thus requiring NEPA compliance and documentation. However, it had been previously demonstrated that the historic practices did not result in any significant environmental impact to any natural resources and that any alternatives considered for a final rule would not lessen the historic environmental restrictions; therefore, there was no real potential for significant impacts to the natural environment. If anything, the measures required of the shipping industry for DCR management under a final rule would strengthen environmental protection measures. Thus, from the standpoint of the natural environment, an EIS was not necessary, and it might have be possible to issue a FONSI based on the analysis done as part of the 2008 EIS. However, the economic impacts on the shipping industry resulting from a final rule with more stringent environmental protection requirements had not been fully analyzed, and there was the potential for significant economic impacts.

Given this history, there were a number of NEPA compliance options available to the USCG. They could promulgate a final DCR management rule that was at least as environmentally friendly as historic practices and did not necessitate expenditure by the shipping industry beyond current levels. This would not have any significant or unknown impact, and reference to the 2008 EIS would satisfy NEPA compliance. However, it would leave open the question of why the mitigation measures and alternatives with potentially fewer impacts on the natural environment identified as part of the 2008 EIS were not implemented, and the option of moving forward on this basis was rejected.

Another option was to address impacts on the natural environment by "inclusion by reference" to the 2008 EIS and embark on preparation of an EA to analyze the economic impacts. Preparation of the EA would begin by addressing the scarcity of economic data as cited in the first environmental impact analysis and initiating an investigation of shipping economic data to fill the gaps of the 2008 analysis. However, there was the real possibility that an economic evaluation using the new data could indicate a significant impact to the economic viability of the shipping industry for one or more of the DCR management alternatives. This could then require a full EIS if an alternative with significant economic impacts was selected, resulting in lost time and expenditure of resources. Thus, this alternative was also rejected.

Another alternative was to conduct a supplemental EIS that included the new economic data and results of the new alternative/mitigation measures' effectiveness evaluations to reevaluate the alternatives in light of the new data. This alternative was very attractive in that it could proceed expeditiously and would be very efficient because most of the information had already been fully vetted by the public. The supplemental EIS would simply describe the extensive effort to evaluate the cost and effectiveness of alternative DCR discharge control methods and then analyze the economic impact to the shipping industry resulting from the implementation of the measures. These costs could then be weighed against the benefits to the natural environment as described in the 2008 EIS, to make a decision regarding DCR management regulations with a full understanding of environmental implications as dictated by NEPA.

Another equally attractive option was to conduct a tiered EIS, accepting the conclusions of the 2008 EIS with regard to impacts on the natural environment and continuing the alternative/mitigation measures' effectiveness and economic evaluation initiated in the first EIS. This had the same advantages as the supplemental EIS because it minimized duplication, was efficient, and made full use of the information and conclusions already presented to the public. A tiered EIS also had the advantage of using compliance with the interim rule as the existing condition, whereas there was some question as to whether a supplemental EIS would be required to have the same starting point as the original EIS to represent existing conditions. The choice between a supplemental EIS and a tiered EIS was difficult, but in practice it made little difference. Both took full advantage of the work that was already done and could be executed efficiently. Both options also provided an efficient and accepted approach to incorporating new information and using the information to evaluate and select a course of action for promulgating DCR management regulations. In the final analysis, the USCG decided to conduct a tiered EIS, although a supplemental EIS could also have accomplished the objective. The tiered EIS represented an approach of moving forward from a decision already made when the time was "ripe" and subsequent decisions could be made without retracting from the one made in the ROD resulting from the 2008 EIS. A supplemental EIS could create the impression of reexamining the 2008 decision, which in fact was not the case. With the tiered EIS, the USCG was moving forward consistent with their earlier decision but just adding another level of detail.

6.4 Strategic Environmental Assessment

Environmental evaluations of proposed actions broader and more comprehensive than a single project are collectively termed strategic environmental assessments (SEAs). At the broadest level, an SEA is conducted on a policy, sometimes developed by a central or regional government but often by an agency or some other major subdivision of a central government. There has been a movement worldwide for SEA as a precursor and guide for project-specific environmental impact assessments with a specific goal of not only broad and long-range analysis but also rectifying past environmental insults and achieving enhancement of environmental conditions (Sadler et al. 2011). The environmental analysis and review of a U.S. federal government enforceable policy to develop energy self-sufficiency that minimizes carbon emissions would be an example of an SEA at the policy level. The U.S. Forest Service's (USFS) commitment to multipurpose use of USFS lands might be another policy warranting environmental evaluation at a level higher than an individual environmental impact assessment. An individual environmental impact analysis, as exemplified by NEPA documents, is most often the analysis of a project or specific action that is finite (and generally relatively small scale) in time and space. Individual EISs and other NEPA documents cannot by themselves actively achieve the NEPA goals; they can only address agency actions to make sure they are not in contradiction to the goals (Morgan 1998).

The development and environmental analysis of a land use or environmental resource plan, although less comprehensive than evaluation of a policy, could also be considered an SEA. The SEA for a plan would go a step beyond the policy evaluation by identifying the actions that would be required to implement the policy. A plan to implement the example policy of energy self-sufficiency that minimizes carbon emissions might identify the range of energy requirements, with and without conservation. It would likely also identify the energy sources available (e.g., solar, wind, geothermal) to satisfy the policy and the SEA for the plan and would at least qualitatively identify the types of environmental effects associated with each type of energy production. The plan and SEA associated with the USFS multiuse policy example might similarly identify the types of uses to be considered (e.g., logging, hiking, skiing, air tours), the types of impacts associated with the use, and the conflicts among the uses.

Generally, the least comprehensive and narrowest type of activity appropriate for an SEA is a program. This would be comparable to a programmatic EIS as discussed earlier. The program and SEA for the U.S. energy self-sufficiency example might identify an energy conservation target and the percentage of required energy that would be obtained from various sources such as nuclear, coal, wind, solar, hydroelectric, and fossil fuel. The program and SEA for energy might also identify on a regional basis not only what percentage of energy would come from each source but also alternative general geographic locations, materials extraction approaches, and processing methods (e.g., nuclear waste management and petroleum exploration and refinement). An SEA for the USFS multiuse example could include identification of alternative target uses for individual forests, areas where each use could occur, and the types of impacts that could occur for each alternative use in each alternative area. Any proposed action less than a program would typically be considered a project and more appropriately evaluated using an environmental impact analysis approach as described throughout this book, rather than an SEA procedure.

Project- or action-specific environmental impact analyses most frequently address infrastructure improvements, water resources development, private development, military activities, or transportation initiatives. These types of actions will almost always infringe in some way on one or more environmental resources in order to deliver a benefit to some societal need or desire. The goal of the individual environmental analysis for one of these actions is to evaluate the impact on the resource and identify the alternative that will cause the least environmental impact so that decision makers can weigh the societal benefits against the environmental costs and make informed decisions. In contrast, a policy, plan, or program (PPP) amenable to an SEA is typically for an action intended to rectify past and/or prevent future environmental harm. In the examples described earlier, the PPP for energy development is intended to prevent environmental (both natural and social) damage from oil extraction and produce environmental benefits by reduction of greenhouse gases and the resulting climate change. Similarly, the USFS example addresses potential infringement on one use of the environment (such as hiking in wilderness areas) by controlling other uses (e.g., air tours or all-terrain vehicle use) such that all uses can be achieved with minimal conflict. In this context, a conventional EA for a project is to enable the project "to do the least environmental harm" as opposed to an SEA where the goal can be considered to "do the most environmental good" (Sadler et al. 2011).

The need for programs to rectify or prevent environmental damage was not included in the development of NEPA, so SEA was not directly addressed in the Act or its implementing regulations. However, there is a growing recognition of the benefits of applying SEA in the United States and many other countries, which are currently formulating and implementing SEA requirements and procedures. In the United States, expansion and enhancement of the tiered and programmatic EISs have the potential to develop a successful SEA approach, and the application of SEA can benefit from experience and supportive case law of successful programmatic and tiered EISs. The European Union has mandated an established general protocol for SEA and requires member countries to develop guidelines and specific requirements. Japan and several other Asian governments have also developed SEA procedures, and in southern Africa there is some progress in this area (Sadler et al. 2011). However, developing frameworks is one thing and demonstrating progress by implementation is another matter. SEA has experienced constrained and limited progress due largely to the lack of tested guidelines, the lead time required to develop SEA, and the magnitude and controversy of most policies, plans, and programs that are most amenable to the SEA approach.

There are numerous advantages to nationwide or even regional SEA requirements and approaches. Probably the greatest advantage is the broad and comprehensive environmental evaluation conducted under SEA. If a PPP is developed and then no environmental review is conducted until individual projects are proposed, there is no mechanism to achieve the greatest environmental benefit or least environmental damage from a comprehensive policy. An alternative with the least environmental damage or greatest benefit could be selected for an individual project implementing a PPP, but this selection could preempt even consideration of a far superior alternative for another aspect of the broader initiative.

A hypothetical example of a regional planning agency developing a long-term drinking water supply program for an entire river basin illustrates the preemption of comprehensive environmental benefits of the program with the lack of an SEA. A goal of the hypothetical example is to maintain ecological characteristics of the water resources within the basin, primarily by not exporting or importing water. Because of available funding and an organizational structure that is firmly in place, a project to develop a large reservoir that would supply the projected water demand for the basin is the first proposal under the river basin water supply program. The EIS for the reservoir considers alternative locations, sizes, construction methods, operating methods, and even includes a somewhat cursory consideration of other sources and conservation methods. But there is no mechanism or requirement to force conservation or implement other water supply management approaches by the implementing entity. Consequently, these are dismissed and a combination of specific reservoir location, size, and operation that meets the projected water supply need and has the least environmental impact is selected.

In this hypothetical example, it would appear that environmental protection has been maximized at the project level, but this is not necessarily true at the program level. After an unavoidable delay, perhaps due to lack of funding, non-acceptance of a new technology, or inability to develop political consensus, the next project to address the policy is a relatively new technology to utilize ground water resources. This technology, aquifer storage and recovery (ASR), withdraws water from either a surface or a ground water source when water is plentiful, treats the water, and stores it in the ground water aquifer (Pyne 2005). ASR has many advantages over reservoir development, both economic and environmental, including:

- The cost of surface water reservoir construction is avoided.
- The loss of a free-flowing water body and associated aquatic community in the stream or river is avoided.
- No homes or other developed land uses are displaced.
- Flow is maintained in the stream or river at natural levels.
- The bordering vegetated wetlands associated with the stream are retained.
- Cultural resources, such as archaeological sites of indigenous people or historic sites (both of which are frequently associated with lands adjacent to streams and rivers), are not destroyed by a reservoir.

ASR does have the limitation of strict geologic requirements for the location of the storage and recovery area. There must be low permeable strata that confine the water injected into the aquifer so that it is not lost. Also, the geochemistry must be suitable so that calcium, magnesium, and other compounds do not precipitate and compromise the transmissivity and yield of the aquifer.

In the hypothetical example, the environmental analysis of the ASR project concludes that it is a viable and low impact approach to satisfy the long-term drinking water supply for the river basin. With a modest and previously demonstrated and achievable level of water conservation, ASR can supply the long-term water supply needs, at an affordable price, with little or no environmental damage and create environmental benefits of keeping the water in the basin. However, the only location with acceptable geochemistry and confining geologic strata is within the area proposed for development under the prior reservoir project and is thus unavailable. The result in this example would be the construction of an expensive reservoir, even though the environmentally preferable reservoir site was selected, the proposed action destroyed a free-flowing stream, inundated the critical bordering vegetated wetland habitat, and flooded cultural resources. The environmentally and economically superior ASR project was not implemented to satisfy the regional water supply program because the only geological site was lost to reservoir construction, and an environmentally friendly ASR approach was not available to meet any future demands.

The hypothetical example described above has many similarities to the King William Reservoir Project proposed by Newport News, Virginia (Army Corps of Engineers, Norfolk District draft EIS 1994, supplemental draft 1995, and final EIS 1997; also *Alliance to Save the Mattaponi, et al., Plaintiffs, v. United States Army Corps of Engineers, et al.*, United States District Court for the District of Columbia, Case 1:06-cv-01268-HHK Document 88

Filed 03/31/2009). The City of Newport News sought to meet the regional water supply need by constructing the King William Reservoir. Although there was substantial project-specific environmental analysis (including several iterations of a NEPA EIS and Clean Water Act alternative analyses) there was no comprehensive SEA of a water supply PPP on a regional basis. After more than 20 years and US $55 million in development costs, the project was canceled. The project had a history of opposition due to such environmental issues as significant impact on migratory fish; loss of aquatic habitat; flooding of Native American cultural sites; loss of substantial acreage of bordering vegetated wetlands, Native American subsistence fishing, and defensible water needs projections. The final blow to the project, which resulted in the City abandoning the proposal, came when the U.S. District Court for the District of Columbia ruled that NEPA and Clean Water Act procedural requirements were not met by the U.S. Army Corps of Engineers, which was the federal agency taking action by issuing the necessary permits for the proposed King William Reservoir. The issues cited by the court in the ruling included accurate and timely water demand projections, evaluation of alternatives, and magnitude and mitigation of wetland impacts. Based on the ruling, the U.S. Army Corps of Engineers estimated that it would take a minimum of two years and US $2 million to issue another supplemental EIS to address the deficiencies. Based on the Corps' projection of time and money plus the history of opposition, the City abandoned the reservoir project.

Newport News is currently in the process of reassessing the regional water supply situation (City of Newport News, Office of the City Manager, 2009). Specifically, they have initiated a new water supply needs assessment and committed to include interviews with all major regulators and stakeholders. They will even evaluate potential consequences of climate change and the City is committed to close coordination with the Regional and State Water Supply Plan, which was under development at the time the City decided to reevaluate their water supply program. Once they have completed the needs assessment, they are committed to evaluating alternatives for both the supply side and the demand side (i.e., water conservation) including a full range of environmentally desirable measures identified during the previous 20 plus years of planning and development for the King William Reservoir. This new approach and commitment by the City of Newport to develop a long-term water supply plan incorporates many of the aspects of an SEA.

In addition to a broader and more comprehensive approach, efficiency and consistency are other major advantages of SEA. As with tiering and programmatic environmental documentation under NEPA, using SEA provides an opportunity to address overarching and common issues once and not repeat the treatment of these issues for every environmental analysis for each individual project. In the example used above for a U.S. national policy of energy self-sufficiency, the alternative methods of surface coal mining common to all or most mining operations could be addressed in the SEA to inform decision makers as they determined the advisability and extent of including surface coal mining in the national energy policy. As the process proceeded, the analysis of individual surface mining environmental issues would not have to be totally redone at the program, planning, and project level. Or at least the treatment of alternatives could be limited to identifying the alternatives applicable to a particular location from the universe of alternatives identified and evaluated in the SEA. Impacts and mitigation of the mining methods could similarly be addressed first at the SEA level and then only as necessary to add specific considerations for other levels down to individual projects. This approach not only adds efficiency but also consistency. Perhaps, more importantly, if conducted properly, each level (policy, program, plan, and project) of analysis incorporates lessons learned from previous levels and allows a revisitation of higher levels in the process, if the evaluation of a plan or project identifies incorrect assumptions or assessments at the planning or program level. Similar to other multilevel environmental analysis techniques, determining which decisions can be made without constraining future options in the overall PPP and making these decisions when the "time is ripe" is at the heart of the SEA approach. Other decisions can be put off until more and site-specific information can be collected. This process can also be a major factor in comprehensive scoping with the SEA identifying the issues and in some cases even the environmental analysis methods to be used for the evaluation of the specific actions implementing a PPP.

Most of the potential disadvantages of SEA can be overcome with the experience gained from more and more successful completion of SEAs and benefiting from lessons learned. One of the most commonly raised objections is that the SEA process delays actions that are perceived as necessary in the short term. This concern can be legitimate because under a properly conducted SEA, specific actions should not be taken until the context of the PPP is fully evaluated. However, as evident in the King William Reservoir example, proposing a project before a PPP and the associated SEA are fully developed can result in decades of wasted effort and money, far exceeding that required for an SEA.

Another shortcoming of SEAs has been the lack of adequate guidance and protocol for conducting the assessments. This has been true in the past and not only did the lack of guidance constrain SEA development and acceptance, but it has also resulted in assessments that are of poor quality, of little use to decision makers, and do not result in the SEA benefits outlined earlier. In recognition of these problems, the United Nations Economic Commission for Europe (UNECE) developed an SEA Protocol that was signed by 36 nations. The SEA Protocol became effective in 2010 and the European Union also has an SEA program on a similar time schedule. As part of the program, UNECE has an ongoing effort to develop SEA guidance and protocol including a resource manual to support application of the protocol, which is available at www.unece.org/env/sea/(Bonvosin 2011). As the benefits of SEA are more widely understood and SEAs are efficiently conducted in a timely manner, there is a likelihood they will be accepted and adopted in many jurisdictions. The advantages and benefits of SEA are not in developing procedures, requirements, or analysis methods but rather in the ultimate environmental benefits derived from implementing the program, and it will take time to make a judgment on the value and effectiveness of a fully developed SEA program. Based on the histories of SEA and NEPA, one could reasonably hope that as SEA matures, both in terms of development and refinement of analysis methods and increased acceptance, SEA initiatives will eventually be fully implemented and accrue environmental benefits in a manner similar to what we have witnessed with NEPA over the last 40 years.

An early precursor of SEA, which has been very successful, is the Coastal Zone Management Act (CZMA), passed by the U.S. Congress in 1972 and is still very much in effect. The CZMA requires every state with a coastal zone (which includes the Great Lake states) to prepare a comprehensive coastal zone management plan. Although the Act does not mandate an environmental impact analysis as part of the plan development, it does call for "management of the nation's coastal resources, including the Great Lakes, and balances economic development with environmental conservation." The CZMA is administered by the National Oceanic and Atmospheric Administration's (NOAA) Office of Ocean and Coastal Resource Management (OCRM). NOAA has delegated much of the responsibility and funds to state CZM offices to develop the coastal plans and administer the Act. Under the CZMA, every federal action within the coastal zone, whether it is funding, construction, land transfers, permitting, or regulation, must be consistent with the coastal management plan applicable to the specific location of the action. NEPA compliance for the federal action in the coastal zone requires full documentation of the action's consistency with the plan on a point-by-point basis. The mission of OCRM and the state CZM offices has a very heavy environmental protection and enhancement component, thus the coastal plans and enforcement of CZMA tend to reflect many aspects of SEA.

Another area potentially amenable to successful implementation of SEA is a regional transportation PPP. At the broadest level the SEA would consider where people live, work, and play and also where manufactured goods and food are produced. The SEA would then identify plans to minimize transport among these uses. The next level of analysis would be to evaluate alternative means of transportation such as mass transit, ships, trains, high mileage cars, and so on. Then the final analysis would address where the roads/railroads would be and what type of transportation modes would be employed. Other possible topics of SEA that could be developed with existing environmental impact analysis tools are biodiversity, carbon emissions, and water resources development. The development of the U.S. Interstate Highway System during the 1950s and 1960s is a great example of a missed SEA opportunity. If an SEA requirement had been in place at the time, the highway alignments would have been very different (e.g., avoiding wetlands) and in all probability there would have been a strong integration of rail and highway.

6.5 Conclusions

Any of the multilevel environmental impact analysis approaches discussed in this chapter can be applied to achieve significant efficiency and keep the analysis focused on the appropriate issues. These approaches have the advantage of simplifying environmental and project implementation problem solving by breaking the issues and decisions into pieces of manageable size and complexity. They also provide an opportunity to examine the broader picture without becoming bogged down in the details and establish a general course of action that maximizes benefits and minimizes adverse environmental impacts integrated over space, time, and multiple environmental resources.

There are also a number of potential negative aspects of multilevel environmental analysis, with the most common being delaying project implementation. Particularly, if the process of preparing the follow-up analysis to a programmatic or first, in a tiered approach, environmental impact evaluation is protracted, the conditions, assumptions, analyses, etc., conducted for the first document could be outdated. This can force reexamination of the original analysis, and in a worst case result in a redo of the original document with a potentially different outcome. The tendency for the inherently detail-oriented nature of many environmental professionals, particularly scientists and engineers, can be another obstacle to successful application of multilevel environmental analysis. If the environmental team tries to project and resolve every detail of every potential environmental issue as part of the most general level of the multilevel environmental analysis, the advantages of the approach are missed.

There are some measures that can be taken to maximize the benefits and minimize the potential pitfalls of multilevel environmental analysis:

- Identify which issues are "ripe" for decisions and full environmental analyses.
- Clearly understand and elucidate which future options are eliminated and which are left open after each level of the analysis.
- Identify at least the conceptual scope of the following level of analysis at each step in the process.
- Clearly state the assumptions made at each level and confirm them in subsequent levels as needed to validate earlier decisions and move forward with the analysis.
- When a general course of action is identified in a first-level analysis, provide enough detail to demonstrate that the course of action can be implemented with the assumed or fewer environmental impacts. It is not necessary to present every possible action and evaluate all possible impacts, which is the scope of the next level of analysis.

- Involve appropriate stakeholders at every level and obtain their buy in or at least acceptance of the process at every level.
- Do not reinvent the wheel at every level and avoid unnecessary detail; it defeats the purpose of multilevel environmental analysis.

References

Bonvosin, N. 2011. The SEA protocol. In: *Handbook of Strategic Environmental Assessment*, ed. B. Sadler, R. Aschemann, J. Dusik, T.B. Fischer, M.R. Partidário, and R. Verheem. London: Earthscan.

City of Newport News, Office of the City Manager. September 16, 2009. Memorandum on 120 day report on the King William Reservoir Project.

Morgan, R.K. 1998. *Environmental Impact Assessment: A Methodological Perspective*. Norwell, MA: Kluwer Academic Publishers.

Pyne, R., David, G. 2005. *Aquifer Storage Recovery: A Guide to Groundwater Recharge Through Wells*. Gainesville, FL: ASR Press.

Sadler, B., R. Aschemann, J. Dusik, T.B. Fischer, M.R. Partidário, and R. Verheem, ed. 2011. *Handbook of Strategic Environmental Assessment*. London: Earthscan.

U.S. Army. 2004. *Environmental Assessment to Analyze Standard Practices for Construction Projects in the Cantonment Area*. Fort Campbell, KY: U.S. Army.

U.S. EPA. 1988a. *Draft Supplemental Environmental Impact Statement for Boston Harbor Wastewater Conveyance System*, Volume I and II. Boston, MA: U.S. EPA, Region I.

U.S. EPA. 1988b. *Final Supplemental Environmental Impact Statement for Boston Harbor Wastewater Conveyance System*, Volume I and II. Boston, MA: U.S. EPA, Region I.

7

Environmental Analysis Tools

7.1 Overview

The preceding chapters identified environmental impact analysis statutory requirements (Chapters 2 and 3), planning (Chapter 4), content (Chapter 5), and streamlining (Chapter 6). With the exception of the case study examples presented in these chapters, they provided little detailed discussion of the tools available to conduct the technical aspects of the analyses required to predict and mitigate environmental impacts. This chapter presents some of the tools that are available to fill in the framework and accomplish the analysis. For many environmental analysis practitioners with technical backgrounds and a focus on problem solving, development, and application of such impact analysis, tools are the most challenging and reward aspects of the entire process. Such practitioners gain satisfaction because technical expertise developed through education, training, and experience can be brought to bear on real environmental problems to develop a solution with a high likelihood of success.

To identify, much less describe, all the impact analysis tools available or even example tools for each discipline is beyond the scope of this book or any single publication. But there are two tools outside the purview of standard environmental impact analysis methods, originally developed to evaluate biological resources, which can be very useful for environmental analysis in cases where ecological resources are the primary areas of concern. These tools also represent concepts and approaches which, with imagination and creativity, can be adapted to address impact prediction and mitigation for other resource areas by specialists in their areas of expertise. The two tools, *ecological risk assessment* and *net environmental benefit analysis* (NEBA), are discussed in this chapter and include real and hypothetical examples of applying them to environmental impact analysis.

7.2 Ecological Risk Assessment

As the name implies, ecological risk assessment is an analysis tool used to determine, and to the extent practical, quantify the threat to ecological resources resulting from environmental stresses. The tool was developed in response to contamination of water, soil, air, and other media from hazardous waste and other uncontrolled releases of toxic chemicals and grew out of the regulatory mandates in the United States to address hazardous waste cleanup, specifically:

- Comprehensive Environmental Response, Compensation, and Liability Act (CERCLA) or the Superfund law (42 U.S.C. §9601 et seq. 1980)
- Superfund Amendments and Reauthorization Act of 1986, which amended and updated CERCLA
- National Contingency Plan (40 CFR 300)

As the investigation and cleanup of hazardous waste sites progressed in response to the new legislation and regulations, new tools were needed for the task. One of these tools was ecological risk assessment, which grew from the roots of environmental impact analysis and human health risk assessment. It has proven effective and efficient in defining the need, extent, and approaches for cleanup of contaminated sites, and the methods have been refined and enhanced since the introduction of the tool in the late 1980s. It has matured and there has been some limited adaptation of ecological risk assessment to environmental impact analysis in cases where the primary resources potentially impacted are ecological. The history leading up to the development of the risk analysis tool is presented in this chapter, followed by a summary of the tool, a case study where ecological risk assessment was successfully applied in the EIS for development of an Alaska gold mine, and a discussion of the relationship between ecological risk assessment and environmental analysis.

The presentation of ecological risk assessment in this book is not a comprehensive description of the tool. There are many references, as cited in this chapter, that present all the details necessary to conduct a risk assessment. The objectives of the risk assessment discussion presented here are to give environmental impact analysis practitioners an appreciation of what can be accomplished with the tool, provide an overview of the process, and encourage practitioners to adapt ecological risk assessment procedures to other environmental resources.

7.2.1 History and Development

Ecological risk assessment was not initially used or even available as a component of the first hazardous waste site or contaminant release evaluations. In the early years of hazardous waste awareness (most of the 1980s) ecological resources were of little concern because the environmentally aware public and regulators were narrowly focused on risk to humans and human health. This focus was fully justified by the real threat to human life and health, driven by the sudden awareness of Love Canal and similar

Environmental Analysis Tools

contamination disasters. Love Canal was an abandoned canal partially constructed to generate electricity by conveying water from upper Niagara Falls in upstate New York. When Hooker Chemical Company took over the abandoned canal section and adjacent land, they converted it to an uncontrolled disposal area for toxic chemicals. When this improvised and poorly managed landfill reached capacity, it was covered with soil and conveyed to the City of Niagara for a dollar. During the late 1950s, homes and a school were built directly over the old disposal area, and then in the late 1970s, the dam broke, so to speak. A record rainfall washed away the soil cover exposing corroding waste-disposal drums and killing the vegetation. This was followed by disturbingly high rates of birth defects, miscarriages, and other adverse human health conditions, both chronic and acute (Beck 1979). The U.S. EPA recognized that this was not an isolated case, similar occurrences would erupt throughout the country, and an immediate legislated attack on the problem was critical, thus the passage of Superfund.

With this background and legitimate risk to human life and health, the focus of Superfund was obviously not on the birds, bees, bunnies, and other ecological resources. Also, because human health was the focus of early hazardous waste site identification and cleanup, the activity tended to be in developed areas where humans were at risk and ecological resources were scarce. With human health as the initial focus of Superfund, the human health risk assessment was one of the key tools enhanced for use at hazardous waste sites to determine the threat and thus the need for remediation (Doyle and Young 1993). The human health risk assessment tool evolved to consist of four steps to characterize the baseline or existing risk to humans from contamination (U.S. EPA 1989a):

- Data collection and evaluation, to determine where the contamination exists and at what concentrations
- Exposure assessment, to identify the pathways of exposure (e.g., ingestion, inhalation, dermal exposure), the population exposed, and the degree of exposure
- Toxicity assessment, determination of safe levels (e.g., the mass of dioxin that could be ingested without causing cancer) for each contaminant present at elevated concentrations
- Risk characterization, a comparison of the level of exposure and safe levels to identify and quantify risk

As the most egregious sites that posed human health risk were investigated, the nature of past hazardous waste practices and typical contamination migration patterns became apparent. One common factor that was revealed by this increasing knowledge and understanding was that water flows downhill, and when the water originates or passes through a hazardous waste site, it transports the contaminants with the runoff. Thus the contaminated water and the eroded contaminated soil particles migrated to water bodies and wetlands. Although the human health risk, except for drinking water supplies, was less in streams, rivers, and wetlands than in developed areas, generally the most sensitive and vulnerable ecological resources are located in wetlands and water bodies. As a result, the biologists and ecologists within regulatory and natural resource agencies became increasingly concerned about long-term and significant impacts to sensitive and critical ecological attributes from the migration of hazardous waste from uncontrolled disposal sites to sensitive ecological resources.

In order to assess the magnitude of the potential threat to ecological resources from contamination, EPA surveyed 250 hazardous waste sites (U.S. EPA 1989b and 1989e). None of the sites surveyed were placed on the Superfund priority list (i.e., the sites with highest priority for investigation and remediation) due to ecological concerns because, prior to about 1989 there was no real consideration of ecological resources in the listing process. Hence there was no ecological information available on approximately 75% of the sites and only limited information on the remaining 25% of the sites surveyed. Based on the limited information that was available, the survey concluded that there were severe ecological threats at 10% of the sites and moderate threats at an additional 80% of the sites. Because water does flow downhill, wetlands and freshwater ecological resources were found to be the most common ecological resources threatened at the sites identified as posing an ecological risk.

With this information in hand, EPA launched a program, first within the Superfund divisions and then agency wide, to address ecological risk at Superfund sites. The agency's attention to the matter resulted in requirements and initial guidance for ecological risk assessment (U.S. EPA 1989f and U.S. EPA 1992). This new emphasis on ecological aspects of hazardous waste investigation and cleanup also prompted a series of peer-reviewed articles and two early text/reference books (Maughan 1993 and Suter 1993) on the newly formed ecological risk assessment tool and process. After a decade of experience, research, academic study, and numerous draft versions and associated public comments, the practice of ecological risk assessment matured, and EPA and other federal agencies issued a final guidance (U.S. EPA 1992 and 1997).

7.2.2 Process and Approach

As referenced above, the ecological risk assessment tool grew out of a combination of the well-established human health risk assessment and environmental impact analysis procedures. As a result, elements of each exist in the ecological risk assessment framework. However, ecological risk assessment differs from the other two procedures in significant ways:

- Human health risk assessment compared with ecological risk assessment:
    - Only one receptor is considered as opposed to multiple species, ecological processes, and habitats.

- A wealth of toxicological research data and funds are available for new research, but only limited data and funds available for ecological response to toxic chemicals.
- Assessments end with evaluation of health effects to individuals, whereas the broader community aspects (e.g., continued survival of population, ecological process, and habitat suitability) are under consideration for ecological assessment.
- Tests of site contaminants on receptors are unethical and not permitted, but laboratory exposure of site media to ecological receptors is a key component of the ecological risk assessment.
- Environmental impact analysis compared with ecological risk assessment:
  - Changes (e.g., impacts) must be predicted as opposed to the effects already in place
  - A multitude of receptors as opposed to a limited set
  - A stand-alone description and justification for the decision regarding the proposed action, whereas the ecological risk assessment is just one component in the ultimate cleanup and remediation decision
  - There is always the possible decision to do nothing or select the no-action alternative as opposed to the mandate for action if risk exists
  - Stakeholder involvement is a combination of technical and social input, but risk assessment stakeholders are primarily highly technical in the fields of ecology or toxicology.

Similar to environmental impact analysis, there is an overall framework for ecological risk assessment (Figure 7.1) and then specific tools and methods can be applied within the framework. This framework can also be presented in a simplified, conceptual form (Figure 7.2) similar to the environmental impact analysis credit card diagram depicting impact as a function of the affected environment and the alternatives (Figure 4.1). There is no firm set framework for ecological risk assessment and different regulatory programs (U.S. EPA 1997 and 1998), types of sites, and preferences of the ecological risk assessment practitioners result in a number of variations. But most variations consist of a number of sequential steps and include some expression of the steps discussed in this chapter and illustrated in Figure 7.1. All of the accepted ecological risk assessment approaches include an initial evaluation (screening-level risk assessment) that is analogous to the National Environmental Policy Act (NEPA) environmental assessment (see Section 3.1.4) because both determine the need for additional analysis to determine the level of effects. Each of the individual steps in the ecological risk assessment process is summarized under a separate heading, followed by a description of critical scientific investigation tools available for ecological risk assessment.

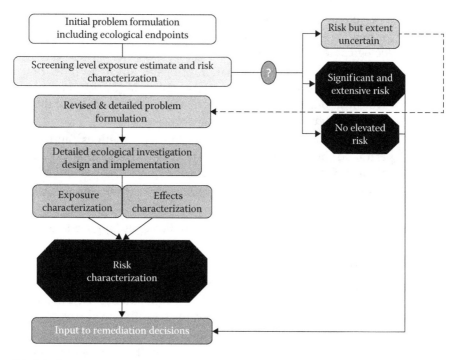

FIGURE 7.1
An ecological risk assessment process summary. (Adapted from U.S. EPA. 1997. Ecological risk assessment guidance for Superfund: Process for designing and conducting ecological risk assessments: EPA 540-R-97-006; U.S. EPA. 1998. *Guidelines for Ecological Risk Assessment:* EPA 630-R-95-002F; Maughan, J.T. 1993. *Ecological Assessment of Hazardous Waste Sites.* New York: Van Nostrand Reinhold.)

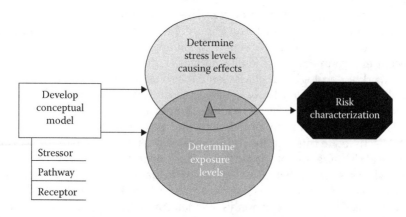

FIGURE 7.2
A simplified ecological risk assessment process.

7.2.3 Initial Problem Formulation

There are two basic elements to the initial problem formulation: a conceptual site model and identification of general ecological endpoints. The conceptual site model must be developed before the endpoints can be determined, and this model is the ecological risk assessment equivalent of the environmental analysis impact prediction conceptual model (see Section 5.3.1). The problem is defined by tracking sources of contamination to potential ecological receptors such as fish, birds, and mammal-breeding dens (Figure 7.3). This requires some background knowledge of the site, including ecological site description, the history of contamination, and contaminant migration pathways (e.g., groundwater transport, soil erosion, aerial dispersion, and food chain transfer).

The ecological site description is similar to the initial description of the affected environment typically prepared as part of the environmental impact analysis scoping effort (see Section 4.3). It is generally accomplished by an experienced ecologist conducting a literature review and more importantly a site reconnaissance to identify the primary ecosystems and communities present within and adjacent to the site. It can be considered the applied equivalent of the observational first step in the scientific method which is to understand the setting before

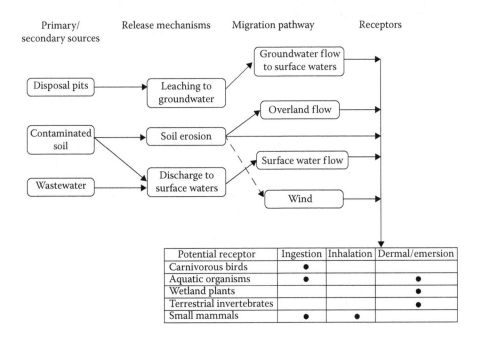

FIGURE 7.3
An example conceptual site model. (Adapted from U.S. EPA. 2011. Environmental cleanup best management practices: Effective use of the project life cycle conceptual site model. EPA 542-F-11-011 July 2011.)

generalizations and hypotheses can be made or procedures designed to test the theories (Keaton 1972). The product of the site description should include:

- A list of potential ecological receptors expected to be associated with the site
- Identification of areas within or adjacent to the site, where the receptors might occur and thus be exposed to contamination
- Their general life history requirements relevant to potential contaminant exposure (e.g., where they breed and their general feeding behavior)
- Potential exposure pathways of receptors given the general characteristics of the site, contamination, and receptors

The site contamination information incorporated into the initial problem formulation is dependent on how far site investigation has progressed prior to initiation of the ecological risk assessment. In some cases, there has been extensive sampling and analysis of site media (e.g., surface water, soil, ground water), and not only is there information available regarding the contaminants present, but the concentrations of the contaminants are frequently known. In other cases, the contaminant information is limited to reports of which chemicals were used or disposed of at the site. At a minimum, the initial conceptual site model must identify the types of contaminants of potential concern (e.g., metals, volatile organic compounds, pesticides) so that ecological receptors, pathways (e.g., ingestion, inhalation, or dermal contact), and types of risk (e.g., acute toxicity, food web magnification) can be included in the model.

The final component of the conceptual site model is a description of the migration pathways present at the site. Identification of potential pathways allows the tracking of the contamination as it moves from a source to an ecological receptor. The most common migration pathways at hazardous waste sites that can affect ecological receptors are groundwater, soil erosion, overland flow, and river/stream flow. Volatilization and aerial dispersion can, in some cases, represent a concern for ecological risk, but it is rare and generally limited to confined conditions, such as beaver lodges or burrows of terrestrial animals. Any of these mechanisms can transport contaminants from their source to critical ecological habitat and result in significant exposure and risk. A simplified example conceptual site model of a common hazardous waste site (buried waste) is depicted in Figure 7.2 and a more complex example is included in the AJ Gold Mine case study (Section 7.2.9).

The second component of the problem formulation step, establishing ecological endpoints, was an innovative breakthrough in the development of ecological risk assessment. In the early years of Superfund and other hazardous waste programs, there was a reluctance to include and address ecological issues in the remediation evaluation and planning due largely to the complexity of ecological systems. Particularly to the geologists and engineers who typically directed the early hazardous waste investigation, the number of ecological resources, interrelationship of receptors, wide diversity of receptor sensitivity, and complexity of ecological exposure pathways were overwhelming and threatened timely site remediation. The concept of ecological endpoints greatly simplified the process of including these receptors and pathways in remediation decisions.

There are two types of ecological endpoints: assessment endpoints and measurement endpoints (Maughan 1993 and Suter 1990). The assessment endpoints represent ecological attributes warranting protection. They are the equivalent of a resource identified during environmental analysis scoping that must be evaluated as part of the impact analysis (see Section 4.3.1). However they differ from the simple definition of an issue because they include both a subject (e.g., salmon population or a pair of bald eagles) and an action (e.g., successful breeding or growth to maturity).

Of course "warranting protection" is a subjective term at best and some might consider it ambiguous. There are some attributes that all would agree warrant protection: a pod of right whales; the salmon run on the Columbia River; or a nesting pair of bald eagles. There are others that most would consider to warrant not the least degree of protection: a stand of the invasive vine kudzu that has engulfed a farmer's barn; a growing mice population in a corn storage silo; or the lamprey infestation in the Great Lakes. A key to the success and usefulness of the ecological risk assessment is determining where between these two extremes an attribute lies and thus should be included as assessment endpoints in the risk assessment.

The assessment endpoint concept works for ecological risk assessment because the primary and influential stakeholders are trained ecologists who have expert experience in determining importance of ecological attributes. The hazardous waste site cleanups are mandated by state and federal regulations and the agencies overseeing the investigations bring their biologists and ecologists to the table during the initial planning of an ecological risk assessment. In fact, many ecological attributes warranting protection have regulations or legislation that mandates protection (and thus simplify determining attributes warranting protection), for example:

- Endangered Species Act and comparable state programs
- Migratory Bird Treaty
- Federal Clean Water Act and comparable state/local programs protecting wetlands
- Marine Mammal Protection Act
- Federal Clean Water Act mandate for states to promulgate water-quality standards to protect aquatic life.

Other ecological resources warrant protection for popular reasons. Many resources have recreational or aesthetic value, such as deer, bear, and many fish species; thus, if they are potentially associated with a site, their continued survival is a preordained assessment endpoint. Others are highly visible and represent a healthy ecosystem even to the nonprofessional, such as a beaver lodge, a pair of river otters, or a majestic stand of old growth forest.

In contrast, there is frequently reluctance on the part of the ecological risk assessor and the overseeing agency ecologist as well, to assign assessment endpoint status to invertebrates, plants, and other ecological units low on the food chain and visibility spectrum. The reluctance is an effort to avoid the criticism and bad publicity: "I'm being forced to spend millions of dollars and sacrifice thousands of jobs to protect some insect?" However, protection of such resources can be incorporated into the ecological risk assessment by other means, and the other ecological endpoint, the measurement endpoint, affords an opportunity to protect resources that do not have the status and publicity of charismatic mega fauna. But these endpoints have to be closely linked to the assessment endpoint so that the need for remediation can be expressed based on a popular ecological attribute warranting protection rather than an unrecognizable ecological characteristic such as the number of seeds produced by an obscure plant species. Guidelines and criteria have been established for the selection of endpoints based on site conditions and contaminants present (U.S. EPA 1989f, Maughan 1993, Suter et al. 2000).

As pointed out above, assessment endpoints typically represent very visible attributes with a high public-interest value. However, directly measuring the health and sustainability of one of these popular attributes can be a daunting task. For example, to measure the density and health of a salmon population in a specific river would require following the fish through several years as they matured and returned to breed. It would also necessitate sampling and survey during multiple seasons and life stages as they breed, mature past the juvenile stage, out-migrate, and return to spawn. This would take years, untold resources, and delay any cleanup action beyond any reasonable expectation. As a substitute to a lengthy and complex evaluation of broadly expressed assessment endpoints, measurement endpoints were developed to indicate, rather than measure directly the status of the assessment endpoints.

Measurement endpoints typically represent single requirements or conditions necessary for the survival and health of the assessment endpoints. These simpler attributes can be easily measured and usually quantified within the scope, budget, and time frame of a hazardous waste site evaluation and cleanup. However in most cases, it is necessary to identify multiple measurement endpoints to gauge the status of the assessment endpoint. In the above salmon population example, at a minimum it is essential that endpoints reflect the chemical suitability of the water (i.e., water-quality standards and criteria) and the condition of the substrate where the fish require gravel nests.

Endpoints would also be required to evaluate the food available to the maturing salmon to ensure their health and continued survival. This is where the less visible and popular ecological attributes enter into the ecological risk assessment, as the health of the aquatic invertebrates and forage fish segments of the aquatic community are designated as measurement endpoints. In the initial problem formulation phase, the measurement endpoints are often general and non-site-specific, but are refined at a later stage of the assessment if the degree of risk is deemed uncertain after the screening-level ecological risk assessment.

Environmental Analysis Tools

The combination of assessment and measurement endpoints serve multiple purposes:

- The measurement endpoints are the primary input to the design of any ecological data collection effort. Any data collected should be directed at quantifying or otherwise evaluating the status of the measurement endpoint at the site; otherwise the use of the data is uncertain at best and frequently useless.
- The status of the endpoint at the site indicates to what degree the contamination has caused an ecological risk.
- The endpoints are a mechanism to evaluate the success of the cleanup following remediation.
- The endpoints focus the scope of the risk assessment making it limited and manageable.

An example of endpoints and associated attributes for a typical hazardous waste site with multiple ecological receptors is illustrated in Table 7.1.

7.2.4 Screening Level Assessment

The objective of screening is to determine, based on limited data and information, whether there is definitely a risk; definitely not a risk; or the risk to ecological receptors is unknown (Figure 7.1). Similar to the screening of alternatives in an environmental impact analysis (see Section 4.5.3), the screening-level assessment identifies the concerns (as indicated by ecological endpoints) that should be the focus of additional investigation and consideration in remediation and other decisions. There are two topics covered by the screening-level assessment: the potential exposure of ecological receptors (included as endpoints) and degree of contamination.

The assessment of potential presence and exposure of receptors is the simpler of the two types of screening. It can frequently be accomplished as part of the ecological site description (see Section 7.2.3) by an experienced ecologist. Based on a simple site reconnaissance, the ecologist can generally determine from the habitat present, which species and communities would be expected on the site, and which potentially contaminated environmental media they may contact to produce exposure. If the site is a fully developed and largely paved oil refinery, the ecological resources would be extremely limited and there would be few, if any, appropriate assessment endpoints. In such a case, the ecological risk assessment might be limited to effects from off-site migration. At the other extreme, if the contamination has reached or has the potential to reach an area comprised of a complex native vegetative cover encompassing both a standing and flowing water habitat, the need for a detailed ecological risk assessment cannot be ruled out based on the potential presence of receptors and their exposure. A common mistake made by nonecologists in screening for receptors is to prematurely conclude that there are no sensitive resources actually present, so a

TABLE 7.1
Examples of Ecological Endpoints and Related Attributes

Assessment Endpoint	Exposure Area	Representative Receptor Species	Exposure Routes	Measurement Endpoints	
				Measures of Exposure	**Measures of Effects**
1. Survival, growth, and reproduction of carnivorous mammal populations	Upland; bottomland hardwood forest	Long-tailed weasel	Direct exposure, ingestion	Measured concentrations of contaminants in surface soil Estimated concentrations of contaminants in vertebrates	Comparison of exposure doses to published values for survival, growth, and reproduction
2. Survival, growth, and reproduction of herbivorous mammal populations	Upland; bottomland hardwood forest	Eastern cottontail	Direct exposure, ingestion	Measured concentrations of contaminants in surface soil Estimated concentrations of contaminants in plants	Comparison of exposure doses to published values for survival, growth, and reproduction

3. Survival, growth, and reproduction of carnivorous bird populations	Upland; bottomland hardwood forest	Eastern screech owl	Direct exposure, ingestion	Measured concentrations of contaminants in surface soil Estimated concentrations of contaminants in vertebrates	Comparison of exposure doses to published values for survival, growth, and reproduction
4. Survival, growth, and reproduction of plant community	Upland; bottomland hardwood forest	Various	Direct exposure	Measured concentrations of contaminants in surface soil	Comparison of concentrations of contaminants in soils and soil Benchmarks.
5. Survival, growth, and reproduction of amphibian and reptile populations	Wetlands; Upland; bottomland hardwood forest	Eastern cottonmouth	Direct exposure, ingestion	Measured concentrations of contaminants in surface soil	Comparison of exposure doses to published values for survival, growth, and reproduction
6. Survival, growth, and reproduction of freshwater fish community	Streams; Cape Fear River	Various including shortnose sturgeon	Direct exposure, ingestion	Measured concentrations of contaminants in surface water and sediment	Comparison of concentrations of contaminants in overlying water to the water-quality criteria

detailed ecological risk assessment is not warranted. This overlooks the possibility, and in extreme cases, the likelihood that there are no resources present because site contamination prevents their presence and survival. Comparison of the site to a reference area with similar physical attributes, except for contamination can correct this fallacy, and in some cases if the sites are structurally similar but sensitive resources are plentiful at the reference site and absent at the other site, it can reveal elevated risk.

The second type of screening addresses the level of contamination and is more quantitative. The first step in the screening-level exposure estimate (Figure 7.1) is to assemble the analytical data for each media (e.g., water, soil, sediment). There should be contaminant concentration data to represent the exposure pathways for each ecological measurement endpoint, and if there is none, additional data collection is necessary to accomplish the screening. The next step is to identify a generic and conservative (i.e., tendancy to overestimate rather than underestimate the risk) ecological benchmark for each environmental media that represents the maximum concentrations protective of the measurement endpoint. For example, if the measurement endpoint is sustainable brook trout population, the benchmark for exposure to surface water would be the ambient water-quality criteria for chronic exposure (U.S. EPA 2013). The benchmarks for each endpoint and each media are then compared with the measured site concentrations to determine which, if any, contaminants pose no risk; pose a severe risk; or warrant more investigation.

These comparisons can be made and presented as calculations of benchmark or hazard quotients. The quotients represent the measured site concentration divided by the benchmark; thus if the measured concentration is equal to the benchmark, the quotient will be 1.0. Values substantially higher than 1.0 indicate significantly elevated risk, and values well below this mark indicate that the particular media, contaminant, and pathway do not pose a risk and can be screened out from further consideration in the detailed ecological risk assessment. Benchmark quotients between these extremes typically warrant additional investigation (see Section 7.2.6). The cutoff quotient value between no elevated risks, need for investigation, and substantial risks are site, receptor, contaminant, and quality of data specific and dependent. If there is a large data set covering the entire site at a high density, a comparison of the benchmark with the mean site concentration revealing a quotient below 1.0 frequently indicates the contaminant can be screened out. In contrast, if the data set is sparse and there is uncertainty in the appropriateness of the benchmark to the specific site, a quotient comparing the benchmark with the maximum site concentration just slightly above 1.0 could indicate the need for additional investigation.

Table 7.2 presents a hypothetical screening of contaminants for a specific media (water) and endpoint (survival, growth, and reproduction of a freshwater fish community). The appropriate benchmark is the chronic water-quality criteria because the endpoint addresses all life history aspects. If survival was the only endpoint, the acute water-quality criteria might be

TABLE 7.2
Example Contaminant Screening for Ecological Risk

Pollutant	Chronic Criteria[a] µg/l	Mean Site Concentration µg/l	Maximum Site Concentration µg/l	Benchmark Quotient for Mean Concentration[b]	Benchmark Quotient for Max Concentration[b]	No Elevated Risk	Significant Elevated Risk	More Investigation Necessary
Cadmium	0.3	0.3	0.8	1.2	3.0			X
Chromium (VI)	11.0	3	6	0.3	0.5	X		
Heptachlor	0.0038	<0.003	<0.003	<1.0	<1.0	X		
Lead	2.5	6	21.6	2.4	8.6			X
Nickel	52.0	240	875	4.6	16.8		X	
Zinc	120.0	238	789	2.0	6.6			X
4, 4 DDT	0.001	0.0001	0.0003	0.1	0.3	X		
Toxaphene	0.0002	<0.0002	<0.0002	<1.0	<1.0	X		
Tributyltin	0.072	0.001	0.004	0.0	0.1	X		
Iron	1000	5200	12890	5.2	12.9		X	
Dieldrin	0.056	<0.002	<0.003	<0.1	<0.1	X		

[a] U.S. EPA 2013.
[b] See text for description.

appropriate. The table also presents quotients for both the mean and maximum site concentrations. This provides additional information so that the risk assessor can make an informed judgment regarding need for more investigation based on site knowledge and extent of available data. Following the ecological resource and contaminant screening, the areas of ecological concerns can be narrowed and the risk assessment can proceed to the next step of detailed investigation focused on the critical endpoints potentially at risk.

7.2.5 Final Problem Formulation and Detailed Ecological Investigation

With the refinement of the endpoints, pathways, contaminants, habitats, and receptors of concern accomplished through the screening-level ecological risk assessment, the detailed assessment can focus on the areas of greatest or unknown level of risk. Often this detailed assessment is termed a *baseline ecological risk assessment* or BERA because the purpose is to define and quantify the level of risk that currently exists before any remediation or additional natural attenuation has occurred. The first activity is to fine tune the problem formulation and associated conceptual site model. The screening assessment has almost always eliminated at least some pathways and receptors of concern and the revised model should reflect the findings. In many cases, the final model can include some degree of quantification or at least relative importance to identify the exposure and migration pathways of greatest concern. Similarly, the initial endpoints identified can be revisited to be more site specific, such as identifying individual species (e.g., osprey) rather than general categories of receptors (e.g., piscivorous birds).

The final problem formulation and conceptual site model is the key tool for identifying what additional studies are necessary to fully characterize the risk and address the final measurement endpoints. A typical goal of the additional investigations is to quantify the extent of ecological damage that has or continues to occur. Another is to relate media concentrations to the extent of damage so an acceptable cleanup level can be established. Each study should address a risk hypothesis related to a measurement endpoint such as:

- Ospreys feeding at the site ingest lead in quantities that impair reproduction.
- DDT levels in soil inhibit microbial remineralization of essential plant nutrients.
- Polycyclic aromatic hydrocarbons in sediments are toxic to benthic invertebrates to the extent there is no sufficient food in the system to support higher trophic levels.
- Eroded soil particles from contaminated areas following a rain event accumulate in stream sediments to toxic levels.
- Volatile organic compounds in soils create air concentration in fox dens acutely toxic to fox cubs.

- Concentrations of all metals in on-site surface waters create synergism resulting in impaired growth and survival of aquatic organisms.

Comparable to technical scoping in environmental impact analysis (see Section 4.3.2), it is critical that the design of the BERA investigation receives full key stakeholder buy-in before the studies are initiated. In the case of ecological risk assessment, it is even more important because the "key stakeholders" are the scientists representing the regulatory agencies that make the final decision on risk and site remediation. If sample collection methods, analytical procedures, quality assurance procedures, or concentration detection limits don't meet the expectations of the regulatory agency scientists, the entire risk assessment will be called into question, and the investigations may need to be repeated incurring significant costs and delays.

Conducting the detailed investigations can be the most revealing and rewarding phase of the ecological risk assessment to the ecologists and other scientists working on the site characterization. These studies represent real science and frequently employ state-of-the-art methods. As pointed out above, each investigation must address a specific question and relate to a measurement endpoint. Examples of detailed investigations conducted to support ecological risk assessments at aquatic sites are presented in Table 7.3. Comparable studies are frequently conducted at sites with risk to terrestrial receptors with the focus on soils and plants as the contaminated media and exposure pathways of concern.

Depending on the degree of contamination, extent of risk, level of uncertainty, and potential remediation costs, very sophisticated, long-term and expensive investigations are sometimes conducted to identify the need/extent for remediation. The investigation can involve testing to track exposure pathways and identifying remediation approaches to disrupt the exposure pathways rather than risk large-scale media cleanup. Also if the studies typically conducted as part of a BERA (Table 7.3) indicate risk to individuals, but the assessment endpoints are at a higher level of biological organization (e.g. populations), ecological modeling may be necessary (Pastorok et al. 2002). The modeling can be conducted on the population, ecosystem, or landscape scales to determine if the loss of some percentage of the individuals would be translated to a broader-scale risk.

7.2.6 Exposure Characterization

There are two integrated components of exposure: 1) if and how a contact is made between a contaminant (or other stressor for a nonhazardous waste risk assessment); and 2) the contaminant concentration (or level of stress) at the point of exposure. Thus the objective of the exposure characterization is to determine or estimate both of these factors for each ecological receptor representing a target in an ecological endpoint. The evaluation of contact is the first of these two components to be conducted and characterization of contact is

TABLE 7.3
Common Investigations Conducted as Part of Detailed Ecological Risk Assessments at Sites with Contamination of Aquatic Systems

Assessment Endpoint	Measurement Endpoint	Risk Hypothesis	Summary of Investigation	Application of Results
Survival, growth, and reproduction of aquatic community	Benthic community density and composition comparable to community in unaffected reference area	Site contaminants have altered the community characteristics of the benthic assemblage at the site	Collect replicate sediment samples from various areas of the site and a similar noncontaminated reference area. Determine species composition, density, diversity, and trophic structure of animals from all samples. Concurrently measure concentration of contaminants in each sample.	Identify areas with degraded benthic assemblages and relation of benthic health to contaminant concentrations. Identify areas at elevated risk and contaminant concentrations associated with elevated risk
Survival, growth, and reproduction of aquatic community	Toxicity of sediment/water to aquatic animals	On-site sediment and water have concentrations of contaminants toxic to aquatic organisms	Collect sediment and water from various locations within the site representing a range of contaminant concentrations. Conduct toxicity tests on media to determine inhibition of survival, reproduction, and/or growth of species potentially occurring at the site. Also measure concentration of contaminants in each sample (can be same samples in benthic investigations described above).	Identification of areas of site with elevated risk and contaminant concentrations associated with toxic effects.

Survival, growth, and reproduction of aquatic community	Ingested dose and response of contaminants by carnivorous fish	Carnivorous fish feeding at the site ingest contaminants at levels that significantly inhibit critical biological processes and organism health	Conduct fish sampling to identify primary carnivorous fish species present or potentially present at the site. Determine primary food sources through examination of gut content and literature review. Collect representative specimens of primary food sources and measure concentration of contaminants in tissue. Calculate ingested dose of contaminants to carnivorous fish and compare to publish information of safe doses.	Quantify risk, and thus need for remediation, to carnivorous fish at the site. Identify contaminants posing the risk.
Survival, growth, and reproduction of semiaquatic and terrestrial populations	Ingested dose and response of contaminants by carnivores feeding on fish or other aquatic animals	Mammals, birds, and reptiles feeding at the site ingest contaminants at levels that significantly inhibit critical biological processes and organism health	Conduct literature review and site observations/surveys to determine what animals potentially at the site rely on the aquatic food chain and their primary sources of food. Collect representative specimens of primary food sources and measure concentration of contaminants in tissue. Calculate ingested dose of contaminants to animals relying on aquatic food sources and compare to publish information of safe doses.	Quantify risk, and thus need for remediation, to mammals, birds, and reptiles feeding at the site. Identify contaminants posing the risk.

accomplished by identifying which of the primary exposure pathways apply: dermal contact, inhalation, total immersion (e.g., fish in contaminated water), or ingestion. The determination of applicability is based on the chemistry of the contaminant (e.g., inhalation for a volatile compound or ingestion for a compound such as methyl mercury that magnifies up the food web), the media involved (e.g., soil, water, air), and the specific receptor (e.g., food items for the ingestion pathway). The characterization of contact identifies not only which pathways but also which media are involved in the exposure.

Once the media producing exposure are identified, the final step in the exposure characterization is to determine the intensity of exposure, which in the case of risk from contamination is the concentration or dose of the contaminant in the environmental media of concern. The determination of concentration is required for inhalation, dermal exposure, and total immersion pathways and begins with a sampling and analysis program throughout the site for each environmental media of concern. Frequently the sampling program is adequate to identify the exposure concentrations for the receptors but sometimes it is necessary to model the media to account for changes in concentration with time or distance (Boucher 1993). Also if the size of the site is large and there is a low density of sampling, a statistical estimate of exposure concentration is necessary to represent uncertainty and distribution of contaminants throughout the site.

The determination of dose for ingestion pathways is a little more complex. For this exposure pathway, a food web model must be constructed to represent the diet of each ecological receptor that is vulnerable to risk from ingestion. First the ingested material, usually a food source, must be identified, quantified, and the percent composition represented by each item is determined or estimated. Then the concentration of contaminant in each item is determined, and the total mass per day of contaminant is calculated as the sum of the concentration times the mass of each food item ingested. This calculation yields exposure via the ingestion pathway, expressed as the total daily dose.

7.2.7 Effects Characterization

As described above, the exposure characterization step in the risk assessment defines which stressors (e.g., contaminants at a hazardous waste site) contact the receptor and at what level of intensity (e.g. contaminant concentration at a hazardous waste site). The effects characterization identifies the effects to the receptor following contact. The ecological benchmarks described above as part of the screening-level assessment (Section 7.2.4 and Table 7.2) are a simple expression of an effects characterization. The screening-level benchmarks are concentrations below which no elevated risk to any life stage of the most sensitive receptors is expected. In some cases, the same benchmarks are applicable in the BERA and represent the final effects characterization for specific receptors, exposure pathways, and endpoints.

However, in many cases the screening-level benchmarks represent concentrations lower than necessary to protect receptors or the designated measurement endpoint. In such cases, site-specific benchmarks can be identified or developed that represent the endpoint species, lifecycle stages of concern, and site specific conditions rather than the broad-based and conservative benchmarks used in the screening assessment. For example, if there is no salmonid (e.g., trout) habitat associated with the site because the natural water temperatures are too high, water-quality benchmarks for warm water fisheries (which are usually higher concentrations) may be more appropriate. Similarly, since juvenile forms of most animals are typically more sensitive to contamination than adults, a higher-effect concentration might be appropriate for sites where there is no potential for breeding or rearing of receptor species.

The effects characterization can also take into account site-specific conditions. For example, the hardness and other natural characteristics of surface waters can affect the concentrations of contaminants that cause toxic effects. In recognition of the phenomena, procedures have been developed to adjust water-quality criteria based on site-specific conditions (U.S. EPA 1994). Also evaluation of the ingestion pathway for a large carnivorous receptor should take into account that they may only derive a portion of their food from the site, because they forage over a large area, thus the effect concentration of on-site prey items would have to be adjusted by the size of the site relative to their forage area (Boucher 1993).

The effects characterization often identifies more than one concentration or dose to represent different degrees of exposure or levels of risk. For example, water-quality concentrations associated with both acute and chronic effects (U.S. EPA 2013) are frequently included in the effects characterization. Thus, if a receptor will only be exposed for short durations, such as swimming by a discharge point, the acute criteria might be the appropriate effects concentration. But if the receptors are subject to long-term exposure, such as within a reach of stream, the chronic criteria will be more appropriate.

The product of the effects characterization is a listing and documentation of concentrations, doses, or other levels of stress that result in various levels of risk. There should be an effects level for comparison to each measurement endpoint and for every media identified in an exposure pathway. Some of these levels will be standard and may even be the same benchmarks used in the screening-level assessment. Others will be adjusted to site conditions and still others will be developed specifically for the site through toxicity testing and other ecological investigations (Table 7.3).

7.2.8 Risk Characterization and Input to Remediation Decisions

Risk characterization is the final technical step in the ecological risk assessment, and because it provides direct input to remediation decisions, it can be considered the most important. However, in concept it is the most simple. At the base level, risk characterization is a simple comparison of the exposure concentrations or doses to the effects levels for each measurement endpoint. If the concentration or other level of stressor exceeds the risk levels identified in the effects characterization for a given measurement endpoint, the conclusion is that endpoint is not achieved. This then results in an input to the remediation decision that some degree of cleanup for that medium and that contaminant is warranted to eliminate an elevated risk. If there are no estimated exposure concentrations exceeding effects levels, then the risk assessment input is that remediation is not necessary to achieve the ecological endpoints.

Of course in practice, nothing in risk characterization or ecology in general is quite that simple. There are almost always complications and additional considerations to ecological risk characterization, such as:

- The exposure concentrations just barely exceed the effects levels and there is a high degree of uncertainty in both.
- No effects level is exceeded by a single compound but the estimated exposure for dozens of compounds are very close to the levels associated with elevated risks.
- After applying a margin of safety, effects levels are marginally exceeded but remediation of the contaminated media would destroy the habitat to the point that the site could not support the receptor potentially at risk (see Section 7.3.1 for a technique devised to deal with such findings).
- Only a small portion of the site exceeds the effects levels but the site mean is well below levels of concern.
- One evidentiary set indicates elevated risk, but another does not.

After the risk has been preliminarily characterized, it is frequently necessary to address the complexities, such as those listed above to provide defensible input to remediation decisions. When the complexity involves conflicting lines of evidence regarding presence or degree of risk, a "weight-of-evidence" approach has been developed to address the conflict (Menzie et al. 1996). This approach uses quantitative and qualitative metrics to rate the various lines of evidence and often requires additional investigations and analyses to address some of the ambiguous finds listed above. Following evaluation of each line of evidence, the weight-of-evidence approach identifies ways to compare the findings and present a unified and defensible representation of ecological risk for remediation decisions.

Another interesting follow-up to initial risk characterization has been developed for cases where the only risk is via ingestion of prey items by upper-level carnivores. In many such cases it is not unusual that prey items residing over most of the site are found to have relatively low body burden of the contaminants of concern, but those residing in the "hot spots" of contamination account for the vast majority of risk to the upper-level carnivores.

Ecological principles, related to NEBA (see Section 7.3), have been used to address this "hot spot and ingestion risk" scenario.

The approach is to reduce the proportion of contaminated prey items in the carnivore's diet by two processes. One is to increase the habitat value of the relatively uncontaminated portions of the site by actions in the clean areas such as:

- Planting and maintaining species that are prime food or cover for the prey items
- Removing invasive plant species that inhibit prey species presence and density
- Creating habitat features, such as nesting areas attractive to prey species
- Providing a limiting habitat requirement such as water or salt.

Increasing the attractiveness of the relatively uncontaminated portions of the site to the prey species will result in higher density food sources in these areas. The carnivores forage and feed preferentially in these areas of relatively uncontaminated prey and thus lower the ingestion dose of contaminants and proportionally reduce the risk.

The area of highest contamination can be rendered less attractive to foraging carnivores and produce the same result of reducing contaminant ingestion and associated risk. This can be done through a combination of very limited and nondestructive removal of contaminated media from the "hot spot" and/or reducing the habitat value of the remaining areas of elevated contamination. In line with the concept of NEBA, the reduced habitat value should be replaced with an improved value in the uncontaminated area such that the food quantity available to the carnivore is at least equal to the preremediation quantity and of better quality. Of course, the two approaches—cleaning and limiting prey quantity in contaminated areas and improving prey density in clean areas—can be used in combination to result in a contaminant dose posing no risk to the carnivore.

7.2.9 AJ Mine Ecological Risk Assessment as Part of Environmental Analysis Case Study

As discussed above, ecological risk assessment was developed as an analysis tool to address ecological impacts associated with hazardous waste sites and contaminated environmental media. This use of the tool continues to be the primary application of ecological risk assessment, but it has been used in a broader environmental impact analysis, particularly if contamination is one of the potential environmental stressors of concern. The tool has been used by the U.S. Army Corps of Engineers for environmental analysis as part of Environmental Impact Statements (EISs) in Mobile, Alabama (U.S. Army Corps of Engineers 2002) and channel dredging programs on the Atlantic Coast.

Ecological risk assessment was also applied in a very controversial EIS for the reopening of the AJ gold mine in Juneau, Alaska. The mine would generate approximately 80 million tons of tailings (i.e., the remaining rock after the gold has been extracted) over its life and the only feasible alternative was marine disposal of the tailings (see Section 10.3 for a discussion of the AJ Mine EIS background). A primary environmental concern related to this method of disposal was the effects of the tailings on the marine habitat within and adjacent to the area of disposal; thus ecological risk assessment was an appropriate tool to predict impacts and compare alternatives (Maughan et al. 1997).

The AJ Mine risk assessment did not exactly follow the process outlined earlier (Figure 7.1). Many of the issues had been defined and the necessary ecological investigations conducted as a result of the environmental analysis process (particularly technical scoping) before the risk assessment was designated as the chosen tool. Thus much of the initial problem formulation and some of the screening-level assessment had been completed. However, the ecological endpoints had not been clearly identified and accepted by the stakeholders. Thus evaluation of endpoints was a critical first step once risk assessment was designated as the primary tool to evaluate the impacts of mine tailings disposal.

7.2.9.1 Ecological Endpoints

The AJ Mine risk assessment development of endpoints was facilitated by a wealth of information collected for the EIS and the highly experienced and dedicated technical advisory committee (TAC). Working with TAC members, the ecological assessment/environmental impact analysis team established a set of comprehensive assessment endpoints based on the following criteria:

- Ecological relevance. The chosen endpoint should be relevant to the ecosystem being examined.
- Societal value. The endpoint should be of value to society.
- Environmental policy goals. Endpoints important to the public and/or regulators should be considered.
- Susceptibility to the chemical or physical stressor.

Based on these criteria, five assessment endpoints were identified (Table 7.4). At least one measurement endpoint was then developed for each assessment endpoint based on the following criteria:

- Readily measured or evaluated. There should be protocols or methods currently available or easily adapted to evaluate the endpoint.
- Consistent with routes of exposure to assessment endpoint. The route of exposure should be the same for interpretive purposes.
- Appropriate to the scale of the site. Impacts on an organism with a large range are harder to tie to the site.

TABLE 7.4

AJ Mine Assessment Endpoints and Their Relation to Selection Criteria

Assessment Endpoint	Ecological Relevance	Societal Value or Policy Goal	Susceptibility to Stressors
Provide food source that does not interfere with sea bird reproduction	Piscivorous birds feed on bottom fish that may be in contact with tailings	Aesthetic value and declining populations	Potential for exposure to bioaccumulation of metals through diet and reliant on food from tailings disposal area
Provide food source at existing levels that does not interfere with fish-eating marine mammal reproduction	Top predator feed on fish that may be in contact with tailings	Aesthetic value; protected species; rights of indigenous peoples	Potential for exposure to bioaccumulative metals through diet
Provide unaltered food and habitat requirements for naturally occurring pelagic fish	Top predator	Commercial and recreational fisheries	Potential for exposure to bioaccumulative metals through diet
Support demersal fish assemblages similar in composition and density to existing conditions during operations and provide for recovery after operations cease	Predator of invertebrates; prey item for predatory fish and birds; feeds and lives on areas potentially affected by tailings deposition	Prey for commercially important fish species and aesthetically important avian species	Potential for exposure to bioaccumulative metals through diet; sensitive to potentially toxic metals in sediments
Maintain suitable conditions for benthic assemblages over most of the prime benthic habitat area of Stephens Passage during operation and provide for recovery after mine closure	Bottom of aquatic food chain; community composition will be affected by tailings deposition; may bioaccumulate chemicals of concern	Prey species for commercial fish	Sensitive measure of water/sediment impact

- Having low natural variability. If there is huge natural variability, it may be difficult to detect responses from the site-related stressor.
- Rapidly responding and sensitive to the stressor. Response will be immediate and results will not take a long time to interpret.

The marine food web was deemed a critical component of the ecological assessment endpoint selection; therefore, impact on the food web, both toxic and habitat alteration effects from tailings disposal, was a major concern. Thus the make-up dynamics, and relationships of the food web were critical components of the environmental analysis impact prediction conceptual model (see Section 5.3.1) and the ecological risk assessment conceptual site model. However, the academic and expansive food web developed for the EIS (Figure 7.4) was unusable for the risk assessment and measurement endpoint approach used in ecological risk assessment. Consequently, a series of simplified models for each major exposure pathway were developed for the risk assessment (Figure 7.5 example for the Pacific herring). These revised and simplified food webs were key inputs to the critical ecological endpoint development step.

Based on these criteria and the adapted food web model, measurement endpoints were developed and their relationship to assessment endpoints documented (Figures 7.5 and 7.6). Also to facilitate risk characterization and input to decision makers, risk thresholds, comparable to environmental analysis impact significance criteria (Section 5.3.2) were developed for each endpoint (Table 7.5).

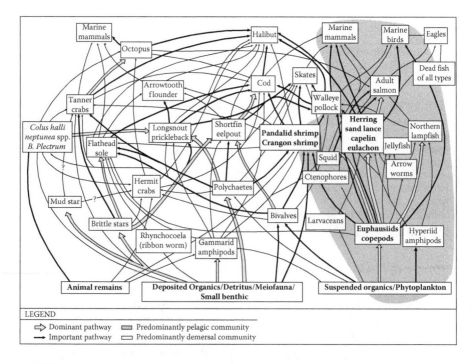

FIGURE 7.4
Initial food web as part of AJ Mine EIS conceptual site model.

Environmental Analysis Tools

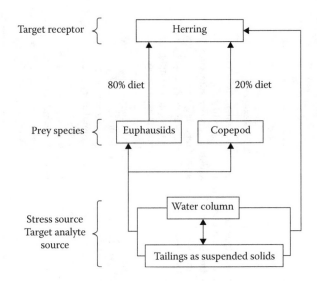

FIGURE 7.5
Herring food web adapted for ecological risk assessment conceptual model.

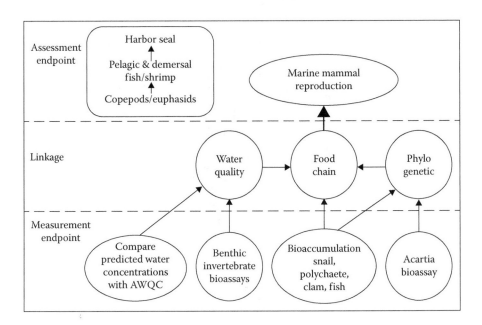

FIGURE 7.6
Marine mammal reproduction ecological endpoints.

TABLE 7.5
AJ Mine Ecological Endpoints and Risk Thresholds

Assessment Endpoint	Measurement Endpoints	Risk Thresholds	Assessment/Measurement Comparison Method & Linkage
Provide food source that does not interfere with sea bird reproduction	Tissue concentrations from clams and polychaetes, bioaccumulation tests; and sole from historic tailings	Fish tissue concentrations less than values known to have no effect on reproduction in seabirds or surrogates; concentrations less than the no observed adverse effect level (NOAEL) if reproduction effects levels are not available	Determine whether feeding in area will alter reproduction by calculating concentrations in seabird diet using measurements from clams, polychaetes, and sole to derive values; and comparing intake to effects levels reported in the literature
Provide food source at existing levels that do not interfere with fish-eating marine mammal reproduction	1. Estimate effects on salmon and herring	No reduction in density of salmon and herring within the study area	
Provide unaltered food and habitat requirements for naturally occurring pelagic fish	2. Tissue concentrations from clams and polychaetes, bioaccumulation tests; snail and sole from historic tailings	Fish tissue concentrations less than values known to have no effect on reproduction in marine mammals or surrogates	Determine whether feeding in area will alter reproduction; by calculating concentrations in marine mammal diet using measurements from clams, polychaetes, snail, and sole to derive values; and comparing intake with effects levels reported in the literature

7.2.9.2 Exposure Characterization

The conceptual site model for the marine disposal of mine tailings identified two types of stress: chemical and physical. The chemical exposure could be determined from a relatively simple analysis of the potentially toxic constitutes (primarily heavy metals) of the mine tailings. This was enhanced by experiments to determine the dissolution of the contaminants into the water column and the sediment pore waters. It was also verified by sampling and analysis of sediments, pore water, and overlying water column of nearby historic mine tailings marine disposal areas. These analyses produced consistent characterization of exposure expressed as milligrams of contaminants per liter of water (mg/l) in the overlying water column and pore water. Similarly, chemical exposure to the mine tailings in the sediments was expressed as milligrams of contaminants per kilogram of sediment (mg/kg). These exposure concentrations could then be compared with the concentrations developed through the effects characterization.

Characterization of physical exposure was not as simple as the chemical characterization. The degree of physical exposure, and thus stress, was dependent on the mass (converted to accumulated depth in centimeters) of tailings deposited per unit area of the seafloor. This was determined by sophisticated hydraulic modeling that factored in the settling rate of the tailings particles, the density of the water, the currents, stability of tailings piles on the sea floor, and several other physical/geological oceanographic variables. The output of the model was a prediction of where the discharged mine tailings would come to lie on the seafloor at the cessation of mining activities and the depth of total tailings accumulation (Figure 7.7). Similar to the chemical exposure conclusions, the degree of physical exposure to mine tailing (both total accumulation and rates of deposition) could be compared with the physical effects characterization.

7.2.9.3 Effects Characterization

Paralleling the exposure characterization, the degree of stress causing adverse effects to ecological receptors had to be determined for both chemical and physical stressors. Similar to the exposure analysis, the determination of effects from chemical effects was relatively straightforward and utilized tried and true ecological risk assessment techniques. The water-quality standards and criteria for the contaminants found in the mine tailings were the starting point for the chemical effects characterization. These were modified by review of recent literature reporting toxicity information on the species representing the ecological endpoints selected for the AJ Mine risk assessment/EIS. There were also adjustments made to account for the chemical behavior of some contaminants under the conditions found in the tailings deposition area (e.g., pH, salinity, and depth). This characterization yielded effects levels or concentration levels above which adverse affects would be expected for all the chemical-related measurement endpoints for both short- and long-term exposures.

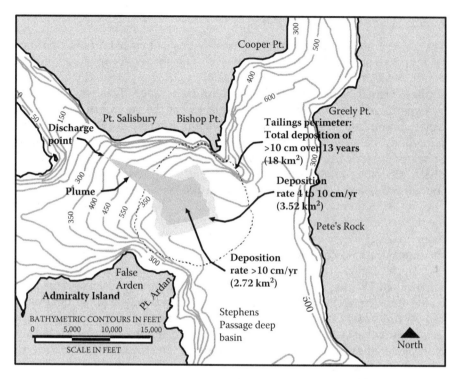

FIGURE 7.7
AJ mine ecological risk assessment physical stress exposure characterization.

The standard chemical characterization of determining literature values reported as effects levels was supplemented with toxicity testing on mine tailings samples. The additional characterization was warranted because of the relative nonbioavailability of the metals in the tailings, the low chemical solubility in the deep, cold, salt, and alkaline disposal areas; and the potential for additive and/or synergistic effects of multiple contaminants in the tailings. After much discussion with the TAC and research on toxicity testing methods, an extensive bioassay program was implemented. It addressed both the solid and dissolved phases of the mine tailings and several benthic as well as pelagic (i.e., living in the water column) species were tested for toxicity (acute and chronic) and bioaccumulation in tissue. When the test results were in, the conclusions were very clear: the contaminants associated with the mine tailings (predominantly heavy metals) were not toxic to ecological receptors of concern and they had not bioaccumulated in the tissue of prey organisms exposed to the tailings. This was consistent with the development of benchmarks which were well above the measured levels of contaminants in the mine tailings; thus no elevated risk was indicated. The results seem to stem from the relatively low concentrations in the tailings, at least partially because most of the metals would be extracted with the gold processing prior to disposal, and the low bioavailability, particularly under the disposal area environmental conditions.

Similar to the physical exposure assessment, the determination of effects from the physical stress of tailings deposition was not straightforward or amenable to standard ecological risk assessment approaches. It was clear to the environmental analysis team that extreme rates and depth of tailing deposition would smother and alter benthic habitat, to the point that little fauna could reside in the sediments. It was also well established that the indigenous benthic infauna and the epibenthic fauna (e.g., flat fish and crabs) could tolerate modest rates of tailing deposition and accumulation. The effects characterization challenge was identifying the break point where the community associated with the bottom was not adversely affected and where it was severely degraded. The literature was sparse on the relationship of solids deposition and its effects on organisms associated with the bottom and at the time the issue was not addressed as part of most ecological risk assessments. However, making use of whatever literature was available, some creativity, and discussions with the TAC, effects levels were established as follows:

- Tailings accumulation less than 4 cm/yr (centimeters per year) would have no adverse effect on the community associated with the seafloor.
- Accumulation between 4 cm/yr and 10 cm/yr would alter the benthic habitat and shift community composition.
- Deposition rates greater than 10 cm/yr would result in virtually uninhabitable benthic conditions.

These rates were derived from studies showing that most organisms could survive instantaneous burial rates greater than 10 cm (Reid and Baumann 1984, Nichols et al. 1978, SAIC 1993) so 10 cm/yr was chosen as a conservative (i.e., over predicts risk) rate above which there would not be a sustainable level of benthic activity. Instantaneous rates below 5 cm did not produce significant mortality (Kranz 1974 and Nichols et al. 1978) so the rate of 5 cm/yr to 10 cm/yr was conservatively chosen to result in a change in the benthic habitat but supportive of a community that could survive. Based on these studies, mine tailings deposition rates less than 4 cm/yr were predicted to have no measurable effects on the benthic or epibenthic communities. These rates were consistent with the natural sedimentation rates (approximately 5 cm/yr) in the adjacent Taku Inlet, which during summer receives the melt from Taku Glacier. With the exposure and effects level characterized, the final step of risk characterization should have been a simple matter of comparing the two. However as stated above, little in ecology or risk assessment is quite so simple.

7.2.9.4 Risk Characterization

Actually, the characterization of risk due to chemical exposure was simple. The concentrations in the mine tailings, both solid and dissolved phases, were below the effects levels. Also the toxicity testing showed no adverse effects from chemical exposure; therefore the risk assessment concluded marine disposal of mine tailings would not result in any elevated risk due to chemical exposure.

The evaluation of risk from physical stressors was a different story altogether. The characterization of exposure demonstrated that tailings deposition would be above the effects levels for the altered and lost habitat in a portion of the study area (Figure 7.7). Approximately 7.5% of the study area habitat would be affected (3.4% altered plus 3.9% lost, Table 7.6); however, risk characterization must go beyond determining how much area or even what percentage of total area would be affected. The risk assessment must address the question of: does the change affect the achievement of ecological endpoints?

Each endpoint was addressed based on the loss of a resource critical to an ecological endpoint, starting with the density of benthic fauna. This was the first endpoint addressed because: (1) the benthic assemblage is the most directly impacted by the deposition of mine tailings and (2) the benthic fauna are near the base of the food chain and virtually every other ecological endpoint is dependent on secondary food production by the benthos. As it turns out, the benthic fauna affected is only about half (1.6% altered and 2.0% lost) of the habitat area affected (Table 7.6). The relatively small effect on the benthic fauna is no coincidence or a random result of the risk characterization. Extensive investigations clearly demonstrated that the slope habitat supported almost five times the faunal density of the basin (5980 animals per square meter on the slope vs. 1350 per square meters in the basin, Table 7.6), and the mine tailings discharge method and location were configured to minimize impact on the slope habitat. The risk characterization demonstrated (and the TAC accepted) that this level of risk to the benthic animals was acceptable and the sustainability of the benthic assemblage would not be threatened.

However, it still left many other ecological endpoints (Tables 7.4 and 7.5), to be addressed for full characterization of the risk. The lack of chemical exposure, bioaccumulation of contaminants, and water column effects resulting from the tailings discharge demonstrated that there would be no risk to the other endpoints (e.g., pelagic fish, sea birds, and marine mammals) related to toxicity. But, nontoxicity related endpoints were dependent to some

TABLE 7.6

Benthic Fauna Risk Characterization of AJ Mine Tailings Discharge

Habitat	Area (km^2)	Benthic Fauna Density (Number/m^2)
Slope: total for study area	25.4	5980
Basin: total for study area	50.6	1350
Other: total for study area	19.3	2550
Total: total for study area	95.3	2827
Total altered	3.4%	1.6%
Total lost	3.9%	2.0%

Environmental Analysis Tools

degree on the benthic infauna and epibenthos as food, and the effects of lost or altered benthic habitat on food availability were calculated to determine whether the measurement endpoints would be met.

A process was followed relating the loss of benthic fauna to the nutritional needs of the other endpoint species through evaluation of diet and abundance of food (Figure 7.8). The diet of each endpoint representative species was determined by a combination of literature review and examination of gut contents of collected specimens. The predicted effect on the diet of each endpoint was determined as the combination of percentage of diet of prey item and loss of that prey item density due to habitat loss or alteration (Figure 7.9). For example if 10% of a halibut's diet was tanner crab and the predicted decline in tanner crab density caused by the tailings discharge was 1.0%, there would be a 0.1% (1% × 10%) reduction in the halibut's available food supply. This was then summed over all dietary items to determine the projected food loss for each ecological endpoint species (Table 7.7).

The risk estimates for all endpoints were deemed by the project team, regulators, and TAC to be within an acceptable range. The loss of food, which was the only area of potentially elevated risk was deemed within natural variability and not of concern for almost all of the representative species, thus the endpoints were achieved. The projected 2.6% reduction in the seal diet was of some concern (Table 7.7). However, when the conservative nature of all of the calculations and assumptions, the diversity of the seal diet, and

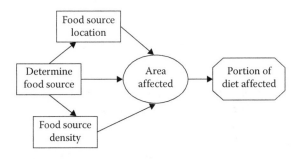

FIGURE 7.8
Food loss prediction process.

TABLE 7.7

Representative Species Food Loss from AJ Mine Tailings Discharge

Representative Species	Food Items Affected	Percentage of Diet Loss (%)
Herring	Meroplankton	0.6
Salmon	Pelagic fish	0.4
Murre	Pollock	1.1
Seal	Pelagic fish, sole, squid, octopus	2.6

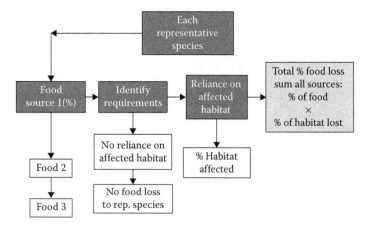

FIGURE 7.9
Calculation of food quantity loss.

typical seal life span (greater than 20 years) were taken into account, this level of risk over the 13-year operation of the mine was deemed acceptable.

7.2.10 Ecological Risk Assessment and Environmental Impact Analysis

Ecological risk assessment concepts can be transferred to impact analysis in two ways. First it can be used as a distinct tool to evaluate impacts to ecological resources as part of a comprehensive environmental analysis, as illustrated in the AJ Mine case study. Second, the concept can be adapted and applied to other environmental resources potentially impacted by the proposed action or an alternative. The adaptation to other environmental resources would incorporate the endpoint concepts, defining the desired state of the resource and then the degree of impact would reflect the achievement of the endpoint. As in the AJ Mine example, this concept can be structured to reflect not just impact versus no impact, but also the degree of impact. The critical hypothesis testing component of the ecological risk assessment related to measurement endpoints can also be applied to other environmental resources to strengthen the objectivity and quantification in impact prediction.

The development of an impact prediction conceptual model (see Section 5.3.1) as part of an environmental impact analysis can also benefit from the ecological risk assessment process. The risk assessment conceptual site model is prepared in multiple iterations with more detail and quantification included at each level. This approach can also be applied in environmental analysis to refine the impact prediction conceptual model as more information on affected environment, alternatives, and impacts is developed.

Ecological risk assessment and environmental impact analysis do differ in one respect. The impact analysis predicts the future condition based on changes in current conditions produced by the proposed action and alternatives. In contrast, the classic ecological risk assessment at a hazardous waste site is faced with a current condition where the impact has already occurred. The challenge of the risk assessment is to determine what the preimpacted condition was and then quantifying the change that has already occurred. As demonstrated in the AJ Mine case study, this difference does not prevent the use of ecological risk assessment in environmental analysis if adaptations are made to both risk assessment and environmental analysis processes. In fact, both can benefit from the integration of the two.

7.3 Net Environmental Benefit Analysis (NEBA)

The national attention to hazardous waste cleanup spawned not only ecological risk assessment as an environmental evaluation tool, but also NEBA. The concept and need to consider the net benefit to the environment arose simultaneously from two aspects of hazardous waste cleanup, one regulatory and the other technical. The regulatory need resulted from a requirement in the CERCLA or the Superfund law [CERCLA §107(a)(4)CC)] and also in the Clean Water Act that parties responsible for causing environmental damage provide retribution to the public for the environmental damage resulting from their actions, intentional or not. As part of CERCLA [Section 301(c)], the president was required to develop procedures for natural resource damages evaluation which President Reagan addressed by Executive Order 12316 signed in August 1981 delegating the authority to the Department of the Interior. The procedures developed called for a Natural Resource Damage Assessment (NRDA) in response to a Natural Resource Damage Claim by the trustees (state, indigenous tribes, or federal resource agencies) of the affected environmental resource. The requirements specified that the retribution had to be an ecological value and could not be a simple payment to the agency responsible for stewardship of the damaged ecological resource for their operating budget or unconstrained use. Thus there was a need, which as discussed below can be filled by NEBA, for tools to determine the extent of damage and the scope of required ecological restoration.

The technical impetus for NEBA was also twofold. One resulted from cleanup efforts that were actually causing more damage to sensitive ecological resources than the damage caused by the ongoing contamination. The classic example is the initial cleanup of the *Exxon Valdez* oil spill in Alaska when large and numerous cleanup teams used pressurized hot water to wash oil from the intertidal zone. The "washing" effort resulted in the combined oil and hot water migrating to the productive and sensitive subtidal zone where the ecological damage exceeded the consequences of leaving the oil in the intertidal area and encouraging gradual natural attenuation (Burger et al. 2003, Efroymson et al. 2004). Thus a tool was needed to assist in determining "how clean is clean" and at what ecological cost, again a need that can often be served by NEBA.

The second technical cleanup issue prompting the need for NEBA was the relationship between cleanup cost and improvement in environmental conditions. In many, if not most, cleanup programs with a goal of reducing environmental risk, the majority (typically 70%–90%) of the contaminated waste site ecological value could be restored by an expenditure of only 10%–20% of the total cost of cleanup up to pristine conditions (Figure 7.10 is a hypothetical representation of the phenomena). NEBA was a tool to assist in getting a better ecological value for each dollar spent.

7.3.1 The NEBA Process

The NEBA process was not a regulatory initiative or even the result of a specific government agency program addressing the ecological concerns associated with the hazardous waste clean up discussed in proceeding sections. The process developed organically as the need was identified starting with a 1990 report addressing the *Exxon Valdez* spill by the National Oceanic and Atmospheric Administration (NOAA) entitled, *Excavation and Rock Washing Treatment Technology: Net Environmental Benefit Analysis* (described in Efroymson et al. 2004). As NEBA matured, there have been methods developed to apply the analysis to specific situations and a number of states, including Texas, Florida, and Washington, have included it in their hazardous waste remediation guidance (Efroymson et al. 2004).

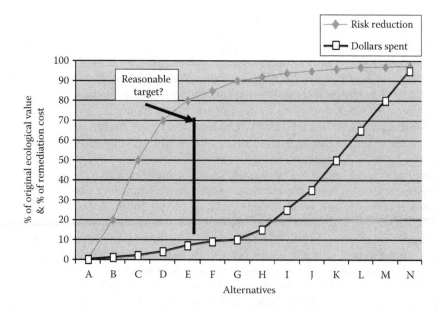

FIGURE 7.10
Hypothetical cost versus ecological value resulting from waste cleanup.

The quantification of environmental value is at the base of NEBA and applies to both achieving maximum environmental benefit for each dollar spent and NRDA addressing an excess ecological damage from hazardous waste remediation. Applying the tool of NRDA, the loss of value from the hazardous waste insult is measured and integrated over time (the debit). Then an environmental improvement is proposed, including at least partial restoration of the damaged resource and the value gained projected into the future (the credit), with the net benefit being the difference in debit and credit measured as the difference in the areas under the curves (Figure 7.11).

A similar process is followed in selecting a remediation alternative that produces the greatest environmental benefit or achieves the greatest environmental value per-remediation dollar expended. The total environmental value following cleanup is calculated for each remedial alternative, including considerations such as:

- Ecological restoration following remediation
- Off-site compensation achieved through habitat creation
- Ecological value gained by purchasing or otherwise ensuring preservation of ecologically valued areas
- Discount for time required for restoration and habitat creation to reach full value
- Diminished value if low levels of contaminants are left on site
- Remaining value of current environmental resources destroyed through soil removal or other remediation activities
- Credits for natural attenuation in areas of little or no active remediation

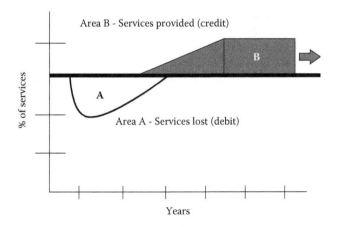

FIGURE 7.11
Application of net environmental benefit analysis to natural resource damage.

Once the net environmental value (credit minus debit) of each remediation alternative has been calculated, they can be compared and provide objective environmental input to remediation decisions.

7.3.2 Quantification of Environmental Value

The summary above and Figure 7.11 make NEBA seem simple and it begs the questions, why did it take so long to develop and why isn't it used all the time? The simple answer is that quantifying the value is no easy task and developing techniques to address the quantification is the key to applying NEBA. In early cleanup efforts, the lack of an approach to quantifying the environmental value of the resources was an impediment to NRDA and ecologically beneficial remediation.

Early in the history of hazardous waste cleanup, some sites with either actual or potential significant ecological value were cleansed to the point of ecological sterility. Soil removal destroyed all vegetation and the material used for replacement caps was unsuitable to support natural vegetation with any ecological value. Also dredging of contaminated sediments totally destroyed the benthic habitat and altered the physical and chemical structure of the sediments, rendering them unsuitable as benthic invertebrate habitat for extended periods of time. Similarly, ecological destruction occurred at contaminated groundwater sites, where pumping of the groundwater lowered the water table to the point that adjacent wetlands dried up and lost all habitat value. Those proposing and implementing the remediation claimed that the site was better off because ecological receptors were not exposed to contamination. But ecologists made the counter point that there were no longer any ecological receptors present; so if there was any change in ecological conditions, it was in the negative. A quantitative process and tool, which NEBA provided, were needed to address and resolve these qualitative conjectures.

Early NRDA efforts often ended in a similar stalemate as attempts were made to calculate a monetary value for the damage inflicted to each resource and then sum the cost for all damaged resources. The approach was then to collect the total calculated lost value from the responsible party and contract the design and construction of replacement natural resources. This monetary determination of environmental value and then collection of the debt was doomed to fail in many, if not most, cases. The debate and frequent failure centered on the "cost of an organism": how much is a larval insect serving as a critical fish prey item worth? Added to this is the complication of how much that insect would be worth today if it had survived the hazardous waste spill or disposal that occurred 10 years ago? Also, there was an endless debate over how the collected funds should be used to compensate the public for the natural resource damage. The process was adversarial, with seemingly endless scientific investigations, economist's calculations, debates over type and location of restoration, and excessive lawyer's fees (sometimes exceeding the calculated value of the ecological damage) producing delays that significantly increased the duration of damage.

Some progress was made when the construction and monitoring costs of restoring the resources to their predamaged condition were used as a basis for negotiating a monetary settlement. However, these efforts were complicated and often derailed by debate and disagreement over the area or intensity of restoration, the projected construction costs, and the specifics of monitoring. But the concept of restoration cost as compensation for natural resource damage did have merit and contributed to development of a more workable approach quantifying environmental value.

If the party responsible for the natural resource damage could restore and replace, rather than make monetary payments for the damaged natural resource, much of the debate and many of the roadblocks to restoration could be avoided. The key was to find nonmonetary currency to measure and confirm that the value of the restoration and replacement equaled or exceeded the natural resource damage integrated over the entire duration of the degradation and recovery. This or a similar currency could also be used as a common environmental benefit measure for various remediation alternatives and the alternatives could be compared. The currency could be used to calculate both the debit from destruction of environmental resources resulting from remediation and the natural resource credit created by removing contaminants and restoring the resource.

For both the NRDA issues and remediation with demonstrable ecological benefits, NEBA could be a very useful approach. However, the value had to be quantified and a common currency developed to express the quantification. Two tools were available to develop site-specific environmental currency and thus enable NEBA to assist in remediation and restoration decisions: Habitat Evaluation Procedure (HEP) and Habitat Equivalency Analysis (HEA).

7.3.3 Environmental Currency

The U.S. Fish and Wildlife Service (USFWS) had developed a technique to quantify habitat value for selected species (U.S. Fish and Wildlife Service 1980), which showed promise as a common currency for determining net environmental benefit. The USFWS HEP was developed to determine the reduction in habitat value for key species as part of their NEPA review of federal agencies' actions. The method was used to determine the Habitat Suitability Index (HSI) of an area for selected species calculated as: Study Area Habitat Conditions/Optimum Habitat Conditions. The existing and optimum conditions could be determined from habitat models created by USFWS for common species of either high societal (i.e., recreational or aesthetic prominence) or ecological integrity value (i.e., performed or indicative of critical ecological functions such as food or nutrient processing). Thus models were developed for such species as small mouth bass, white tail deer, or bald eagle. The product of HSI times the total acreage of habitat in the study area yields the number of habitat units (HU) which could be determined for the affected environment and predict conditions after project implementation.

The difference between the two then represents the level of impact for each alternative under evaluation for satisfying the purpose and need. The difference before and after HU could then be compared among alternatives as input to decisions incorporating environmental considerations.

HEP could easily be adapted to evaluate natural resource damage by comparing HUs of a prior impact or reference conditions to HUs following a hazardous waste release or other sources of damage. The shortcoming of applying HEP to natural resource damage restoration or comparison of remediation alternatives taking into account damage done by hazardous waste removal was the focus on a single species. Consequently, compensation could be provided or habitat preserved for one species at the expense of another. Multiple species could be used but this would not only complicate a process in need of simplification, but it would also raise the question of relative value of one species versus another (e.g., Do you want more deer and fewer bass?). Thus the concept of HEP could be very useful, and it could serve as a tool, but it would not fully satisfy the quantification requirement for NEBA.

In their role as a Natural Resource Trustee, NOAA, a sister agency of USFWS, built on HEP to develop an HEA, to compensate for natural resource damage [National Oceanic and Atmospheric Administration (NOAA) 1997]. This approach took a broader view of environmental currency, which could be applied to both NRDA and remediation evaluation in a NEBA approach. HEA centers on ecological service provided by a resource as the currency instead of single species habitat and restoration, so comparisons can be made on a "service-to-service" basis. HEA takes the concept of evaluation and restoration beyond simple-in-kind habitat replacement or even consideration of a single species, as prescribed in HEP. Service provided by the habitat is the focus and currency of HEA, which achieves integration of multiple benefits provided by a natural resource. Also through a "service-to-service" comparison, HEA enables flexibility in comparisons among different resources and also for restoration of historic damage or future proposed actions.

The service-to-service approach takes into account one of the first principles of ecology that all habitats provide multiple ecological functions, but the value varies greatly among functions and individual habitats. For example, wetlands provide multiple functions including faunal food and shelter, sediment stabilization, nutrient cycling, primary production, fishing, bird watching, and water-quality improvement. Most of the healthy wetlands demonstrate all of these functions to some degree, but the level of service varies greatly among various wetland types, stages of succession, hydrology, dominant plant species, etc. While applying HEA to wetlands, one must first recognize this variation and select the service to be replaced or used as the currency for comparing alternatives or measuring compensation. Sometimes, the primary service of the habitat under consideration is selected and other times a service deemed most desirable or in shortest supply in the general area is used as the currency. It is also possible to develop a currency that integrates multiple services, such as primary production or plant density if the habitat is relatively uniform, such as a mangrove swamp or *Spartina* spp. salt marsh.

Selecting a currency that is indicative of the health of a habitat is another approach that can be useful for some habitats. For example, the density of grass shrimp (*Palaemonetes pugio*), considered a key species by local experts, is highly correlated with many and diverse services provided by marsh habitats (NOAA 1997). Thus, grass shrimp population density can serve as the currency in HEA because it is indicative of the services provided by the environmental resource under consideration and can ultimately be used to calculate ecological debit and credit in a NEBA approach.

An advantage of HEA is that the service can be replaced without replacing the actual habitat lost. For example, at the Blackbird mine site, the salmon population was severely damaged by water quality and habitat alteration resulting from the disposal of waste in Panther Creek, a tributary to the Snake River in Idaho (Ray 2007). The damage to the fish had occurred over a long time, and compensation for years of damage could not be achieved by simply restoring the river to the pre-waste disposal value. The value had to be increased over baseline conditions to achieve the target of supporting at least 200 naturally spawning salmons each year to make up for years of loss productively. The value was increased by a number of enhancements including: creating a new off-channel habitat, establishing a salmon hatchery to restock the creek, fencing off several kilometers of land along the creek to keep livestock out of riparian habitat, and restoration of the natural meandering of the creek. The HEA demonstrated that these measures increased the value to a level that offset the decades of damage created by the waste disposal.

The selection of the currency to be used is frequently a combination of environmental attributes and negotiation among stakeholders. A proposed currency can be strongly defended if it demonstrates several of the following attributes:

- Indicative of multiple environmental functions such as the grass shrimp discussed above
- Integrates multiple functions such as nutrient cycling that depends on a suitable chemical and physical structure and the presence or absence of a full suite of organisms involved in decomposition and remineralization
- Indicative of high social and/or recreational value such as hunting/birdwatching species density or habitat requirement
- A service in limiting or short supply for the area/regional ecosystem (e.g., food, shelter, and water)
- A service that would make a substantial contribution to a threatened or endangered species, such as isolating and protecting nesting sites for bald eagles

- Measurable within the constraints (e.g., time and money) of the specific remediation/restoration

Even if a proposed currency meets every one of the above attributes and more, it must be acceptable to the key decision makers. This typically includes the overseeing regulatory authority, the proponent who will ultimately have to pay the bill, and those directly affected by the natural resource damage (e.g., an indigenous population losing a source of subsistence food).

If a single currency cannot be identified to reflect all aspects of a remediation or NRDA settlement, there is potential to develop multiple currencies. Use of multiple currencies does compromise some of the advantages of NEBA, but sometimes it is necessary to reach agreement among stakeholders and implement a comprehensive comparison or identify just compensation for damage. If multiple currencies are necessary, a system of scaling (i.e., assigning the worth of one's currency in relation to the other) comparable to a monetary exchange rate must be part of the process. This is typically done through stakeholder negotiation, but analytical tools such as determining the relative abundance of critical habitat attributes are available to facilitate the negotiations.

Once the currency has been established and accepted by the stakeholders, NEBA can proceed. The first step is the design and implementation of studies and predictions addressing the currency metric(s) used to determine the value of the credit and debit of the remediation, damage, restoration, or the alternatives, in the case of environmental impact analysis. Once completed, the results of the studies can be entered into the ledger and the difference between debit and credit calculated. In the case of remediation or environmental impact analysis alternatives evaluation, the delta between debit and credit can be a critical input to the comparison and selection of an alternative. In the case of natural resource retribution, the proposed restoration and habitat creation can be adjusted to achieve the target net benefit.

7.3.4 NEBA Examples

Two examples of NEBA represent different applications of the approach related to environmental impact analysis. One example is of a contaminated sediment site where the original preferred remediation was dredging and disposal of all the contaminated material. Although an actual example, it was done in confidence and thus is presented without specific details. The other example is hypothetical but reflective of trade-offs that water resource and wastewater management professionals have to face in many real-world cases.

7.3.4.1 Contaminated Sediment Remediation

Spills and operation of ongoing industrial and marine transport over decades had resulted in significant sediment contamination adjacent to a chemical facility. The originally proposed and standard remediation approach was

Environmental Analysis Tools

to dredge the contaminated sediment and dispose of the material in a containment facility to prevent release and further environmental damage. The responsible party saw this as a bad idea because: (1) the natural resource was diminished in environmental value prior to the contamination due to historic use and continued future industrial/commercial use (it was estimated at only 60% of the value of a similar resource in a less industrialized area); (2) it would be a severe drain on the available funds allocated for environmental issues (estimated cost was $25 million) and addressing more immediate and serious problems would be delayed; and (3) the physical dredging in similar situations had been shown to cause extensive habitat destruction and release to adjacent clean areas. However, if no direct actions were taken, with only reliance on natural processes to address the issue, there would be a significant net environmental deficit of 4000 service acre years (SAYs) which was the chosen and negotiated NEBA currency (Figure 7.12).

Based on NEBA currency and the extensive studies and predictions used to calculate the benefit, the preferred alternative of dredging all contaminated sediment would only worsen the situation (Figure 7.13). There had been substantial historic damage and even an extensive dredging operation could not remove all the contamination, thus there would be a 2000-SAY debit from this alternative. In addition, the dredging would alter the benthic physical habitat, requiring a period to recover, and some contaminated material would be released to adjacent habitat. These two activities would result in an additional 1600-SAY debit for a total net benefit of negative 3600 SAY. Given this loss of benefit and the high price tag, the responsible party was in search of a better way.

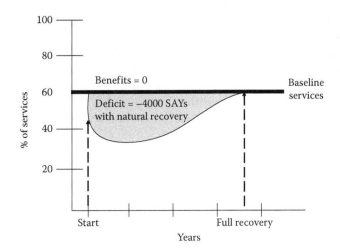

FIGURE 7.12
Environmental deficit from contaminated sediment with no active remediation.

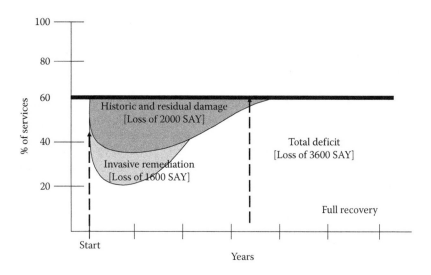

FIGURE 7.13
Environmental deficits from contaminated sediment dredging remediation.

NEBA provided a tool to define a superior approach to remediating the contamination. A remediation alternative was formulated to dredge a limited area of only the most contaminated sediment ("hot spot"), which had a large benefit because almost 90% of the contaminant mass was in only about 20% of the contaminated area. This activity resulted in a net debit of 2000 SAYs (Figure 7.14). However, the alternative included the restoration of a rare and ecologically important hardwood wetland habitat that had been severely damaged by past unrelated activities. This produced a 2500-SAY credit resulting in a net environmental benefit of 500 SAYs. The responsible party was happy because they actually achieved a benefit and the cost of the NEBA-derived alternative (approximately $3 million including the hardwood wetland restoration) was only a fraction of the conventional dredging alternative. The other stakeholders were also pleased with the plan because of the hardwood wetland restoration that had been an unfunded priority in their regional plan for years and they could demonstrate to their constituents the removal of approximately 90% of the contamination.

7.3.4.2 Pollution Abatement in the Green River

In this hypothetical (but reflective of common issues in water resources management) example, the pollution problems in the Green River resulted from nutrient overloading from a variety of sources. The responsible state environmental agency addressed the problem in this impaired water body through the preparation of a Clean Water Act–mandated program by conducting a total maximum daily load (TMDL) evaluation. This evaluation

Environmental Analysis Tools   305

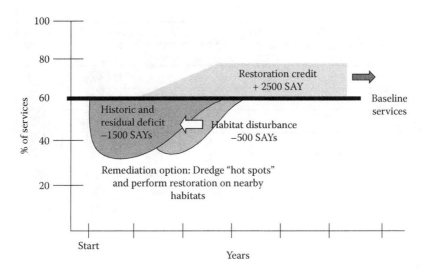

FIGURE 7.14
NEBA-derived hot spot removal with restoration contaminated sediment remediation.

identified nutrient (primarily nitrogen) enrichment causing noxious algal blooms as the primary impairment in the Green River. The sources of nitrogen identified and the associated loads were:

- Regional wastewater treatment plant (WWTP) – 2 kg N/day
- Urban storm water – 3 kg N/day
- Agricultural runoff – 4 kg N/day
- Feed lots – 6 kg N/day
- Total load – 15 kg N/day

The TMDL evaluation determined that a daily load of 12 kg N/day, a 3 kg N/day reduction from the current load, was all the river could assimilate without producing noxious algal blooms that ultimately compromised the beneficial uses of the river.

The initial plan was to apply best practical treatment (BPT) requirements to each nitrogen source as part of their Clean Water Act–mandated National Pollutant Discharge Elimination System (NPDES) permit (Figure 7.15). However this was very expensive ($13 million) with the bulk of the cost ($10 million) falling on the regional WWTP (Figure 7.16). Yet the reduction in nitrogen load from the WWTP (1 kg N/day) was only half the total reduction. The operators of the WWTP felt it was not equitable that they should pay 77% of the cost ($10 million out of $13 million) when they only represented 13% of the problem (2 kg N/day out of 15 kg N/day). Thus, they sought a

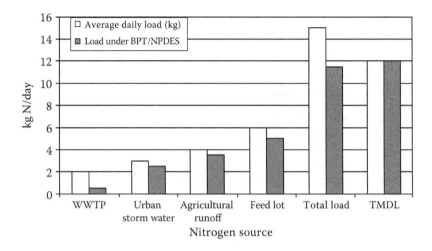

FIGURE 7.15
Nitrogen loads to green river under existing conditions and the best practical treatment under the national pollutant discharge elimination systems (BPT/NPDES).

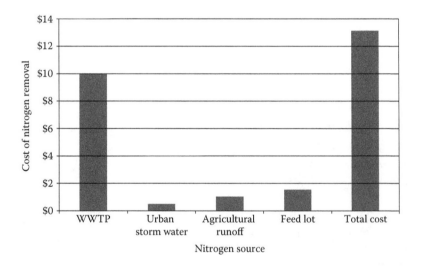

FIGURE 7.16
Costs for nitrogen removal for each source for the best practical treatment under the national pollutant discharge elimination systems.

NEBA approach using nitrogen load and cost of nitrogen control as the environmental currency.

The net environmental benefit sought was a reduction in the current 15 kg N day to the TMDL specified load of 12 kg N/day (Figure 7.15) and the approach should address cost per benefit received. Nitrogen removal costs

Environmental Analysis Tools

varied widely over various sources, ranging from $1.5 million to $6.6 million total cost per kilogram nitrogen removed on a daily basis (Figure 7.17). The NEBA alternative was formulated to take advantage of this variation, with the maximum reduction coming from sources with the least expensive removal costs (feed lots) and none coming from the most expensive (WWTP) to arrive at the specified 12 kg N/day TMDL (Figure 7.18). This formulation does take into account nitrogen reduction from some of the most egregious specific cases, even if the costs are elevated. Urban runoff falls into this category, and although addressing the runoff is expensive, it is included in the NEBA alternative. The result is an alternative that removes more than the required mass of nitrogen (Figure 7.18) at less than half the cost of the classic BPT alternative (Figure 7.19).

As is often the case, the impediment to this approach is that several entities (e.g., the municipality, the feed lot operators, and the regional wastewater authority) must work closely together and share costs. Even though it is much more efficient than the classic alternative, under the NEBA approach the portion of the costs incurred by one entity (e.g., feed lot operators) must be borne by another entity (regional wastewater authority). Experience has shown that this is a tough sell, but the objective and analytical approaches inherent in NEBA and HEA make these tools useful in presenting the case

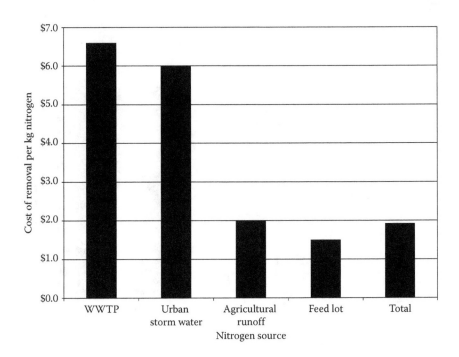

FIGURE 7.17
Cost of nitrogen removal per kg for sources to the Green River.

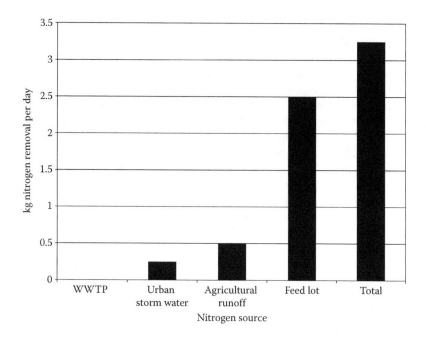

FIGURE 7.18
Green River nitrogen removal (kg per day) from each source under the NEBA alternative.

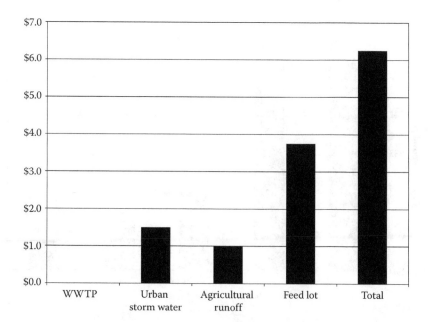

FIGURE 7.19
Nitrogen removal total costs for the NEBA alternative.

and thus increase the likelihood of reaching consensus and achieving environmental benefit at a reduced cost.

7.4 Adapting Tools to Environmental Impact Analysis

Although ecological risk assessment and NEBA were developed for different purposes, they can readily be adapted to more generic environmental impact analysis. NEBA can be directly applicable because the currency and the calculation of total benefit can be used to objectively and quantitatively compare alternatives. It is also an excellent tool for development and evaluation of mitigating measures. Since there is a single currency, or scaled multiple currencies, measures to increase the currency metrics can be targeted for mitigation (e.g., if density of osprey is the currency, mitigation can be creation of more nesting sites). Also if the net benefit has been determined for an alternative, the same calculation can be made with mitigation and then an informed decision made as to the relationship of cost of mitigation and environmental benefit derived.

Ecological risk assessment is also adaptable to broader environmental analyses, as demonstrated by the AJ Mine case study (Section 7.2.9). Adjustments must be made to the methods because a classic risk assessment is primarily concerned with impacts that have already occurred, whereas a classic environmental impact analysis projects future impacts. But the risk assessment concepts of conceptual models, assessment and measurement endpoints, and application of acceptable levels of risk/impact are directly transferable to environmental analysis.

Finding a tool to use in the analysis can be a rewarding challenge to the environmental practitioner. Just as ecologists were able to reach into their bag of methods and pull out ecological risk assessment, HEA, and NEBA, practitioners from other fields should be able to find tools common in their profession to apply in the environmental analysis. Both example methods were developed to address biological impacts. But many of the concepts presented can be adapted to other types of environmental resources by applying knowledge of the methods, the specific situation under consideration, and some creativity. For example, if the purpose and need are to expand a school system, the educators on the analysis team should be able to identify tools to compare alternatives. Common metrics such as test scores, graduation rates, job placement, and college acceptance might be developed into a tool similar to NEBA to compare the alternatives of expanding a school into an environmentally sensitive area versus more online courses.

Typically, team members have tools that they use in their area of expertise as standard practice to measure and evaluate new ideas, programs, infrastructure improvements, etc. It can be a relatively easy task to adapt these methods for use in an environmental analysis to compare alternatives and also to compare impacts to different resources (e.g., traffic congestion vs. lost business to local retailers from diverting traffic from commercial areas). The concepts of ecological endpoints and NEBA can be used in different fields, but it takes the knowledge of the environmental resource at risk, understanding of the different alternatives, and a degree of creativity and initiative to identify and successfully apply tools in environmental analysis and provide meaningful environmental inputs into decisions.

References

Beck, E.C. 1979. The Love Canal tragedy. *EPA Journal*. Available from: www.epa.gov/history/topics/lovecanal/01html. (accessed 7 February 2013).

Boucher, P.M. 1993. Middle March ecological assessment: A case study. In: *Ecological Assessment of Hazardous Waste Sites*. ed. James T. Maughan. 294–342. New York: Van Nostrand Reinhold.

Burger, J., T.M. Leschine, M. Greenberg, J.R. Karr, M. Gochfeld, and C.W. Powers. 2003. Shifting priorities at the department of energy's bomb factories: Protecting human and ecological health. *Environmental Management* 31(2): 157–167.

Doyle, M.E. and J.C. Young. 1993. Human health risk assessments. In: *Ecological Assessment of Hazardous Waste Sites*. ed. J.T. Maughan. 113–143. New York: Van Nostrand Reinhold.

Efroymson, R.A., J.P. Nicolette, and G.W. Suter II. 2004. A framework for net environmental benefit analysis for remediation or restoration of contaminated sites. *Environmental Management* 34(3): 315–331.

Keaton, W.T. 1972. *Biological Science*. 2nd edn. New York: W.W. Norton & Company, Inc.

Kranz, P. 1974. The antistrophic burial of bivalves and its paleontological significance. *Journal of Geology* 82: 237–265.

Maughan, J.T. 1993. *Ecological Assessment of Hazardous Waste Sites*. New York: Van Nostrand Reinhold.

Maughan, J.T., R. Whitman, and J. Gendron. 1997. *Marine Community Risk from Submarine Mine Tailings Disposal Induced Reduction in Food Availability*. Society of Environmental Toxicology and Chemistry, San Francisco, CA, November 17–21.

Menzie, C., M.H. Henning, Maughan, J.T., et al. 1996. Special report of the Massachusetts weight-of-evidence workgroup: A weight-of evidence approach for evaluating ecological risk. *Human and Ecological Risk Assessment* 2: 277–304.

National Oceanic and Atmospheric Administration (NOAA). 1997. *Habitat Equivalency Analysis: An Overview*. Washington, DC: NOAA Damage Assessment and Restoration Program, NOAA. Available from: http://www.darrp.noaa.gov/library/pdf/heaoverv.pdf.

Nichols, J.A., G.T. Rowe, C.H.H. Cliffird, and R.A. Young. 1978. In situ experiments on the burial of marine invertebrates. *Journal of Sedimentary Petrology* 48: 419–425.

Pastorok, R.A., S.M. Bartell, S. Ferson, and L.R. Ginzburg. 2002. *Ecological Modeling in Risk Assessment: Chemical Effects on Populations, Ecosystems, and Landscapes*. Boca Raton: Lewis Publishers.

Ray, G. L. 2007. *Habitat Equivalency Analysis: A Potential Tool for Estimating Environmental Benefits. EMRRP Technical Notes Collection* (ERDC TN-EMRRP-EI-02). Vicksburg, MS: U.S. Army Engineer Research and Development Center.

Reid, B.J. and J. Baumann. 1984. Preliminary laboratory study of the effects of burial by AMAX/Kisault Mine tailings of marine invertebrates. *Canadian Manuscript Report of Fisheries and Aquatic Sciences*. No. 1781.

SAIC (Science Applications International Corporation). 1993. *Environmental Impact Statement for Farrolones Disposal Site*. California, USA: SAIC.

Suter, G.W. II. 1990. Endpoints of regional ecological risk assessments. *Environment Management* 14: 9–23.

Suter, G.W. II, R.A. Efroymson, B.E. Sample, and D.S. Jones. 2000. *Ecological Risk Assessment for Contaminated Sites*. Boca Raton: CRC Press.

Suter, G.W. II. 1993. *Ecological Risk Assessment*. Boca Raton: Lewis Publishers.

U.S. Army Corps of Engineers. 2002. *Environmental Impact Statement for Proposed Restoration of Areas Adjacent to Arlington Channel-Garrows Bend Channel, Mobile County, Alabama*. Mobile, AL: U.S. Army Corps of Engineers.

U.S. EPA. 1989a. *Risk Assessment Guidance for Superfund, Volume I: Human Health Evaluation Manual (Part A)*. Office of Emergency Response. EPA/540/1-89/002.

U.S. EPA. 1989b. Summary of ecological risks, assessment methods, and risk management decisions in Superfund and RCRA. EPA 230-03-89-046.

U.S. EPA. 1989c. *Ecological Risk Assessment Methods: A Review and Evaluation of Past Practices in the Superfund and RCRA Programs*. EPA 230-03-89-044.

U.S. EPA. 1989d. Ecological risk management in the Superfund and RCRA programs. EPA 230-03-89-045.

U.S. EPA. 1989e. *The Nature and Extent of Ecological Risk at Superfund Sites and RCRA Facilities*. EPA 230-03-89-043.

U.S. EPA. 1989f. *Ecological Assessment of Hazardous Waste Sites: A Field and Laboratory Reference*. Corvallis, OR: Office of Research and Development, Environmental Research Laboratory: EPA 600-03-89-013.

U.S. EPA. 1992. Framework for ecological risk assessment EPA 630-R-92-003.

U.S. EPA. 1994. *Interim Guidance on Determination and Use of Water-Effect Ratios for Metals*. EPA Publication Number: 823B94001.

U.S. EPA. 1997. Ecological risk assessment guidance for Superfund: Process for designing and conducting ecological risk assessments: EPA 540-R-97-006.

U.S. EPA. 1998. *Guidelines for Ecological Risk Assessment*: EPA 630-R-95-002F.

U.S. EPA. 2011. Environmental cleanup best management practices: Effective use of the project life cycle conceptual site model. EPA 542-F-11-011 July 2011.

U.S. EPA. 2013. Aquatic life criteria table. Most current version. Available from: http://water.epa.gov/scitech/swguidance/standards/criteria/current/index.cfm. (accessed 12 February 2013).

U.S. Fish and Wildlife Service. 1980. *Habitat Evaluation Procedure (HEP) Manual (102 ESM)*. Washington, DC: U.S. Fish and Wildlife Service. Available from: http://www.fws.gov/policy/ESM102.pdf.

8

International and Individual State Environmental Impact Analysis Programs

8.1 Introduction

Many countries and U.S. states followed the National Environmental Policy Act (NEPA) model by developing their own environmental policy statements, enabling legislation, and ultimately introducing regulations to address the growing environmental concerns. All of these programs drew from the NEPA experience to some degree, sometimes embracing the concepts, and at other times rejecting some of the components, but always considering NEPA and adapting to local situations. Presented below are example programs for several countries and two states with well-developed comprehensive programs and long histories. The summaries are presented in part as comparisons to NEPA, reflecting the history of development and "lessons learned." Also, once you are familiar with NEPA, after assimilating Chapters 1 through 7 of this book, a description of other programs in relation to the first and most mature program is more meaningful.

8.2 Canadian Environmental Program

Canada has promulgated both federal and provincial environmental legislation and the two are complementary. The federal legislation (originally passed as the Canadian Environmental Assessment Act, S.C. 1992, and recently revised as CEAA 2012) and implementing regulations, as summarized below, draw heavily from the U.S. NEPA experience. This summary is based on the original Act, as the new Act is not designed to result in substantive changes. New regulations under CEAA 2012 have not been fully promulgated and there is no history with the revisions. The summary of the 1992 CEAA is followed by comments on anticipated changes based on the 2012 Act and associated regulations. The Canadian approach to environmental protection enhances and adds to the U.S. NEPA experience, recognizing conditions in the large open spaces of Canada, and the strong provincial governments.

The Environmental Assessment Act applies only to actions by the federal government of Canada and is "triggered" when a federal authority:

- Proposes to implement a specific project.
- Provides direct funding, loans, or other financial assistance to a proponent to enable project implementation.
- Involves transfer or use by others of federal land in any way (e.g., leasing, sale, or other forms of conveyance and resource extraction activities on military bases and First Nations Reserves).
- Provides any approval (e.g., license or permit) allowing a proponent to carry out a project. Department of Fisheries and Oceans or Transport Canada (Navigable Waters) actions are often the triggers to this provision. Where a project may cause a harmful alteration, disruption, or destruction to a fish-bearing watercourse or perhaps where navigation of a water body may be impacted, an approval from the Navigable Waters division of Transport Canada is required.

The CEAA has overarching goals and specific requirements for critical aspects of environmental analysis. Similar to NEPA, the Act calls for achieving sustainable development and anticipating and preventing environmental degradation. Taking a step beyond NEPA and drawing from guidelines of the Council on Environmental Quality (see Section 2.3), the Act is specific in stating the mechanism of promoting high quality environmental assessment (EA) to achieve these goals. The Act also recognizes the importance of public participation and early consideration of environmental factors in the planning and decision-making processes for government projects and actions to achieve the specified environmental goals. A process for determining the need for environmental analysis is set forward in the Act and summarized in Figure 8.1. This process includes decision points to determine not only the necessity to conduct environmental analysis but also the level of analysis required based on the type of action. The Canadian provision for different levels of environmental analysis, based on the complexity of the action and potential for significant impact, has a parallel approach to NEPA as summarized in Table 8.1.

The focus of the Canadian Act is to ensure federal authorities consider environmental costs and enhancement opportunities when taking any action. However, in comparison to the U.S. NEPA, there is a stronger emphasis on preventing environmental damage. The Act also provides a mechanism to resolve conflicts over environmental issues during the process rather than the provision in NEPA to elevate to the Council on Environmental Quality after the environmental analysis has been completed. The resolution mechanism specified in the Canadian Act is mediation, which is a voluntary process using an independent and impartial mediator to assist interested parties resolve their issues. After the Minister of the Environment consults with the interested parties and the responsible authority, the minister appoints the mediator.

International and Individual State Environmental Impact Analysis

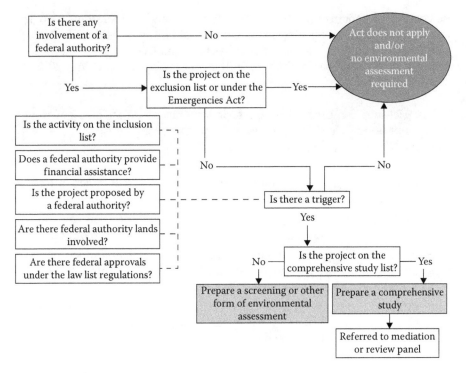

FIGURE 8.1
A schematic diagram summarizing the process to determine whether the Canadian Environmental Assessment Act applies.

TABLE 8.1

Levels of Environmental Analysis for Canada and the United States

Canadian Environmental Assessment Act	U.S. National Environmental Policy Act
Exclusion List or Emergencies Act	Categorical exclusion (see Section 3.6)
Screening and class screening	Environmental Assessment (see Section 3.1.4)
Comprehensive study	Environmental Impact Statement (see Section 3.1.3)

Also similar to NEPA, Canadian regulations and common practice encourage coordination of federal and provincial environmental requirements. There are a number of federal–provincial/territorial EA cooperation agreements, federal–aboriginal agreements, and even federal–international agreements. These agreements set the framework for the cooperation and coordination between the various levels of government within that province or territory. These agreements can be found at: http://www.ceaa-acee.gc.ca/default.asp?lang=En&n=CA03020B-1#fp. In both countries, the processes can be conducted concurrently to achieve efficiency and cooperation. Also, the coordination encourages active participation of both parties during the process to avoid conflicts after a decision has been made.

Some of the anticipated changes in the process based on CEAA 2012 are summarized below:

- Strict and compressed time restrictions placed on various government reviews and other assigned responsibilities.
- Referral by the Minister of the Environment of a project to a review panel with revised responsibilities and authority under CEAA 2012. After receiving public comments, the panel makes recommendations to the minister regarding further environmental analysis.
- Additional public participation requirements, including a new comment period, when the agency is determining whether an EA is required.
- Access to the public of key project information and documents related to an EA on the website of the Canadian Environmental Assessment Registry.
- Emphasis on cooperation and communication with aboriginal people on such topics as health and socioeconomic issues, cultural heritage, land use and resources for traditional purposes, and items that are culturally significant.
- Follow-up programs to verify the accuracy of the impact predictions and to determine if mitigation measures are working as intended are mandatory.
- Limitation on intervening parties to interested parties that are directly affected by carrying out the designated project or have relevant information or expertise.
- Elimination of potential multiple assessments by various federal authorities.
- Creation of more severe enforcement mechanisms.

8.3 Australian Environmental Program

Application of the primary Australian environmental legislation (Environmental Protection and Biodiversity Conservation Act 1999 or EPBC Act) is broader than NEPA. It applies to three categories of actions with the first covering any potential effects on specific environmental resources and includes:

- World heritage properties
- National heritage places

- Ramsar wetlands of international significance. These significant wetlands are those identified on the *List of Wetlands of International Importance*, which originated during the Ramsar Convention on Wetlands (Ramsar, Iran, 1971) and has since resulted in an international treaty with 162 signatories
- Listed migratory species
- The Commonwealth marine environment

The second category subject to EPBC is actions taken in a self-governing territory or a state where there is agreement between the appropriate State or Territorial Minister and the Commonwealth (i.e., national government) Minister. The third category of actions addresses action on Commonwealth land and the final category is actions by any Commonwealth agency taken anywhere in the world.

The Australian protection of specific resources parallels U.S. legislation and regulation, such as Section 404 of the Clean Water Act (33 U.S.C. 1251 et seq.) for wetlands, the National Historic Preservation Act (16 U.S.C. 470 et seq.), which encompasses heritage properties and places, and the Marine Protection, Research, and Sanctuaries Act (16 U.S.C. §1431 et seq. and 33 U.S.C. §1401 et seq.). The provision in the Australian EPBC Act for activities on state lands reflects NEPA "piggyback" provision (see Section 6.2.4) and the EPBC Act coverage of Commonwealth lands and actions taken by the Commonwealth is identical to the scope of NEPA's limitation to federal actions.

Australia's environmental process has matured and expanded beyond original NEPA provisions. Coordinated with the national law, all Australian provinces and territories have set procedures or put legislation in place requiring environmental impact assessments (EIAs) for all activities they propose which might affect the environment. Some states, such as Western Australia, have expanded requirements to cover monitoring of post-implementation impact and other significant features (Wood and Bailey 1994).

8.4 European Union Environmental Program

The European Union began addressing environmental issues in the early 1970s, shortly after the passage of NEPA (Hey 2006). However, they did not enact legislation requiring environmental analysis until 1985, with the passage of their original Environmental Impact Assessment Directive (EC 1985). The directive has been amended several times (1997 and 2003), with recent consolidation and clarification in 2011 (Directive 2011/92/EU). Just as NEPA placed environmental requirements on each federal agency and mandated each agency develop implementing regulations, the EU Environmental Impact Assessment Directive called for similar requirements for each member state. Thus, each individual country had to develop specific environmental analysis procedures addressing the EU Directive requirements.

The EU Directive is also similar to NEPA in that it is focused on procedure, full public disclosure, and ensuring environmental considerations are incorporated into decisions. It is also similar in the basic requirements for the environmental impact analysis report, calling for prediction of impacts, analysis of alternatives, consideration of mitigation measures, and a nontechnical presentation understandable to all stakeholders. There is a requirement, almost identical to the NEPA Record of Decision mandate, calling for public disclosure of:

- Contents of decisions
- Reasons for the selection of the proposed action
- Response to public comments received on the process, EIA, and comparison of alternatives
- A description of measures considered and adopted to avoid impacts

The EU environmental analysis process does differ from NEPA in both procedures and scope. The largest procedural difference is the hierarchical and strictly prescribed process to determine when an analysis is necessary. There are project types that automatically require a full environmental analysis and they are listed as Annex I in the Directive (examples presented in Table 8.2). If the project type falls within the Annex II list, the member state must decide whether an analysis is necessary. Annex II lists numerous activities under the following major topics:

- Agriculture, silviculture, and aquaculture
- Extractive and energy industries
- Production and processing of metals
- Mineral, chemical, and food industries
- Textile, leather, wood, paper, and rubber industries
- Infrastructure projects
- Tourism and leisure

The Directive calls for the member states to determine the necessity of an environmental analysis on each of these types of projects either on a case-by-case basis or by applying pre-established criteria identified in Annex III:

- Characteristics of projects
- Location of projects
- Characteristics of potential impacts

There are two major differences in scope between the EU Directive and NEPA and both result in very different programs. The first is the application

TABLE 8.2

Example Projects Subject to Environmental Assessment under European Union Directive 2011/92/EU (Annex I)

Annex I Designation	Project Type
1	Crude-oil refineries and gasification installations
2.a	Thermal power stations
4.a	Facilities for initial smelting of iron and steel
5	Large size installations for extraction of asbestos or the processing and transformation of asbestos and products containing asbestos
6 (a through f)	Integrated chemical installations (listing specific products)
7 (a through c)	Construction of rail, airports, and motor highways of certain sizes and lengths
9	Waste disposal installations
11	Groundwater abstraction or recharge systems with a capacity of 10 million cubic meters or greater
13	Wastewater treatment plants with a capacity exceeding 150,000 population served
15	Dams designed for a storage capacity of 10 million cubic meters or more
17 (a through c)	Installations for rearing poultry or pigs (of varying capacity based on the type of animals)
19	Quarries and mines of over 150 hectares
21	Petroleum, petrochemical, or chemical storage facilities with a capacity of 200,000 tons or more

of the Directive to both public and private actions as opposed to NEPA, which only applies to federal agencies. This greatly expands the scope of the European program and has the potential to have far-reaching environmental implications.

However, the second major difference between the EU environmental program and NEPA lies with compliance and enforcement, which limits the Directive's potential for addressing environmental issues and degradation. Adherence with the EU Directive is relegated to the member states that have a history of compliance inconsistency and variation among countries (Scheuer 2006; Hey 2006). Also, history has demonstrated quality of analysis inconsistencies among members. As compliance tends to degrade to the lowest common denominator, the program has decreased in effectiveness since its original inception. Enforcement differs greatly between the two programs because meaningful litigation is lacking in the EU Directive compared with NEPA (Ekmetzoglou-Newson 2006). This has resulted in a more streamlined and expedited program in Europe, but also one that frequently fails to take a "hard look" at the impacts as the courts have mandated under NEPA.

There has been concern within the EU and among stakeholders over the shortcomings of the Directive and the attempts made to strengthen the overall environmental program in Europe. As summarized by Scheuer (2006), media-specific pollution and prevention control directives may be the most important measures taken to address environmental degradation. For example, through the permitting procedure, industrial installations are checked against EU's "best available techniques" for waste treatment and environmental sustainability. Some specific waste streams have been addressed with collection and recycling targets for batteries, packaging waste, and electronic wastes. Also, strict media-specific standards (e.g., water and air) were identified to address pollution issues, and legislation has been passed to address specific areas of environmental concern such as the objective to halt biodiversity decline specified in EU's nature and environmental media protection legislation. The result was that for much of the 1990s and 2000s, the European environmental program was more focused on the permit and approval side of environmental protection than on an NEPA-like environmental analysis approach (see Section 9.2 for a discussion and comparison of the approaches).

There has been some movement back to analysis as an important environmental management and protection tool in Europe. The consolidation and simplification of the various Environmental Impact Analysis Directive amendments into the Directive 2011/92/EU are in recognition of the need for improvement. The Directive points toward a renewed effort to institute consistency and adherence to environmental analysis as a force in European environmental policy. Also, the emphasis in the new directive and associated efforts on strategic environmental assessment (see Section 6.4) recognizes the importance of not only controlling but also fully understanding environmental implications and benefits of comprehensive environmental policies and programs.

8.5 Japanese Environmental Program

Japan's history in environmental legislation differs from that in most countries in that there have been several attempts abandoned and new approaches attempted. In contrast, most other countries have had a progression of the environmental analysis process, with each step and refinement typically broadening the previous legislation and regulations that provide greater environmental analysis and protection. Initially, there were two Japanese laws (Basic Law for Environmental Pollution Control 1967 and the Nature Conservation Law 1972) governing specific resources and actions but included no policy or environmental analysis component. The first attempt requiring environmental analysis in Japan was a 1972 rule addressing only public works projects, which included little guidance and few specific requirements. The rule was expanded in 1980 to also cover the major areas of development underway in Japan at the time: ports and harbors, super-express train system, power plants, and reclamation. The inconsistent and often superficial environmental analysis persisted with the addition of these areas to the rule, so the Diet (the national legislative body) proposed a more structured and integrated approach to environmental analysis in the early 1980s, but the bill failed to pass. The lack of legislative approval resulted in an administrative process of "Implementation of Environmental Impact Assessment" in 1984. This process suffered from inconsistencies and the same lack of guidance and mandates as earlier efforts such that its effectiveness was limited.

Recognizing the limitations of the environmental control, pollution prevention, and comprehensive environmental analysis in force at the time, in 1990 Japan began to update their approach. In 1993, they passed the "Basic Environmental Law." This law set the national environmental policy and goals and required an annual national review and reporting of environmental conditions. It also directed the government to establish standards, such as water- and air-quality permissible levels of pollutants, and set goals for industry. The law does have a very general mention of the need for environmental analysis (Article 20) but there is no clear requirement or guidance. However, the Basic Environmental Law set the stage for the passage of the "Environmental Impact Assessment Law" in 1997, which is currently the mandate, guidance, and authority for environmental analysis of activities that could affect the environment in Japan.

Reflecting the policy established in the Basic Environmental Law and echoing NEPA, the purposes of the Environmental Impact Assessment Law are to ensure that proper consideration is given to environmental protection issues relating to major projects, and ultimately, to ensure that the present and future generations of the nation enjoy healthy and culturally rewarding lives. The law prescribes the process to first identify the need for environmental analysis and then the procedures to produce, review, and finalize the analysis. The law includes a very specific list of categories and subcategories of projects that require a full environmental impact analysis (Class 1) and others in which the decision is made on a case by case basis (Class 2) are covered (Table 8.3). The law then proceeds to describe what level of activity in each category warrants environmental analysis and review. For example, the number of lanes and length of roadway that require a full EIA are specified. For projects with a size above a threshold but below the requirement for a full EIA, the need for an EIA is made by the authorizing federal entity and the project proponent on a case-by-case basis. The law provides for local environmental ordinances and an assessment system that can add projects requiring an EIA to specific conditions and situations occurring in their jurisdiction.

After the need for an EIA is determined, the scope of the analysis is established through a public comment and input scoping procedure. Based on the input received during scoping, the project proponent circulates the resulting scoping document to the appropriate local governmental officials and the public at large. The proponent has the opportunity to revise the scope based on this second round of public input and then distributes the final scoping

TABLE 8.3
Categories of Projects Subject to Japan's Environmental Assessment Law

Category	Subcategory	Class 1 Project (EIA is always required)	Class 21 Project (EIA required on case-by-case basis)
Roads	National expressway	All	
	Metropolitan expressway	4 lanes or more	
	National roads	4 lanes or more and 10 km	4 lanes or more and 7.5–10 km
	Large-scale forest roads	2 lanes or more and 20 km	2 lanes or more and 15–20 km
Waterways	Dams and weirs	Reservoir > 100 ha	Reservoir 75–100 ha
	Diversion channel	Land altered > 100 ha	Land altered 75–100 ha
	Lake development	Land altered > 100 ha	Land altered 75–100 ha
Railway	Super express train railway track	All	7.5–10 km
		> 10 km	
Airport		Runway > 2.5 km	Runway 1.8–2.5 km
Power plant	Hydraulic power plant	> 30 MW	22.5–30 MW
	Thermal power plant	> 150 MW	112.5–150 MW
	Geothermal power plant	> 10 MW	7.5–10 MW
	Nuclear power plant	All	
Waste disposal		> 30 ha	25–30 ha
Landfill and reclamation		> 50 ha	40–50 ha
Land readjustment		> 100 ha	75–100 ha
New residential development project		> 100 ha	75–100 ha
Industrial estate development project		> 100 ha	75–100 ha
New town infrastructure development projects		> 100 ha	75–100 ha
Distribution center complex development project		> 100 ha	75–100 ha
Residential or industrial land development by specific organizations		> 100 ha	75–100 ha
Port and harbor planning		> 300 ha	NA

Source: Adapted from Japanese Ministry of the Environment, Environmental impact assessment in Japan, www.env.go.jp/en/policy/assess/pamph.pdf. Accessed July 19, 2013.

document to the appropriate government entities and includes a summary of comments received and how they were addressed in the final document. The law provides no guidance or requirements for the analyses, the information to be gathered, treatment of alternatives, or structure of the assessment. Instead, it relies on the project proponent, who knows most about the project and the stakeholders, who have the greatest potential to be affected, to develop an EIA scope on a project-specific basis. Once the scoping is complete and the investigations and analyses called for during scoping are completed, a draft EIA is produced and distributed to the public.

The project proponent then responds to all comments on the draft EIA and redistributes the document to all stakeholders. This document includes summaries of all comments on the draft and the project proponent's response to the comments and suggestions. At this stage, the document is submitted to the Minister of the Environment who then expresses an opinion on the project and the results of the environmental analysis. The proponent cannot take any action on the project until the final Environmental Impact Statement (EIS) is completed and made public. Under this procedure, the decision on project implementation and impact mitigation rests with the project proponent and the authorizing agency, who are most often closely linked, but the law does require the Ministry of Environment to comment on the project. This process is somewhat analogous to the draft EIS, final EIS, and Record of Decision (ROD) requirements under NEPA.

The Japanese approach to environmental analysis has both advantages and disadvantages compared with NEPA. After several false starts, the Environmental Impact Assessment Law requires extensive input and review from local government entities and the public, and even allows them to be heavily involved in the design of the environmental analysis. Similarly, the input to two versions of the document (comparable to the draft EIS and EIS) is required and the project proponent must respond to all suggestions and comments expressed. This can result in a very site-specific environmental analysis that focuses exclusively on the critical issues and does not become overburdened with red tape. The Japanese law lacks specific procedural requirements such as required evaluation of alternatives, an authorizing agency to take a "hard look" at environmental impacts, and a complete evaluation of mitigation. This can sometimes be a disadvantage if no stakeholder raises these issues, or the project proponent decides not to include them in the analysis. The law does make provisions for the decision makers to have the necessary information to fully understand the impacts of any proposed action and also to receive and address opinions from stakeholders, including the Minister of the Environment. However, it relies on the commitment of the authorizing government agency and the project proponent (which are frequently one and the same or at least closely tied by common objectives) to consider opinions and comments of others and make decisions that are environmentally protective and sustaining.

8.6 U.S. State Environmental Analysis Programs

All U.S. states have some form of environmental policy and requirements but there is a wide range among the 50 states based at least in part on the wide variety of development density and land use among the states. The more densely developed areas with the most intense land use (e.g., California and New York) tend to have the most comprehensive requirements, while states with more open space (e.g., Utah, New Mexico, and Wyoming) have less intense requirements. However, many of the states with low development density and extensive open space (often in excess of 90%), including the examples above, are dominated by federal lands and activities. Thus, any proposed action on these lands fall under NEPA and state requirements would be somewhat redundant.

There is a wide variability in the reported number of states with comprehensive NEPA-like environmental analysis requirements, with some reports of only 15 states (Montana Legislative Environmental Quality Council 2000). Others have reported a considerably higher number (Bass and Herson 1993) and direct contact with several states has identified states with NEPA-like legislation that appeared on neither list. The discrepancy seems to lie in the definition of a "comprehensive" environmental analysis requirement with some listings including only those requirements applicable to all actions that are public (including subjurisdictional units of the state), private (with or without government approval), and all programs (e.g., water, wastewater, air, solid waste). Review of published lists and direct contact indicates there are more than 30 states with some form of comprehensive environmental impact analysis mandate.

Of the states that do not have comprehensive analysis mandates under the most restrictive definition, most do have such requirements under many of their individual programs. These requirements include specific environmental performance standards, requirements to identify and mitigate impacts, and program-specific comparison of alternatives based on environmental impact analysis. For example, while Alaska does not have comprehensive environmental policies or requirements covering all programs, they build analysis and public outreach requirements into their permit issuing process. Also, most states that do not have their own legislated environmental analysis programs do have procedures for state compliance with NEPA for all programs receiving federal money, approval, or other federal agency involvement.

Both Vermont (wastewater management) and Tennessee (transportation) have detailed and comprehensive environmental analysis requirements for specific programs. Although Vermont does not have a comprehensive law or regulation requiring environmental analysis for all proposed actions, if the state issues a grant or loan for wastewater management programs to a municipality, they are required to prepare a comprehensive environmental review procedure. The guidelines they have issued mirror the federal NEPA compliance guidelines very closely. This is not a coincidence because the U.S. EPA initiated the wastewater grant and loan program, and since providing funds to municipalities for wastewater management was a federal action with potential environmental impacts, NEPA fully applied. In Tennessee, the Department of Transportation's NEPA Documentation Office is responsible for the preparation of the environmental documents for transportation projects. The requirement includes collecting, compiling, and analyzing information on social, economic, and environmental areas for all transportation projects, including purpose and need; the natural, cultural, social and economic environment; farm land; land use; energy; conservation; hazardous waste; visual concerns; and construction impacts. The office conducts public meetings on the projects consistent with NEPA public disclosure requirements and guidance. Similar to the Vermont wastewater example, this was initially driven by the inclusion of federal funds and NEPA requirements that accompany the funds for highway projects.

Massachusetts and California have long-standing comprehensive environmental analysis programs legislated in 1972 and 1970, respectively. They have been amended several times, most often to expand jurisdiction and provide more detail based on lessons learned. Two programs are briefly summarized and compared with NEPA in the following sections.

8.6.1 Massachusetts Environmental Policy Act

Under the Massachusetts program, the project proponent first makes a determination whether the project falls within the jurisdiction of the Act (M.G.L. c. 30, §§61-62H: Massachusetts Environmental Policy Act; 301 CMR 11.00). This is achieved by determining whether the project requires a state license, permit, funding, or any other approvals by a state agency. If the action is proposed by a state entity, the Massachusetts Environmental Policy Act (MEPA) automatically applies and if it is proposed by a private concern, it is up to the proponent to determine whether any state approval or action (and thus MEPA compliance) is necessary for implementation. This jurisdictional definition is very similar to NEPA on the surface because it only applies to agency action; however, in practice it is much more inclusive. Almost any proposed development beyond that of a single family house will require water, sewer, access to a highway, etc., and in most cases these require approval by the state. Also in Massachusetts most, if not all, municipal projects (schools, library, highway improvement, water, sewer, etc.) include at least partial state funding, and thus MEPA applies. Similarly, an alteration in the use of any large parcel of land requires local planning approval, which is typically accompanied by a traffic management and highway improvement plan, requiring the Massachusetts Highway Department's approval, and thus MEPA applies.

MEPA expands and codifies the NEPA concepts of categorical exclusion (CATEX) and adds categorical inclusion. The categorical exclusion or inclusion requirement is stratified and refers the project proponent to threshold screening criteria. At the lowest level, the regulations specify types of actions where MEPA does not apply:

- Lawfully existing structure, facility, or activity
- Routine maintenance
- Replacement project
- Project consistent with a special review procedure previously subject to MEPA review

At the next level, if the proposed action is less intense than a lower threshold, no action or filing under MEPA is required. If it exceeds the lower threshold, an Environmental Notification Form (ENF) must be filed and if it also exceeds the upper threshold, a full Environmental Impact Report (EIR) must be prepared (Figure 8.2). Thresholds are identified in a number of different categories:

- Land
- State-listed species
- Wetlands, waterways, and tidelands
- Water
- Wastewater
- Transportation
- Energy
- Air

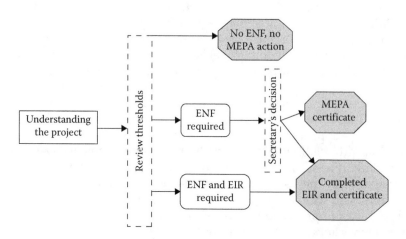

FIGURE 8.2
A threshold screening process under the Massachusetts Environmental Policy Act (MEPA).

- Solid and hazardous waste
- Historical and archaeological resources
- Areas of critical environmental concern
- Regulations

In each category, upper and lower thresholds are specified to determine the need to submit a MEPA ENF application or the need for a full EIR. The type of action generally determines the applicability and the magnitude of action is a major consideration mandating need for an EIR. Table 8.4 gives example thresholds for selected categories and the full list is included in Section 11.03 of the MEPA regulations.

If the proposal exceeds a threshold, the required ENF serves as the scoping document. The applicant prepares the ENF, which includes at a minimum the project description, alternatives, and the environmental resources potentially at risk. The ENF is not only submitted to the office of the state Secretary of Environmental and Energy Affairs but also noticed in the state *Environmental Monitor*, which is regularly perused by the local and statewide environmental advocacy groups and municipalities. The public and all state agencies that might be involved in approval, funding, and review of the action provide their comments (similar to NEPA scoping, see Section 4.3) to the secretary. If the proposed action falls between the upper and lower thresholds, the secretary takes the comments received on the ENF into account and issues a certificate either finding that the ENF fulfills MEPA (i.e., analogous to a NEPA CATEX) or an EIR is required. If an EIR is required, either through decision on the ENF or if the proposal exceeds the upper threshold, the secretary considers all the comments received and issues a certificate specifying the scope of the environmental review, comparable to a NEPA scoping statement. The statement under MEPA differs from the one under NEPA in that in the first case an independent regulatory entity, the secretary, determines the scope, whereas under NEPA the proponent (the federal agency proposing the action) makes the decision on the scope of the analysis.

There is also a provision for a "Fail-Safe Review" in cases where an ENF is not required under any of the criteria, but the project raises significant concern. One or more agencies or ten or more persons can petition the secretary to require the project proponent to file an ENF if the proposal does not meet one of the thresholds established in the Act, provided the following apply:

- It is subject to MEPA jurisdiction.
- It has the potential to cause damage to the environment.
- An ENF filing is essential to avoid or minimize damage to the environment.
- Filing will not result in an undue hardship for the project proponent.

TABLE 8.4
Examples of Massachusetts Environmental Analysis Thresholds

Category	ENF and Secretary Review	ENF and Mandatory EIR
Land	Direct alteration of 20 or more hectares Creation of 4 or more hectares of impervious area	Direct alteration of 10 or more hectares Creation of 2 or more hectares of impervious area Conversion of land held for natural resources purposes Conversion of existing agricultural land to nonagricultural land Release of interest in conservation, preservation, agricultural or watershed preservation purposes Approval of new urban redevelopment projects
State-listed species	None	Alteration of designated significant habitat Disturbance of 2 acres of priority habitat
Wetlands, waterways, and tidelands	Alteration of 1 or more acres of salt marsh or bordering vegetated wetlands; or 10 or more acres of any other wetland type Variance of Wetlands Protection Act requirement Construction of a new dam Structural alteration or expansion of an existing dam Issuance of Chapter 91 License for non-water dependent use of tidal lands greater than 1 acre	Provide a permit for major actions in jurisdictional areas New infrastructure to barrier beach Dredging of more than 10,000 m^3 Disposal of more than 10,000 m^3 of dredged material Issuance of Chapter 91 License for non-water dependent use of tidal lands less than 1 acre Construct or reconstruct a solid-fill structure of 1000 or more square feet
Water	Withdraw more than 2,500,000 gallons per day of surface water Withdraw more than 1,500,000 gallons per day of groundwater Construct new water main 10 miles or longer Provide water across a municipal boundary (except in emergency)	Withdraw more than 100,000 gallons per day requiring new construction Withdraw more than 500,000 gallons per day without new construction Construct new water main 5 miles or longer Construct or expand a drinking water treatment plant of 1,000,000 gallons a day or more Variance of Watershed Protection Act Non-bridge forest-harvesting stream crossing 1000 or less feet upstream of public drinking water supply

The project proponent has a number of options when preparing and submitting the ENF. If their project falls between the two thresholds, they can prepare an expanded ENF, which goes into great detail addressing the potential for impacts for each environmental resource possibly at risk. If they make a strong enough case that there will be little or no adverse impact, the secretary can find no need for an EIR and require the mitigation proposed in the expanded ENF in the certificate. If the proposed action exceeds the upper threshold and an EIR is mandatory, an expanded ENF can be used to dismiss certain areas of concern and thus limit the EIR scope issued by the secretary. An expanded ENF can also include a proposed scope for an EIR with the objective of focusing the scope issued by the secretary on the critical issues, and thus an attempt to structure and limit the EIR.

If the secretary determines an EIR is required based on the ENF and comments received, the certificate issued will include a very specific scope for the EIR. In cases where the state action is approving a permit, the agency responsible for issuing the permit generally issues a proposed EIR scope as part of their comments on the ENF. The secretary then uses the agency's proposed scope as a starting point, and adds to it issues raised in other comments. The required scope of environmental analysis is broad when the action is undertaken directly by an agency or involves state funding, but the scope is typically limited when the only state involvement is approval. Also, if there is only limited jurisdiction by a state entity, the scope normally only applies to the aspects of the project subject to some form of agency action.

In many cases, the EIR scope is focused on just a few issues such as traffic, and other environmental resources need not be addressed in the environmental analysis. Typically, a draft EIR and a final EIR are specified in the scope, but under certain circumstances where the scope is limited to just a few issues, the proponent can make a case for a final EIR only. Unless there are reasons uncovered during the ENF comment process, the EIR scope generally requires a standard format similar to that specified in Council on Environmental Quality guidelines for a NEPA EIS:

- Project description
- Alternatives to the project
- Existing environment
- Assessment of impacts
- Statutory requirements
- Mitigation measures
- Response to comments

The ROD required by NEPA is mirrored by the MEPA requirement for a "Section 61" finding. This finding specifically and concisely identifies the adverse impacts identified in the EIR and more importantly the project proponent documents their mitigation commitment for each identified impact. All state entities are then obligated to include and enforce the commitments in actions they take (e.g., issuing a discharge permit or approving a highway connection). This goes a step beyond the ROD and has proven to be an effective environmental protection measure.

MEPA institutionalizes the concept of enhanced scoping and public outreach (see Section 4.3.3) with the establishment of a Special Review Procedure. Under this procedure, there is a project-specific schedule with interim submittals, public hearings, review procedures, public participation program, and final submittals. The procedure also mandates a citizens advisory committee (CAC) with assigned roles and responsibilities. The intent of a mandated CAC is involvement of the critical stakeholders at every decision point and critical juncture of project development and environmental analysis. With the concurrence of the CAC, the normal requirements of MEPA can be waived and project-specific requirements adopted for such items as: submittal of final documents, only a single ruling by the Secretary of Environmental and Energy Affairs, review time constraints, etc. The Special Review Procedure is designed for large, complex, and potentially controversial projects. It allows the proponent and key stakeholders to reach an agreement on interim decisions before time, money, and energy are wasted in pursuing and analyzing an option or project element that has no public support. It also creates an environment of cooperation and can result in a project that is jointly designed and supported by the community and the proponent. The downside of the Special Review Procedure is the time and resources that must be committed by both proponent and stakeholders to achieve the benefits, so the pluses and minuses must be weighed carefully before embarking on this expanded and enhanced public outreach procedure.

The redevelopment of the closed Weymouth Naval Air Station southeast of Boston, Massachusetts is an example of both the Special Review Procedure and the interaction of state and federal environmental programs. The station was one of the last large tracts of undeveloped land in metropolitan Boston that was not designated for permanent open space preservation, and it was closed during a period of high demand for residential housing. Also, the infrastructure (including transportation, water, and sewer) in the surrounding area was already over capacity, so additional development created added pressure and much public concern. Thus, the situation was ripe for controversy with the competing interests of the pro-development community and the citizens of Weymouth, who were concerned about crowding, traffic, and a drain on municipal services. A Special Review Procedure and establishment of a CAC under MEPA were keys to addressing the potential for controversy before public opposition totally derailed the project. It also contributed to what ended up as a project with significant benefits to the community as well as the proponent, including:

- Enhanced and expanded commuter rail transportation
- Procurement of potable water from the excess capacity in the adjacent regional water purveyor

- Expanded bus service
- Environmentally sustainable storm water management system
- State-of-the-art wastewater treatment system
- In perpetuity, a guarantee of substantial open space and preservation of critical habitat
- Expedited cleanup of hazardous waste
- Proponent-supported infrastructure improvements
- State-supported traffic improvements

The coordination of environmental analysis at the state and federal level is another aspect of environmental analysis illustrated by the South Weymouth Naval Air Station case study. When the Navy closed the station, they were obligated to comply with NEPA and prepare an EIS because the closure was a federal action and had the potential to significantly affect the human environment. However, the ultimate use of the land following the Navy's cessation of operation was unknown, thus the EIS was very generic in their prediction of impacts, incorporating many assumptions with a high degree of uncertainty. As the entire situation was controversial and there was significant opposition to any proposed new use, either a public group in opposition or another federal agency with natural resource responsibility could have made a strong case that the Navy's EIS did not take a "hard look" at the impacts due to the necessity for broad assumptions and a high degree of uncertainty. However, any proposed action on the site required numerous state approvals, and the land was to be initially transferred to the state before it was turned over to private developers. Thus the Navy, other federal agencies (particularly EPA who was heavily involved in the hazardous waste cleanup and the MEPA process), and citizens groups understood there would be an extensive and highly scrutinized MEPA process before any action was taken. They understood that the MEPA process would fully meet the intent of NEPA, and challenging the action under NEPA would most likely be a waste of time and instead focused their efforts on MEPA. The result was consensus (not unanimous but very strong) and ultimately a much better process with benefits for all parties.

Enactment and implementation of MEPA reflects much of NEPA and the lessons learned. It builds on and expands the public outreach, scoping, and screening multilevel approaches (i.e., CATEX, EA, and EIS) of NEPA. It embraces the concepts of affected environment, impact prediction, and mitigation developed over decades of NEPA experience. It does differ in one very important aspect. MEPA is administered by an independent entity (Secretary of Environmental and Energy Affairs) with staff and authority to not only define the scope but also approve or reject the environmental impact analysis and ultimate the proposed action. In practice, the secretary pays close attention to comments by agencies such as the state Department of Environmental Protection, which must ultimately issue approvals or permits for the proposed action.

8.6.2 California Environmental Quality Act

The requirements, structure, and procedures for the California Environmental Quality Act (CEQA) (Public Resources Code 21000–21177 and the CEQA Guidelines, California Code of Regulations, Title 14, Division 6, Chapter 3, Sections 15000–15387) are modeled very closely on those of NEPA and it was enacted in 1970, just shortly after passage of the national policy. Similar to NEPA, the California equivalent applies to any action taken by any government entity with the action including approval or funding of a predominately private undertaking. CEQA also is based on a policy of encouraging environmental protection and integrating environmental considerations into any decision by a government entity that could potentially affect the environment. Also similar to NEPA, CEQA is based on an overall policy of harmony between humans and their environment, but CEQA is more specific and direct in the policy statement, which includes a requirement for all government agencies to take all action necessary to:

- Provide the citizens of California with clean air and water.
- Ensure enjoyment of aesthetic, natural, scenic, and historic environmental qualities.
- Provide freedom from excessive noise.
- Prevent the elimination of fish or wildlife species due to human activity.

Unlike most other state programs, under CEQA, local and regional governments and all their subdivisions (e.g., planning boards, fire departments, water districts) must comply with provisions of the Act. This provision greatly expands the number and range of responsibilities exercised by government agencies and other entities that are covered and thus the number of environmental analyses that are conducted every year. It also increases the breadth of activities covered by the Act and integrates environmental consideration into a much greater spectrum of the day-to-day life of California residents, which in turn increases their environmental awareness. For example, development of local zoning ordinances is covered by CEQA, which directly affects even small-scale development at the local level. The crafters of California's environmental policy broadened the scope of CEQA, compared to NEPA when it was first enacted, in 1970 and with subsequent amendments in anticipation of the exponential population and development growth predicted for the state. Their concern was justified, as California's population doubled, from 15 million to 30 million, during the first 15 years CEQA was in effect (Medvitz and Sokolow 1995). Proceeding unchecked, this growth and its associated impacts on open space and other environmental resources could have been devastating.

8.6.2.1 CEQA Process

Compliance with CEQA is the responsibility of each California agency or other government entity. Similar to NEPA (but different from MEPA, Massachusetts equivalent) there is no command and control agency policing compliance with the regulations and guidance. Thus litigation and the court of public opinion are the only checks on agency compliance. However, somewhat similar to NEPA's Council on Environmental Quality, the California Office of Planning and Research develops, reviews, and periodically revises the general guidance for CEQA compliance. Each government agency is then required to develop specific procedures compliant with the general guidance, which is adapted to their mission and specific conditions. The Office of Planning and Research has been diligent in preparing and updating the guidance, including the level of detail and specificity, which is unparalleled in any other state or national environmental analysis program. As a result, most other agencies have not seen the need to develop separate and agency-specific procedures or guidance and deferred to the Office of Planning and Research guidance.

The CEQA process starts when a government agency anticipates directly taking action, or becomes aware that they will have to issue an approval of a privately initiated action resulting in physical alteration of an environmental resource. The first step in the process is for the public agency to determine whether the anticipated activity is a "project" under CEQA, which they do by careful review of the 450 plus pages of CEQA and associated guidance (Association of Environmental Professionals 2012). If the activity meets one or more of the general criteria for coverage, the next step is to determine whether an exemption applies to the specific type of activity, location, or category of activities. This is no small task, as there are over 25 pages of exemptions and more are added every year. For example, the Olympics were exempted, and recent efforts have made the Act lenient on activities such as geothermal energy, which are considered environmentally sustainable. Also, there are a number of initiatives in California, such as thermal generation of electricity, which are covered by parallel programs with their own environmental analysis requirements and thus are exempted from CEQA in a manner similar to the Comprehensive Environmental Response, Compensation, and Liability Act (CERCLA) hazardous waste projects exemption from NEPA. Also similar to NEPA, there are provisions for categorical exclusions of certain actions from further CEQA action based on absence of significant impact as determined from past agency experience. If the activity is determined to be a "project" under CEQA, is not categorically excluded, and it is not exempt, they must then make a general determination as to the possibility of significant effects on the environment. If the activity clears this hurdle, then the CEQA process starts in earnest.

The government agency or other entity with the greatest anticipated role in the activity must be determined and designated as the lead agency. If it is a direct agency action or only one state approval is required, the designation of the lead agency is automatic. However, if multiple approvals by several agencies are required, a determination of the approval with the greatest potential impact must be made and the agency responsible for the approval assigned as lead agency. Alternatively, an agency with a lesser approval responsibility but expertise with the environmental resource at greatest risk could be designated as lead. Occasionally there is not overall consensus as to the appropriate lead, but the regulations specify a mechanism for resolution. The lead agency manages the CEQA process through all steps, up to and including project approval, similar to the lead federal agency preparing the EIS and issuing the ROD under NEPA.

The agencies with lesser authority or approval implications for the action are assigned the role of "responsible agency." These agencies have a vested interest in full and complete compliance with CEQA because they must ultimately issue a permit for the proposed action and demonstrate compliance with the Act if they are challenged. Thus, they must determine what approvals they must render and the information needed to make the decisions for their approval. They then weigh in during all steps in the process, as summarized immediately below, to ensure that the lead agency's compliance with CEQA meets these needs so that they can issue the permits under their jurisdiction.

The lead agency's first task is to prepare an "Initial Study" with the primary objective of determining whether a full EIR is necessary. The study is conducted with the information available by comparison with a checklist or matrix to determine whether there are sensitive environmental resources present or components of the proposed action have historically been associated with potentially significant impacts (this is a perfect opportunity to develop an Impact Prediction Conceptual Model, as described in Section 5.3.1). The lead agency does have the option of bypassing the Initial Study if it is obvious based on the magnitude of the action or the historical record that the action will ultimately require a full EIR. However, an opportunity to begin scoping and focusing the environmental impact analysis on the critical issues is not realized if the Initial Study is not prepared and reviewed. If the checklist approach identifies potential for impact, but there are accepted techniques to mitigate the impacts, they can be identified in the Initial Study. The Initial Study under CEQA is similar to the EA under NEPA in that it is used to determine the next step in the process (i.e., full analysis or authorization to proceed) but the limited level of analysis and preparation time for an Initial Study differs from the more comprehensive scope of a NEPA EA.

Completion of the Initial Study leaves the lead agency with one of two decisions: issue a negative determination if there is no significant environmental impact or prepare a full EIR. The negative determination reaches closure by determining that no further action is required under CEQA and the proposed action can proceed to implementation. It is comparable to a NEPA Finding of No Significant Impact (FONSI) following completion of an EA. Similar to a FONSI, the CEQA negative determination can be conditioned by imposing mitigation that must be accomplished when the action is implemented.

The lead agency issues the negative determination, including any mandated mitigation, but it is incumbent on all the responsible agencies to concur. The other agencies must base their issuance of approvals on the finding and are liable to litigation if all CEQA requirements and conditions are not satisfied by the determination. Thus, they have a vested interest in review and validation of decisions associated with the issuance of a negative determination.

If the action could cause significant impact, the lead agency makes and publicizes the decision to prepare an EIR. The lead agency is responsible for the preparation of the EIR, but they may request information, data, evaluations, or even funding from the project proponent if the agency action is approval of a private action. They may also adopt analyses done by others after comprehensive review and affirmation that they accurately reflect conditions applicable to the project under review. The CEQA requirements and procedures for an EIR follow very closely those for a NEPA EIS. Like NEPA, but unlike the Massachusetts equivalent (Section 8.6.1), scoping of the environmental analysis is the responsibility of the lead agency. They are required to actively involve other agencies that must ultimately issue approvals (which is similar to the Massachusetts program) and other stakeholders in the scoping process. But the lead agency makes the final decision on scope and under CEQA guidance must consider and include, or dismiss with cause, a laundry list of environmental resources and types of effects that are specified as part of CEQA.

The CEQA EIR follows very closely the structure and environmental impact analysis process described in this book. It includes a description of the affected environment, prediction of impacts, evaluation of alternatives, consideration of mitigation, and a decision at the end of the process. There are provisions for multilevel environmental analyses (see Chapter 6) including programmatic, tiered, master EIR, supplemental, focused EIR, and joint EIS-EIR (i.e., "piggyback"). There is some variation from NEPA in the areas covered, treatment of alternatives, and the basis for the decision and these are discussed in the following section.

8.6.2.2 Unique Features of CEQA

CEQA requires consideration of environmental quality in public decision making, like most other environmental policies, but California takes it a step further. Before the lead agency can approve a project and responsible agencies can issue permits, the EIR must demonstrate that there are no other alternatives that meet the purpose and need and have less impact. This requirement results in a substantially different perspective on alternative analysis than in other programs because development of alternatives can follow determination of impacts and focuses on ways to avoid or minimize impacts identified for the proposed action. Similarly, if the approved action is predicted to result in adverse impacts, there is a requirement that mitigation must not only be evaluated but imposed on the proposed action. The lead agency can impose restrictions on any area of impact, but responsible agencies are restricted in imposing limitations to areas directly under their jurisdiction. There are exceptions based on the practicality of the other alternatives and mitigation measures, but legislative history has placed a heavy burden on lead agencies to select the alternative with the least impact and impose stringent impact mitigation. If the lead agency approves a project that will result in significant environmental impacts, it must be accompanied by a written statement specifying the rationale for approval and supporting the decision by substantial evidence. The courts have placed a heavy burden on the agencies to satisfy this condition of approval.

This mandate for the alternative with least impact and significant mitigation except in extreme circumstances has produced a much more command-and-control approach to environmental sustainability than experienced in other jurisdictions. This in turn has created on occasions both a high degree of environmental sustainability and significant opposition to the regulations. There is currently a move in the California legislature to "streamline" the regulations, which environmental advocacy groups interpret as weakening the protection provided by CEQA. In light of the 2008 economic downturn, new procedures were enacted to lessen CEQA requirements for projects that generated significant economic stimulus and associated employment as long as specific thresholds are satisfied.

CEQA also contains a provision that somewhat narrows the scope of environmental review compared with other programs. The California environmental regulations only apply to "physical alteration" of environmental resources, thus excluding social and economic conditions. This provision has the greatest influence when the lead agency reaches a conclusion as to significant impact and issuance of a negative determination. If changes in social or economic conditions could generate secondary impacts causing physical alteration (such as development pressures encroaching on environmentally sensitive open space), they can be considered by the lead agency in making a decision regarding significant impact.

Not only does CEQA require mitigation of significant impacts as discussed earlier, it mandates a monitoring or similar program to ensure that the measures are effective. The EIR and associated notice of determination must not only identify the mandatory monitoring program but specify that issuance of any environmental approval or permit must include the monitoring requirement. Many other comprehensive environmental programs accomplish the same objective with the integration of analysis and permits (see Chapter 9), but CEQA institutionalizes the process and ensures that commitments made to placate the opposition during the analysis phase are realized.

One of the most striking and differentiating features of CEQA compared with all other programs is the length and detail of the legislation and implementing regulations. Over 450 dense pages, which seem to cover any potential specific situation that could be encountered, sometimes seem overwhelming. For example, there are specific regulations and procedures if a project affects an oak woodlands system (§21083.4). Among other requirements, if the project could affect such an ecosystem, the affected county must require one or more of the following mitigation alternatives or measures:

- Conserve oak woodlands through the use of conservation easements.
- Plant an appropriate number of trees, including maintaining planting and replacing dead or diseased trees.
- Planting shall not fulfill more than half the mitigation requirement.
- Contribute funds to the Oak Woodlands Conservation Fund.

California has legislated one of, if not, the most demanding and restrictive environmental policies and associated regulations. This has produced enhanced environmental protection and sustainability but also opposition to CEQA. The major complaint is the potential to slow development and economic growth both through the constraints contained in the Act and complexity and expense of the required procedures. However, a case can certainly be made that these complaints are exaggerated because California's population and associated economic growth in the first decade since the passage of the Act doubled, which would be the envy of any other development-oriented state.

8.7 Summary

Most international and U.S. state (with California as an exception) environmental legislation or regulatory efforts began with some version of "NEPA Light." They generally maintained the NEPA concept of limiting jurisdiction to actions by the government, including funding, construction, land use, and permitting. Many countries and states observed the frequent litigation and resulting significant delays in project implementation resulting from NEPA in the early years and purposefully crafted their environmental controls to preempt litigation or failed to enact comprehensive environmental analysis policies. Avoiding litigation was sometimes accomplished by administration procedures, rather than legislation, to require environmental analysis. In other programs, environmental review was included in existing legislation to minimize litigation and project delays. Where this occurred, the approaches frequently limited enforcement, consistency, and effectiveness resulting in incomplete analysis or noncompliance. Also, incorporation of environmental review under existing, often agency-specific, legislation and environmental review as part of an agency's administrative procedure with no broader requirement, created potential for conflict of interest. With the same entity proposing an action and having sole responsibility for environmental review, there can be pressure to avoid a hard look at environmental concerns that could delay, alter, or prevent project implementation. While NEPA and state programs like California inherently have these same limitations, the practice of public disclosure and litigation largely negate these constraints. With time, many countries recognized the weakness and marginal achievement of environmental protection using existing legislation and administrative procedures (Clark and Canter 1997). The resultant experience from their own environmental regulations and the maturation of NEPA has led many countries and states to adopt modified and refined EIA systems, and it is remarkable how much they have come to resemble the full NEPA model.

One difference compared with NEPA is that the environmental analysis requirements in many situations identify specific thresholds (e.g., length of roadway, volume of water withdrawn, size of development) and specific types of action (e.g., power plant, airport expansion) requiring various levels of environmental analysis. NEPA does not specify such thresholds or project types but relies on a more ambiguous likelihood of a "significant environmental impact" caused by an agency action to determine whether and what type of environmental analysis is required. However, U.S. federal agencies do take advantage of the CATEX provision in NEPA guidance (see Section 3.1.2) to accomplish what other countries have achieved by including thresholds and specific projects mandating analysis in their legislation.

The inclusion of environmental analysis and environmental approval/permits is another concept that is integral to most environmental policies and programs. The integration is institutionalized in many programs, like CEQA, and less formal in others, such as NEPA. But the combination of the two provides multiple tools for addressing environmental protection, even if there is opposition or inattention to one of the two as is in the EU at times.

California probably has the most comprehensive and environmentally protective environmental impact analysis policy and regulations. Approval based on no alternative with less impact, stringent mitigation requirements, and monitoring of mitigation has produced significant environmental sustainability. However, it has come at a cost of extreme complexity in regulations and procedures, which in turn has generated significant opposition.

References

Association of Environmental Professionals. 2012. California Environmental Quality Act statute and guidelines. Available from: http://leginfo.ca.gov.

Bass, R.E. and A.I. Herson. 1993. *Mastering NEPA: A Step-by-Step Approach*. Point Arena, CA: Solano Press Books.
Clark, R. and L. Canter. 1997. *Environmental Policy and NEPA*. Boca Raton, FL: St. Lucie Press.
EC. 1985. Council Directive of 27 June 1985 on the assessment of the effects of certain public and private projects on the environment (85/337/EEC). *Official Journal of the European Communities*, No L 175/40, 05.07.1985.
Ekmetzoglou-Newson, T. 2006. The EIA directive. In: *EU Environmental Policy Handbook: A Critical Analysis of EU Environmental Legislation*. ed. S. Scheuer. 228–236. Brussels: European Environmental Bureau.
Hey, C. 2006. EU environmental policies: A short history of the policy strategies. In: *EU Environmental Policy Handbook: A Critical Analysis of EU Environmental Legislation*. ed. S. Scheuer. 17–31. Brussels: European Environmental Bureau.
Medvitz, A.G. and A. Sokolow. 1995. Can we stop farmland losses? Population growth threatens agriculture, open space. *California Agriculture* 49: 11–17.
Montana Legislative Environmental Quality Council. 2000. Final report to the 57th Legislature of the State of Montana. Environmental Quality Council, Helena.
Scheuer, S. 2006. Introduction. In: *EU Environmental Policy Handbook: A Critical Analysis of EU Environmental Legislation*, ed. S. Scheuer. 7–11. Brussels: European Environmental Bureau.
Wood, C.M. and J. Bailey. 1994. Predominance and independence in environmental impact assessment: The Western Australia model. *Environmental Impact Assessment Review* 14: 37–59.

9

Coordinating and Managing the Environmental Impact Analysis Processes

9.1 Introduction

Environmental impact analysis is not conducted in isolation but must be coordinated with other activities in order to meet the stated purpose and need. Following environmental analysis, including the selection of an alternative, the proposed action proceeds to detailed development that requires engineering (or policy formulation) details and environmental approvals. Acknowledging the need for these details and integrating them into the environmental analysis at an early stage in the process expedites the process and ensures successful implementation. The coordination with final development, such as engineering design or policy and plan details can often be simplified in the environmental analysis by specifying the commitment to environmental impact mitigation and requiring that they be fully included in project design and other implementing procedures, similar to the Section 61 Finding process in Massachusetts Environmental Policy Act (MEPA) described in Section 8.6.1. However, the environmental approvals are a different matter and must be fully integrated and addressed in the environmental analysis. The importance of this integration has been stressed in Chapters 4 and 5 but is revisited in this chapter, building on the approach and implementation of environmental impact analysis presented in the previous eight chapters.

Another challenge that can derail the proposed action before it even approaches the implementation phase is the management of the environmental analysis process. One of the most common and often fatal impediments to successfully managing the process is taking too long and costing too much. There are basic project management concepts and methods that can be applied to environmental analysis, which greatly increase the efficiency of the analysis, and support on schedule and on budget completion. Some tools that have proved useful in making environmental analysis efficient are summarized below under separate headings to give the environmental practitioner an appreciation of effective project management.

9.2 Integrating Environmental Approvals and Analysis

Environmental protection under the National Environmental Policy Act (NEPA), as well as other programs as discussed in Chapter 8 has evolved into a two-pronged approach. The first is the analysis phase, as described in the preceding chapters, and is aimed at understanding the environmental implications of a proposed action before they occur. This phase, when conducted properly and comprehensively, can do much to understand the impacts and lay the groundwork for environmental protection. But just analyzing and understanding the impacts alone does not guarantee full environmental protection. The second phase, environmental approvals and permits, is a companion to environmental analysis, and when fully coordinated, the two can maximize environmental protection.

The primary goal of environmental analysis is to understand the consequences of an action prior to implementation, but the intent of environmental permits is to protect the resource. The European Union (EU) has almost by default developed a system of both analysis and resource protection (see Section 8.4) with a tiered approach to analysis and a series of resource-specific policies, goals, and protection. However under the EU program, the resource policies and regulations are only weakly enforced and there is no strong incentive or mechanism to closely link the analysis and environmental approval process. Thus the potential environmental protection and impact avoidance have not been maximized under the EU system, but there is a strong movement in that direction. Under NEPA and individual state programs in the United States, there has been more success in integrating the two approaches to maximize protection and ensure adherence to environmental policy. The integration starts with the analysis phase, but prior to a discussion of integration approaches, a summary of concepts and examples of environmental permits and approvals is presented here to put the integration in perspective.

9.2.1 Environmental Permits and Approvals

Environmental legislation defining the need for permits or other approvals not only establishes the policy for resource protection, but it also mandates the development of regulations. These regulations lay out specific procedures to follow before an activity potentially affecting an environmental resource can be approved and moved to implementation. In contrast to environmental analysis, which is broad, comparative, 100% completed prior to implementation, and generally does not mandate specific actions to achieve environmental protection, the environmental permitting process:

- Establishes specific conditions that must be met to protect an environmental resource
- Sets rules that must be followed

- Includes monitoring requirements to ensure the rules and conditions are met
- Establishes monetary and sometimes criminal penalties for violations
- In most cases must be periodically renewed or at least reviewed based on monitoring results
- Ensures environmental protection, usually in a very narrow area, such as dissolved oxygen levels or square meters of wetlands resources that can be altered

Some environmental approval programs are resource specific and some are activity specific, but all are directed at protecting a specific environmental resource. The Endangered Species Act (7 U.S.C. §136, 16 U.S.C. §1531 et seq.) and the Migratory Bird Treaty Act (16 U.S.C. §§703–712) are examples of environmental approvals that are resource specific, and the implementing regulations for such Acts specify the level of effects allowable on these resources from any proposed action. In contrast, activity-specific permitting regulations have established through scientific research, engineering feasibility, practicality, or some combination a given intensity level that will not threaten an environmental resource. The Clean Water Act (33 U.S.C. §1251 et seq.), Section 404 is an example of an activity-specific environmental permitting program addressing activities in wetlands. After research, public input, compromises, and evaluation of practicality, the U.S. EPA and Army Corps of Engineers who jointly administer wetlands protection under the Act have established maximum allowable activities affecting wetlands (e.g., square meters of fill allowed). Similarly the U.S. EPA has established secondary treatment of municipal sanitary waste as the minimum requirement under Section 401 of the Clean Water Act (see Section 5.3.3 for a full discussion of the limitation).

Whether an environmental approval addresses an activity or resource, it is based on an objective and typically scientific consideration of the resource's requirements and sensitivity. For example, approval under the Endangered Species Act is based on research determining the population levels necessary to ensure the species survival and the number of individuals that can be "taken" without threatening the population. For an individual proposed action, the information on the local population is developed through a location or activity-specific Biological Assessment, but the final approval is also based on comprehensive research determining sustainable population density, habitat characteristics, food consumption, nutrition requirements, etc.

The concept of developing standards protective of the environment and then basing permit approvals on meeting those standards has evolved into an effective environmental protection system. Many diverse environmental approval procedures including the Safe Drinking Water Act (42 U.S.C. §300f et seq.), Clean Air Act (42 U.S.C. §7401 et seq.), Historic Preservation Act (16 U.S.C. §470 et seq.), and as cited above the Endangered Species Act follow the model to some degree, but the Clean Water Act is probably the best example of this system. Section 304 of the Clean Water Act requires the U.S. EPA, which administers the Act, to research the effects of pollutants on the aquatic community and promulgate water quality criteria. These criteria define the level of pollutants (e.g., metals, pesticides) aquatic organisms or aquatic ecosystems can tolerate or the level of essential characteristics (e.g., dissolved oxygen) necessary for the community to thrive. These water quality criteria are based on decades of research evaluating toxicity and species requirements for over a hundred different compounds on dozens of different organisms, in numerous environmental settings. Section 303 of the Clean Water Act requires states to promulgate standards based on the scientifically developed criteria, and in many cases, the toxicity tests and resulting water quality criteria developed by the U.S. EPA are adopted directly by the individual states as enforceable standards. Sections 301, 302, and 402 of the Clean Water Act establish the National Pollutant Discharge Elimination System (NPDES) discharge permit program, which controls water pollution by regulating sources that discharge pollutants into waters of the United States.

Before Section 304 and State water quality standards were fully developed, the NPDES permitting approach was to specify the level of wastewater treatment necessary to protect the waters of the United States. For example, municipal wastewater treatment systems were required to achieve a minimum of secondary treatment (i.e., physical followed by biological treatment) with specific discharge limits such as a maximum of 30 mg/l biochemical oxygen demand or concentration of suspended solids. However, as the water quality necessary to protect aquatic life was defined, the reissued NPDES permits became increasingly directed towards meeting the water quality criteria and standards in the receiving waters. Thus discharges had to develop a combination of waste elimination and wastewater treatment processes such that once the discharged treated effluent was mixed with the receiving waters, the water quality standards were not exceeded.

Not all environmental approvals link the resource protection requirements and approval so closely to scientific based research, but to some extent all are based on resource protection. The basic philosophy behind environmental protection via permits and approvals is that an environmental resource can withstand some degree of alteration without compromising the value, service, sustainability, or other benefits provided by the resource. However, once that level of alteration is exceeded, the resource can no longer remain self-sustaining and provide the intended value and service. Thus a precursor to any set of environmental permit requirements is to establish the acceptable level of effect that will not compromise the sustainability of the resource, such as the water quality standards required under the Clean Water Act. These are then incorporated into the NPDES permit issued under the same act to achieve the desired protection of the resource.

Frequently a piece of environmental protection legislation (including the Acts discussed above) is directed at protecting a single environmental resource. Under the legislative umbrella there can be many approvals and permitting requirements to protect different aspects of the resource. Again using the Clean Water Act as an example, there are approvals required for multiple activities such as dredging and filling, waste water discharges, and storm water management. Similarly, under the Safe Drinking Water Act (42 U.S.C. §300f et seq.) there are requirements for disinfection, storage of finished water, and acceptable levels of chemicals and pathogens in potable water.

9.2.2 The Role of Environmental Permits in Environmental Analysis

A primary role of permits in environmental analysis is the establishment of impact significance criteria (as discussed above in Section 5.3.2). Most environmental permits and approval procedures include very specific standards or requirements and these can form metrics against which to judge an alternative's degree of impact during the environmental analysis. Thus, identifying what approvals may be required and the standards that must be met for approval must be an early step in the analysis and built into the scope and impact prediction steps in the analysis. For example, if achievement of the purpose and need requires a wastewater discharge or air emission approval, the requirements of the permit can form the basis for the impact significance criteria for aquatic and air-quality environmental resources. If wetlands are involved, the allowable square meters of fill allowed can define the threshold of a significant impact.

Other areas of analysis and environmental approval integration are screening, development, and comparison of alternatives. If an alternative directly affecting an environmental resource protected by legislation does not meet the criteria established for an approved permit, it can be screened out early in the process. Similarly, the permit requirements can be used to establish broader screening criteria. For example, if the purpose and need requirement is for sanitary wastewater treatment for a municipality, and the NPDES permit requires secondary treatment, all wastewater treatment sites too small to construct the required level of treatment can be eliminated at the screening stage. The permit requirements can also be major factors in developing alternatives by defining the level of performance, degree of treatment, and other requirements.

The comparison of alternatives can benefit from integration of permit requirements into the environmental impact analysis in two ways. The measurement of an alternative against the impact significance criteria based on permit requirements puts the comparison on a common basis across different alternatives and environmental resources (Figure 5.12). Also, the complexity, duration, and likelihood of success of permit approval can be a major selection factor in the alternative comparison.

Identification of key stakeholders is the other area where environmental permitting comes into play during the impact analysis process. If a permit is ultimately required, representatives of the issuing authority should be contacted directly during scoping to get their early input on scope of the analysis and identification of alternatives. It is also standard practice to have such agency representatives on the technical advisory committee to ensure methods satisfactory to subsequent permitting activities are employed and get an early read on the permitability of various aspects of alternatives. Of course, having full environmental analysis participation and cooperation of those who will ultimately issue the permit will expedite and simplify the permitting process for the proposed action.

9.2.3 Environmental Analysis in Support of Environmental Permitting

Many of the requirements and procedures for permit approvals are based on environmental evaluations in addition to meeting standards, and the comprehensive environmental impact analysis conducted before selecting a proposed action can fulfill these requirements. Thus, early identification of the permits is necessary and any evaluations required for issuance of the permits included in the comprehensive environmental impact analysis process is essential. Similarly as pointed out earlier, involvement of the permitting personnel in scoping and other aspects of the analysis process is critical.

Again the Clean Water Act provides perhaps the best example of the overlap in analysis and permit activities. Section 404 of the Act identifies wetlands protection, and Section 404(b)(1) identifies some of the specific conditions that must be met to obtain a permit for activities potentially affecting wetlands. Under Section 404(b)(1) no discharge of fill material into wetlands shall be permitted if:

- There is a practicable alternative to the proposed discharge which will have less adverse impact on the aquatic ecosystem.
- It violates state water quality or toxic effluent standards.
- The fill jeopardizes the continued existence of an endangered or threatened species.
- It violates the protection of any marine sanctuary.
- It will cause or contribute to significant degradation of the waters of the United States.
- Appropriate and practicable steps have not been taken which will minimize potential adverse impacts on the aquatic ecosystem.
- The proposed action, as determined by the U.S. Army Corps of Engineers, is water dependent and requires siting in waters of the United States.
- An upland alternative does not exist which will involve less wetlands impacts.

Most, if not all, of these provisions are embodied in the methods and procedures which must be conducted for a technically sound and defensible environmental impact analysis. The primary goal of environmental analysis is to compare the impacts of alternatives and incorporate environmental consideration into decision making; this is mirrored by the 404(b)(1) requirements to consider any "practicable alternative... which would have less adverse impact" and an "alternative... which will involve less... impacts." The requirements to avoid violation of state water quality standards or threaten special status species are prime examples of impact significance criteria employed as part of a well-planned and executed environmental impact analysis. Also the assessment concept of mitigation is included in the 404(b)(1) requirements and should be accomplished during the analysis phase rather than after the fact during application for the wetlands permit. Thus, including these concepts and procedures in the environmental analysis and making the required demonstrations for the proposed action is a tremendous head start on acquiring the 404 wetlands permit. Alternatively, neglecting these critical thresholds and demonstrations in the environmental analysis can lead to the necessity to revisit the entire environmental impact analysis process as part of the 404 permit procedures.

The necessity to integrate permits, particularly Clean Water Act permits and analysis is demonstrated by the King William Reservoir EIS/404 Permit in Newport News, Virginia (see Section 6.4 for additional discussion of the project). Development of the reservoir required both an environmental impact statement (EIS) and a 404 wetlands permit because it involved federal action (i.e., issuance of a permit) and affected approximately 160 hectares of valuable wetlands. The analysis and permit application processes were decoupled because the purpose and need changed between the issuance of the EIS and the permit application. The volume of water needed decreased by approximately half between the EIS and permit application because water conservation practices were above predictions and population growth was below predictions presented in the EIS. Thus alternatives were dismissed in the EIS because they could not supply the projected water need, and as ruled by the courts (*Alliance to Save the Mattaponi, et al. v. U.S. Army Corps of Engineers, et al.*, U.S. District Court for the District of Columbia, Case 06-01268 HHK; March 2009), the EIS analysis did not support the alternative analysis and other considerations required under 404(b)(1). Thus an entirely new alternative development, impact analysis, and comparison were required for the wetlands permit, and if it reached a different conclusion, a supplemental EIS would be required. Since the project proponent had already spent over 25 years and $50 million on the planning and environmental analysis, they took this court decision as the death knell and abandoned the project.

The "Big Dig" in Boston, Massachusetts, represents an interesting, although extreme, example of environmental analysis and permitting integration. The project was the largest public works project at a single location that was ever envisioned in the United States. The planning, environmental analysis, design, and construction spanned three decades, and the final cost was approximately $16 billion. The purpose for the project was to replace the aging and deteriorating elevated highway through the center of Boston and relieve congestion in the two tunnels under Boston Harbor that serviced Logan Airport. The need was to relieve traffic on the major artery entering the city from the north, mitigate the division of the city by the elevated highway, and create open space in downtown Boston.

The environmental analysis, which was an EIS under NEPA because of the large percentage of U.S. Department of Transportation funding, took several years and was forced to address a largely conceptual project description. The description lacked detail because much of the engineering and construction was cutting edge and the exact implementation methods, and in some cases, exact location could not be determined until well into engineering design. The analysis considered a full suite of alternatives, including mass transit, various alignments, and conceptual construction techniques but was short on details, such as traffic management during construction and specific description of excavated material handling and disposal.

The proposed action resulting from the analysis and comparison of alternatives consisted of two primary components. The first and largest was the destruction and replacement of approximately three kilometers of elevated highway that divided the city and isolated the neighborhoods. The replacement was to be a depressed central artery constructed by an open cut trench through the center of the city, hence the affectionate name Big Dig. The second component was a new, Third Harbor Tunnel connecting the Massachusetts Turnpike directly to Logan Airport. The proposed action also included significant mitigation:

- Expanded mass transit
- A linear park over the depressed artery with an in-perpetuity guarantee of no structures over the artery
- Connections to walking/biking trails in the city
- Very specific constraints addressing disruption of surface traffic during construction

Not unusual for an endeavor of this scope and magnitude, several dozen environmental permits were required to implement the Big Dig and these were identified in the EIS. However, in many cases the permitability of many of the individual elements of the project could not be specifically evaluated due to the lack of engineering detail at the environmental analysis phase. To compensate for this lack of detail and permissibility analysis, the EIS included commitment to a permitting plan and approach which was supported by the state, local, and federal permitting agencies. Part of the plan was to have a dedicated permitting staff as part of the Big Dig engineering/construction team and an assigned staff at the permitting agencies to work closely with the project proponent. The approach worked well and eventually the project was issued all the necessary environmental permits. There were conditions built into the permits to largely address the mitigation and impact reduction commitments made in the EIS.

The head of the Big Dig environmental permitting team was addressing a graduate class on Methods in Environmental Impact Analysis specifically on the topic of integrating permitting and environmental analysis. After over two hours of describing the project, the EIS, and the permitting effort, he asked the class to comment on whether the environmental impact analysis or environmental permitting effort afforded greater environmental protection and enhancement. The class was equally divided with those favoring the analysis for environmental protection citing the inclusion of open space, improved mass transit, parks, and trials as benefits that would never have been included in the Big Dig as part of permitting. Those favoring permitting identified the impacts avoided from individual actions because of the imposed permit limits as the major environmental protection mechanism. For example, strict testing and disposal specifications were imposed on the excavated and dredged material generated by the project. Also there were very specific conditions put on the management of exhaust from the tunnel as a result of the Air Quality Act permits.

After listening to and commenting on the student's consideration of permitting versus analysis, he gave his own strong opinion, formulated over more than five years of living the project every day. Although he acknowledged both were important and the integration of the two was perhaps more important, his conclusion was that permitting achieved far and away the greatest environmental protection. His rationale was:

- Not enough detail was available during analysis to specify environmentally sound construction methods, and a detailed description of the methods was required for permit applications.
- Many of the permits required monitoring following construction to ensure environmental standards were achieved, as opposed to the EIS which made predictions but there was no follow-up.
- As the project was implemented, there were many changes from the original conceptual plan and potentially significant environmental impacts resulting from these actions, as well as new impacts generated from specific design and construction methods that were not identified in the EIS. The permitting process included provisions for simple revisiting and "project modification," whereas the EIS could not readily accommodate changes.
- Since the permitting process was working well, there was not a reverberating cry for a supplemental EIS when change was necessary, which would have further delayed the already protracted and over-budget Big Dig.

The parallel permitting and analysis processes must be integrated as part of comprehensive environmental protection as well as successful

TABLE 9.1

Environmental Analysis and Permits: Comparisons and Contrasts

Environmental Impact Analysis	Environmental Permits and Approvals
What to Do?	*How to do it.*
Decisions	Control
Agency actions (under NEPA)	Agency approvals and enforcement
No penalties for violation	Adherence dependent on threat of penalties
Open and flexible technical requirements (dependent on scoping)	Strict standards and requirements must be met
One time only, prior to implementation	Monitoring and renewal following implementation
Unrestricted scope	Limited to resource or activity
Relation to permits	*Relation to environmental impact analysis*
Must list and describe permits needed	Frequently must rely on alternative analysis and data collected during the environmental impact analysis
Permitability as decision criteria	Permitting agency exerts influence on decisions
Permit limit as maximum impact level	Incentive for impact mitigation
Vehicle for conflict resolution	Proponent knows requirement in advance

project, policy, and plan implementation. The two complement each other, yet have important differences (Table 9.1). The relationships between these two prongs of environmental policy must be identified early in the planning process, considered in the scope of the environmental impact analysis, and periodically reviewed as the analysis progresses. Also the environmental impact analysis document should be at hand and consulted closely during the permitting and approval process for the proposed action.

9.3 Managing the Environmental Impact Analysis

One of the largest and often most deserved criticisms leveled at environmental analyses is that they take too long and cost too much. As discussed throughout this book (particularly Section 4.3 and Chapter 6) much has been done from a regulatory and guidance standpoint to address this issue and there has been marked success. However, the environmental analysis practitioners are on the frontline conducting the day-to-day management and implementation of the analysis, and their activities directly and profoundly affect schedule and resource expenditure.

A first step for practitioners in maximizing efficiency is to recognize some of the common failings and flaws that have historically been at the center of delays and budget overruns for the analysis. The environmental analysis by necessity and design is an interdisciplinary approach, typically dominated by scientists and planners. These scientists (particularly biological scientists) have a well-earned reputation of curiosity and a desire to know every detail about how things are and how they work, which is a laudable attribute in research but can run amuck in environmental analysis. If the scientists are managing the environmental analysis, they must develop basic project management skills and apply them to the effort, or the direction and control of the analysis will be assigned to engineers or managers with the required training. When this happens, important career opportunities can be lost and the environmental practitioners with the skills and understanding of the environmental issues do not have a seat at the table when important decisions affecting the environment are made and then implemented. Thus developing management skills should be included in the career development path of environmental analysis practitioners with appropriate skills and a desire to increase the influence of their work.

It is far beyond the scope of this book to convey proficiency in project management. But summary descriptions of some of the basic management tools and processes applicable to environmental impact analysis can highlight the importance, benefits, and rewards of an efficiently conducted analysis. The summary description should also make it obvious that successful project management is 90% common sense and thus not the exclusive purview of business and engineering professionals. Experience has demonstrated that many basic management tools are particularly helpful in conducting environmental analysis and their use within the capabilities of environmental analysis practitioners is summarized below.

Probably the most important axiom in project management is adherence to the scope of work. If the effort strays beyond the stated, authorized, and funded scope and the project exceedance of budget and schedule is assured, it is just a question of how much. Most of the following sections describe tools and techniques to ensure the environmental analysis effort stays within the scope, but prior to employing these techniques, there are other efforts which can be used successfully to stay on schedule and increase resource efficiency. These have to do with focusing and limiting the scope of work for the environmental impact analysis before work even starts in earnest, and these tools are discussed first.

9.3.1 Focused Scope of Work and Truncated Analyses

Limiting the scope at the outset is an obvious, important, and useful tool in reducing the budget and time required for the environmental impact analysis. Or alternatively, if there is a fixed budget and schedule, limiting the scope allows the focus and concentration of time and effort on the issues that are most critical. As described in Section 4.3, scoping is a key tool in limiting and focusing the scope. Understanding, both from a social and technical standpoint, which issues are of greatest concern or have the potential for the most significant impact, should receive the greatest effort and time. If a topic or environmental resource, such as community services or cumulative impacts, has not been identified as a concern during scoping, any proposal to spend more than a few hours on the topic must be met with skepticism, and the rationale to address the issues in any detail should be fully justified.

Another technique to focus the effort is maximizing the use of senior consultants. For many standard resources and sources of impact, individuals with years of experience can be recruited to provide advice early in the project planning. They have most likely dealt with the same issues in numerous similar situations and know in advance, or think they know, within a relatively narrow range, the full impact prediction and the environmental analysis results. Sharing this prediction with the environmental analysis team can provide very useful shortcuts and increase efficiency because starting with an impact prediction to prove or disprove is very often an easier task than starting with a blank sheet of paper. However, this technique must be used with extreme caution and the following pitfalls must be avoided:

- The team members must still do the full analysis to confirm or modify the predictions and not rely blindly on the senior consultant's prediction.
- The senior consultants are not being asked to do the analysis (which is sometimes the tendency of technically oriented team members) but just share their past experience. If the typically higher paid senior consultant does the analysis, the drain on the budget will be more severe and any efficiency gained from their prior experience could be lost.
- Senior consultants can make wrong predictions, and proving or disproving them is an important and efficient outcome of this technique. The manager must avoid "bending" the full analysis to match the early prediction intended to give guidance.

Conventional alternative (Section 4.5.3) and resource (Section 5.2.1) screening can be maximized early in the process to significantly reduce the scope of evaluations and predictions required for the environmental analysis. It thus constitutes another effective tool in increasing the efficiency and decreasing the time required for the environmental analysis. Also establishing criteria constituting a fatal flaw (such as economic viability or proximity to active endangered species habitats) is a quick and efficient method of eliminating an alternative before any environmental analysis effort or time is expended. Another technique to enhance screening is an early side-by-side alternative comparison, and when one alternative is found to be less advantageous than another in at least one category and not more advantageous in any other category, it can often be eliminated from further consideration. For example, for the U.S. Coast Guard dry cargo residue management EIS, use of street sweepers was found to be more expensive and operationally complex than using a broom and shovel and provided no advantages. Thus it was eliminated from detailed consideration before any effort or time was expended determining exactly how much they would cost to operate and maintain, the efficiency of the machines, on-deck storage requirements, safety considerations, etc.

Since each resource is evaluated for each alternative, eliminating a resource or alternative from consideration has cascading efficiencies. For example, a detailed environmental impact analysis of six alternatives and eight environmental resources of concern could require a total of up to 48 separate descriptions each of affected environments, impact prediction, and impact mitigation (for a total of 3*48 or 144 sections in the document). A screening process that eliminates just two alternatives and a similar number of environmental resources of concern will require only half the descriptions and sections in the analysis document (4*6*3 or 72). This can have a major positive effect on the budget and time required for the analysis or alternatively allow more resource availability for focus on the critical issues.

Frequently, during the course of the environmental analysis, the predicted impact, acceptability, or fatal flaw of an alternative will become apparent before the analysis is 100% complete, and this information can be used to truncate the rest of the analysis for the alternative, at least for some areas of impact. If preliminary impact prediction for one alternative demonstrates that it is an order of magnitude (i.e., 10 times) above or below all other alternatives or significance criteria, it may not be necessary to further refine the analysis. For example if a rough estimate demonstrates that the a wastewater treatment alternative will result in copper, cadmium, and lead concentrations between 100 and 500 µg/l in the receiving waters, and the significance criterion (i.e., water quality standard) for these heavy metals is between 1 and 3 µg/l, there may not be any need to further refine the analysis. It may also be possible to eliminate the alternative from further consideration on this single and simple analysis without expending any additional effort. It would defy common sense to spend dozens of hours to calculate the energy required for a wastewater treatment process that could not achieve acceptable standards and thus could never be implemented.

9.3.2 Project Plan and Delegation of Work by Discipline

Another important axiom of project management is that if you don't know where you are going and if you do not have a plan to get there, meeting your goal is just a random chance. So a plan for the environmental impact analysis is critical, and the scoping statement (Section 4.3.3) for the analysis is the first step in developing the plan. The scope identifies all the environmental analysis tasks and the methods to be employed, and the project plan takes management of the environmental analysis to the next level. The plan identifies not only what and how but also who and when. The plan lays out all the tasks, identifies their start and completion dates, assigns a responsible person, and specifies the level of effort for each task (Figure 9.1 presents a simplified example).

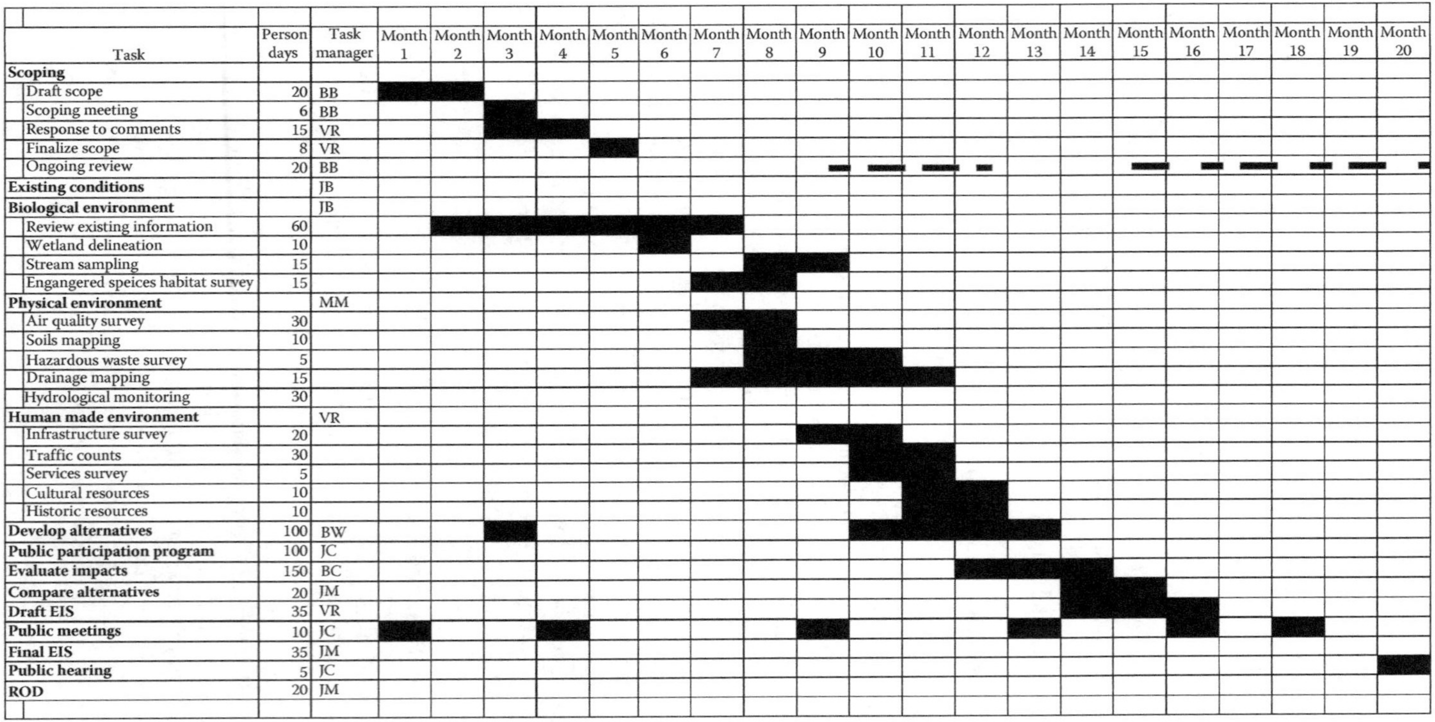

354   Environmental Impact Analysis

Task	Person days	Task manager	Month 1	Month 2	Month 3	Month 4	Month 5	Month 6	Month 7	Month 8	Month 9	Month 10	Month 11	Month 12	Month 13	Month 14	Month 15	Month 16	Month 17	Month 18	Month 19	Month 20
Scoping																						
Draft scope	20	BB																				
Scoping meeting	6	BB																				
Response to comments	15	VR																				
Finalize scope	8	VR																				
Ongoing review	20	BB																				
Existing conditions		JB																				
Biological environment		JB																				
Review existing information	60																					
Wetland delineation	10																					
Stream sampling	15																					
Engangered speices habitat survey	15																					
Physical environment		MM																				
Air quality survey	30																					
Soils mapping	10																					
Hazardous waste survey	5																					
Drainage mapping	15																					
Hydrological monitoring	30																					
Human made environment		VR																				
Infrastructure survey	20																					
Traffic counts	30																					
Services survey	5																					
Cultural resources	10																					
Historic resources	10																					
Develop alternatives	100	BW																				
Public participation program	100	JC																				
Evaluate impacts	150	BC																				
Compare alternatives	20	JM																				
Draft EIS	35	VR																				
Public meetings	10	JC																				
Final EIS	35	JM																				
Public hearing	5	JC																				
ROD	20	JM																				

FIGURE 9.1
A simplified example of a project plan tool.

The plan represents the directions and road map for conducting the environmental analysis, and once developed, all environmental analysis team members must be fully aware of the details of the plan. Thus it is incumbent on the manager of the analysis to share it with each team member, ensure they are fully aware of what is expected, and each practitioner must commit to fulfilling the assignment within scope and budget. If they feel this is not possible they must raise the issue at the outset and not after the time is gone, money spent, and their section unfinished. One approach in giving this assurance is to not only provide them with the relevant sections of the scope and plan but also briefly describe the expected product of their work. Or better yet, ask each team member to describe what he or she expects to produce including what questions will be answered. If they cannot do this, they probably do not understand the scope and plan to successfully implement the scope of work.

One critical aspect of successful management is assigning responsibility for each work task and subtask in the scope. The environmental analysis team member assigned the management of the task is responsible for producing a product that fully satisfies the needs of the analysis, as defined in the scope, and completing it within schedule and budget. It is also the task manager's job to identify the resources, including staff, senior technical consultants, software, etc. required for the task, but it is the responsibility of the overall environmental analysis manager to make them all available to the task manager.

Experience has shown that assigning responsibility for a task is not only critical for maintaining the schedule and budget, but conveying ownership to the task manager also fosters a superior product. With ownership comes dedication; it also encourages creativity. Ownership not only produces a better product but results in commitment and full productive participation of the task managers in the interdisciplinary team. Aligning tasks to disciplines (e.g., ecology, cultural resources, energy, public health) and assigning task management to a team member with interest and expertise in the discipline are very useful in fostering ownership. The staff member will not only be familiar with the technical aspects of the discipline but also care about its preservation and enhancement, which provides an incentive to ensure it is fully considered and addressed in the environmental analysis. One way to convey ownership is to address each discipline as a technical appendix to the environmental analysis. That way the task manager and discipline specialist have a distinct element of the analysis within their control. Within guidelines established at the outset and consistent with the scope, they have latitude to research, organize, and produce the appendix. They can then summarize the salient points of the discipline for inclusion in the body of the environmental analysis document.

9.3.3 Monitoring Progress and Bad News

Establishing a good project plan and having buy-in from the full environmental analysis team are critical to managing the process, but if the progress against the plan is not monitored, much of the benefit is lost. The basic monitoring consists of each task manager estimating the current percent complete and the estimated total time and budget required to complete each task (estimate to complete, etc.) at regularly scheduled intervals. These estimates are then compared with the percent of the budget spent for the work completed and the total budget for completion. The estimates are reviewed by the overall environmental analysis managers as a quality control, but having the task managers perform the reviews adds to their sense of ownership and forces them to pay attention to progress and schedule. A wise manager will keep time and funds in reserve as a contingency as issues, problems, and unanticipated complications arise as they always do. There is then a backstop to shore up a task until corrective measures to bring it back in line are put in place.

Another form of monitoring to keep the analysis within budget and schedule is periodic task review meetings. At these meetings the task managers present their findings and work status to the overall manager, senior consultants, and other task managers. This meeting serves multiple purposes including:

- Forces the task managers to realistically assess the status of their work.
- Produces feedback and identifies opportunities for streamlining the work, particularly if task completion and budget spent are out of line.
- Creates a midstream opportunity for senior consultants or others to identify opportunities to truncate the analysis on a specific resource or alternative.
- Informs each task manager what the others are doing which lends consistency and quality to the final product in line with a multidiscipline approach.
- Provides informed opportunity to revisit the project plan and make adjustments as necessary such as focusing on critical issues and truncating effort that has little benefit.
- Provides a form to nurture a truly interdisciplinary approach to environmental impact analysis.

One important aspect of project management and monitoring is to identify and address bad news as soon as it is uncovered. If the description of an affected environment, alternative description, or impact prediction develops into a more complicated, and thus resource demanding, effort than originally envisioned, early identification and acknowledgment provides an opportunity to address the situation. It also allows the environmental analysis manager to inform the project proponent (i.e., the one who is paying for the project) before the situation snowballs and allows the project proponent to participate in addressing the problem. Ignoring anticipated schedule or budget problems almost never makes them go away, and knowing about the possibility before it happens usually makes the shock less severe.

9.4 Environmental Impact Analysis Critical Success Factors

In summary, paying close and frequent attention to a few critical elements can greatly improve the chance of success and environmental enhancement of an environmental impact analysis. These include:

1. Identify critical stakeholders and actively solicit their input.
2. Clearly identify the purpose and need. In the process, make sure the critical stakeholders and environmental analysis team are aware and on-board with the purpose and need.
3. Pay attention to the scope, make sure it:
   a. Is distributed and acknowledged early in the process
   b. Addresses stakeholder issues
   c. Is focused on key issues
   d. Guides all analysis work (i.e., do no work outside of the scope)
   e. Provides all the information needed for affected environment description, impact prediction, alternative development and comparison
   f. Is compared to work completed and ongoing to monitor progress and make adjustments as necessary
4. Make sure the intended use of any data collected or engineering studies conducted is clear and needed for the analysis before the collection or study is undertaken.
5. Develop an Impact Prediction Conceptual Model early in the process and check/revise it often to help guide the environmental analysis.
6. Establish impact significance criteria for use in designing studies, predicting impacts, and comparing alternatives.
7. Maximize use of existing information for the description of affected environment and, if necessary, design of analysis-specific investigations.
8. Make all analyses are objective and quantitative to the extent possible. Use or modify proven techniques from each discipline (e.g., ecological risk assessment, net environmental benefit analysis, traffic studies) to maximize objectivity, quantification, and acceptance.
9. Develop and analyze in detail a full range of alternatives that reflect the issues and impacts to resources of concern raised during scoping.
10. Involve the public and key stakeholders throughout the process.
11. Make the comparison and selection of alternatives transparent, objective, well documented, and analytical to the extent possible given constraints of specific situations.

12. Integrate the environmental analysis and subsequent approval/permitting effort to the maximum extent possible.
13. Approach the environmental analysis as a problem-solving challenge to achieve the purpose and need with:
    a. Minimum of adverse and maximum of beneficial environmental impact.
    b. Stakeholder acceptance.
    c. Fully implementable and permittable proposed action.
    d. Maximum use of creative, objective, quantitative, and elegant environmental analysis tools.

10

Background on Case Studies

Several case studies are used throughout this book to illustrate specific points, and when cited, only aspects of the project, policy, or plan relevant to issue under discussion are presented. This chapter gives a broader background on the case studies used. It also addresses some aspects of case studies that are based on experience, hindsight, and not necessarily fully addressed in the public record.

10.1 Boston Harbor Cleanup Environmental Impact Statement

Boston Harbor has a long history of receiving discharges from human activity, beginning even before the Sons of Liberty threw East India Company tea into the harbor to protest against British taxes on December 16, 1773. Since then until the late 1900s, Boston Harbor and its tributaries have served as the sanitary waste conveyance and disposal system for the Boston metropolitan area. Prior to the late 1800s, the sanitary waste largely flowed on or under the streets and discharged either directly to the harbor or to one of the tributaries. Recognizing the relationship of poor sanitation (e.g., raw sewage on the streets) and public health, the state commissioned the Boston Main Drainage System to collect sanitary waste and convey it to Moon Island in Boston Harbor. At Moon Island, it was held in a series of large granite lagoons to enable some solid removal and then discharged to the harbor on the outgoing tide.

The Moon Island's method of waste disposal represented an advanced approach to sanitary waste management at the time of implementation but by the mid-20th century it was viewed as rudimentary and creating public health and aesthetic issues. After much debate, both socially and technically, a new sanitary system was put in place beginning in the 1950s. This system consisted of a primary wastewater treatment plant on Nut Island, which received the waste from the southern metropolitan area and a similar facility on Deer Island received the waste from the northern part of the system. Primary treatment, which removed most of the solids by the physical process of settling, was considered the standard at the time for large systems discharging to open waters. The two systems also included disinfection before discharging the treated effluent to outer Boston Harbor, adjacent to the wastewater treatment plants. The solids removed from the process (sewage sludge or preferably referred to as "residuals") were stabilized by anaerobic digestion and then discharged to Boston Harbor. The design was to only discharge the residuals on the outgoing tide, similar to the Moon Island system, and thus flushing them out into Massachusetts Bay where dilution and dispersion would transport them away from shore and render them harmless. But the operators of the system were not always able to adhere to limiting the sludge discharge to the ebb tide.

The wastewater treatment system was eventually turned over to the Massachusetts Metropolitan District Commission (MDC) as part of the mid-20th century regionalization trend in municipal government. The MDC operated not only the regional wastewater system but also water supply, major highways, parks, recreational facilities, and other infrastructure for the cities and towns surrounding Boston (up to 42 for at least some of these services). They even had their own police department and were a major state entity and source of patronage jobs. The system worked well for a while, but during the late 1960s and early 1970s, similar to many other U.S. cities, Boston experienced urban flight and an associated decay of the urban environment, including the infrastructure. With this decay came a severe financial stress similar to, but less extreme than, that seen during the period when New York City had to be bailed out financially. Thus the MDC budget was severely stretched, and wastewater treatment took the brunt of the budget cuts because of its public standing compared with the high-profile water supply, parks, roadways, and police.

With the budget cuts the MDC could not keep up with the operation and maintenance of the wastewater system and it deteriorated. The collection of raw wastewater became an issue and there were numerous and large "overflows" to the surface waters in and around Boston. Also the operation and maintenance of the two wastewater treatment plants declined and there were times when all of the pumps conveying wastewater to the Deer Island treatment plant were inoperable and raw sewage was discharged to surface waters throughout the system. The MDC could not adequately keep up with day-to-day needs; much less prepare for the much more stringent municipal wastewater management requirements of the 1972 Clean Water Act. As a result, the Boston Metropolitan Area fell behind most of the rest of the country as other municipalities were taking advantage of Clean Water Act funding to plan, design, and construct upgraded collection systems and secondary wastewater treatment plants.

In the late 1970s, MDC grasped at a final straw as they applied for a waiver of secondary treatment requirement under Section 301h of the Clean Water Act. After several years, engineering studies, and scientific investigation of environmental conditions, the MDC was denied the waiver of secondary treatment because among other factors they could not convincingly demonstrate, even with an extended discharge 15 km offshore, that continuation of primary treatment would meet all water quality standards (particularly dissolved oxygen concentration) and support a balanced indigenous population of marine organisms. As a result, MDC was behind the curve and had not

Background on Case Studies 361

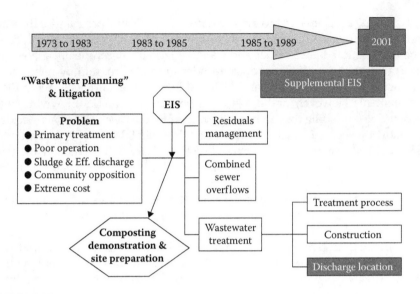

FIGURE 10.1
Boston Harbor cleanup program overview. *Abbreviations:* EIS, Environmental Impact Statement; Eff, Effluent.

completed the planning to implement full secondary treatment. Thus there were continuing violations of water quality standards, and noncompliance with the Clean Water Act.

By the early 1980s, state legislators representing municipalities on Boston Harbor, environmental advocacy groups, and even the U.S. EPA could not tolerate the conditions in Boston Harbor and lack of progress in meeting the Clean Water Act (Figure 10.1). They filed a suit in federal court to address noncompliance by MDC and force action. It did not take the "sludge judge," Paul G. Garrity, long to clearly realize and rule that not only was MDC in violation of the law; there did not seem to be any viable scenario within the existing organizational structure that could bring the situation under control within a reasonable time frame. He thus ordered the State of Massachusetts to not only move expeditiously to rectify the situation and meet Clean Water Act requirements but also to form a new independent agency to address the situation. The new agency, Massachusetts Water Resources Authority (MWRA) was assured of adequate funding because under their charter, they collected sewer service fees directly from the user communities, and the money could not be used for any purpose but the wastewater system. Also they were not conflicted with other more visible and popular responsibilities such as parks, highways, and police.

Following the court's ruling, the MWRA immediately proceeded with wastewater facilities planning to address the Boston Harbor Cleanup following Clean Water Act regulations and procedures. The planning was also done in compliance with the state's Department of Environmental Protection and Massachusetts Environmental Policy Act (MEPA) requirements, and as discussed in Section 8.6, many federal agencies defer to mature state environmental programs if the state entity has primary responsibility. In such cases either a "piggyback" (i.e., combined state and federal analysis and documentation) environmental analysis or incorporating the analysis conducted under the state program by reference is conducted to meet both state and National Environmental Policy Act (NEPA) requirements. However, for the Boston Harbor Cleanup the U.S. EPA, which was responsible for issuing the discharge permit under the Clean Water Act, decided to conduct a separate and independent environmental impact analysis under NEPA. This decision was largely based on the history of the situation and past problems, inactivity of the state agencies involved with wastewater management, and oversight.

As described in detail in Section 6.3.1, the cleanup was too complex and wrought with controversy to address at one time with a single comprehensive solution. Thus the process was a multilevel environmental impact analysis, with the first step being an original EIS addressing the key issue and past primary stumbling block of where to site the wastewater treatment facility. This EIS also delegated to subsequent tiers the other environmental analyses and decisions including:

- Type of wastewater treatment
- Conveyance of raw wastewater
- Location of effluent discharge
- Residuals management, both interim and final
- Combined sewer overflows (CSOs)

The first of these three were addressed by a supplemental EIS (U.S. EPA 1988), which as described in Section 6.3.1 was actually a tiered EIS under the current use of the term. The next issue (residuals management) was addressed by an interim plan and then another supplemental EIS. The environmental impact analysis of the final issue, CSOs, was delegated to the state program (MEPA) because by that time Massachusetts and MWRA had demonstrated their competence and commitment to the cleanup of Boston Harbor.

The major elements of the Boston Harbor Cleanup proceeded to completion in 2001 (Figure 10.1). The sanitary wastewater from the metropolitan area was reliably collected, treated, and discharged to Massachusetts Bay with continuous monitoring under the Clean Water Act National Pollutant Discharge Elimination System (NPDES). Almost two decades of monitoring has demonstrated attainment of water-quality standards and environmental sustainability. It has also confirmed the prediction of impacts in the EIS. The wastewater residuals program has been similarly successful, achieving near full reuse of the solids as an agricultural product, and the discharge of sludge to Boston Harbor has ceased, hopefully for all time. There have been major achievements and advances in CSOs, but it is an ongoing program with improvements added every year on an as-needed basis.

10.2 U.S. Coast Guard Dry Cargo Residue Management EIS

The natural resources and shipping lanes in the Great Lakes basin create a "perfect storm" for manufacturing and other commercial/industrial activities, thus creating one of the major economic engines of the United States. The abundance of easily extractable iron ore at the western end of the Great Lakes, the coal resources in both the eastern and western ends and the limestone present in the center of the basin provide the necessary materials and an optimum opportunity for steel manufacturing. In the nineteenth century this opportunity was recognized and it has been realized through the efficient, inexpensive, and timely transport of the material via Great Lakes carriers (large lake-going cargo vessels up to 300 m long) between the sources of materials and manufacturing centers. These Great Lakes carriers are also heavily involved with the transport of other dry cargo originating in the Great Lakes Basin including grains, stone for construction aggregate, and even salt (there is a large salt deposit under Lake Erie offshore from Cleveland, Ohio).

The Great Lakes carriers load, transport, and unload an annual average of over 100 million tons of iron ore (which is preprocessed to marble size pellets called taconite), coal, limestone, and other materials (U.S. Coast Guard (USCG) 2008). During the process some of the material inadvertently spills on the deck and within the bowels of the ship. Historically this spilled dry cargo residue (DCR) has been swept overboard in order to maintain a clean and safe working area for the ship's crew. This has resulted in an estimated 500 tons per year of DCR discharged to the Great Lakes since the late 1900s (USCG 2008). This common and historic practice of DCR sweeping and discharge did not change with the passage of the Clean Water Act in 1972 or the negotiation of treaties between the US and Canada addressing Great Lakes environmental protection even though the practice was arguably a violation of both.

Recognizing this potential conflict with regulations and treaties, in the 1990s, the USCG developed and issued interim regulations governing the discharge of DCR and the U.S. Congress authorized them to develop permanent regulations that balanced environmental protection and the continued viability of the Great Lakes shipping industry. Since promulgation of the regulations was a major federal action potentially affecting the environment, the USCG prepared an EIS under NEPA. The process was tiered because following the completion of the original EIS (USCG 2008), the USCG felt there was inadequate information regarding the economic impact on the shipping industry and the effectiveness of various DCR discharge management options was not well documented. Thus they issued an interim rule and commitment to continued study of the issues, preparation of a tiered EIS (currently under review), and promulgation of a final rule.

Preparation of the EIS for DCR management on the Great Lakes presented an opportunity not typically encountered in an environmental impact analysis. The practice of virtually uncontrolled DCR discharge within the major Great Lakes shipping lanes had occurred for over a century, which created the possibility of actually measuring rather than predicting the impacts resulting from the discharge. Since the degree of impact was generally proportionate to the mass of the DCR discharged, alternatives that reduced the discharge could be assumed to have lesser impact scaled to the anticipated reduction in DCR discharge.

The DCR environmental impact analysis focused on the physical, biological, and chemical conditions in the Great Lakes shipping lanes and represented both the affected environment and impact from historic DCR discharges. The chemical impacts were addressed by analysis of the concentrations of potentially toxic components of the various cargo types, measurement of sediment chemistry in areas of heavy DCR discharge, and toxicity tests using indigenous aquatic organisms exposed to DCR. All of these tests unambiguously demonstrated that no chemical effects on the aquatic system were anticipated regardless of the DCR discharge rate or mass.

The combined analysis of biological and physical stressors from DCR discharge indicated a potential for effects. The possible effect was on the soft bottom invertebrate community because the addition of hard substrate provided opportunity for attachment of different species, but the impact was not seen as significant or with consequences reaching beyond the immediate area of discharge. However, the addition of DCR and associated changed substrate conditions was seen as a concern for the expansion of invasive mussel populations, and a detailed scientific investigation was initiated to fully evaluate the effects and understand the implications (see Section 5.3.4).

The analysis demonstrated that some effects related to invasive mussels could occur in certain locations. Specifically, Lake Superior would not be affected because the water chemistry (e.g., low-calcium concentrations) prevented establishment of mussel populations and DCR discharge would not alter this situation. Mussel populations in Lakes Erie and Ontario were already at maximum levels and the addition of more DCR would not result in a higher mussel density. The populations in Lakes Michigan and Huron were not at maximum density, at least partially due to lack of suitable substrates and the discharge of DCR could result in an expansion of the invasive mussel populations. However the effects would not be immediate or irreversible and based on the analysis, the USCG decided to issue an interim rule without substantial changes to current practices. The analysis supported the position that no significant environmental harm would occur while the USCG more closely evaluated the mussel situation in relation to economic impacts on the shipping industry resulting from more stringent DCR discharge control requirements.

The Record of Decision and interim regulation included requirements for each Great Lake carrier to submit to the USCG information on DCR discharge. This included estimated location and mass of DCR discharge and the measure used to minimize the spillage and discharge. The information along with observations and measurement of DCR handling is being used in support of the follow-up tiered EIS to fully evaluate the effectiveness and economic implications resulting from more stringent regulations and issue a final rule.

10.3 AJ Mine Tailings Disposal EIS

Juneau, the capital of Alaska is a gold mining city. It was established to exploit the rich mineral deposits in the mountains surrounding what is now the coastal city of Juneau, and much of the city land is the result of using mine tailings (waste rock resulting from mining) as fill. The AJ mine, which is directly adjacent to the city, was productive for a long time but the increased costs of extracting gold from the less productive formation combined with the lack of resources to continue mining associated with the outbreak of World War II forced the shutdown of the AJ Mine. But in the 1990s, advanced mining techniques and economic conditions created an interest in reopening the mine.

One of the major impediments to reinitiating mining operations was the management and disposal of the 80 million metric tons of mine tailings. No land was available for land-filling the tailings and other alternatives, such as creating a dam in one of the valleys to retain the material, were dismissed as far too environmentally damaging. This left marine disposal of the tailing as the only viable option available if the mine was to reopen. The marine disposal was regulated by the U.S. EPA under the Clean Water Act, and thus EPA was required to comply with NEPA by preparing an EIS as a precursor to issuing a discharge permit under the Act.

Early attempts by the mining company to address environmental concerns were met with skepticism by the stakeholders, including the Juneau municipal officials, state environmental regulatory personnel, and employees of several federal agencies. As a result, the mining company and U.S. EPA agreed to conduct a "third party EIS," whereby a third party contractor was hired to conduct the environmental impact analysis. The mining company paid for the analysis but the third party took direction from the EPA, thus satisfying stakeholders that the analysis would be unbiased and the level and detail of scientific studies to understand environmental effects relatively unconstrained.

This approach and the mining company's willingness to do "whatever it takes" for EPA to approve the marine discharge resulted in elements of the environmental analysis which went beyond typical EISs, including:

- Extensive Technical Advisory Committee (TAC) and Citizen Advisory Committee (CAC) involvement:
  - Formation of numerous subgroups based on different environmental resources (e.g., water quality, fisheries, marine mammals, sea birds, and benthos)

- Meetings at key decision points (e.g., study design, date interpretation, impact prediction, and design of alternatives) sometimes occurring monthly
 - Funding of TAC and CAC key members to facilitate timely review
- Year-long studies of critical resources, including:
 - Submersible observations of benthos
 - Radio tagging of tanner crabs to determine home range
 - Gut analysis and food chain studies of major predators
 - Physical oceanographic measurement programs
- State-of-the-art techniques for impact prediction:
 - Toxicity testing
 - Bioaccumulation studies
 - Marine geochemistry modeling
 - Physical oceanographic modeling, including sophisticated prediction of sediment transport
- Use of the newly developed ecological risk assessment technique to evaluate impacts and alternatives (see Section 7.2.9)
- Greatly expanded scoping and public outreach programs
- Mitigation of potential impacts on water supply by construction of a municipal water treatment plant

The result of the process was a renewed and improved relationship among the regulatory, environmental resource, mining, environmental advocacy, native peoples, and municipal government communities. However, when the process was 99% complete, the mining company decided to terminate the project based on economic and mining engineering considerations; the EIS was not completed (thus there is no published reference for the document but much of the scientific results were presented in Maughan et al. 1997); and a marine mine tailings discharge permit was never issued. But the process and extensive date collection program did produce consensus among stakeholders on the impact analysis and environmental sustainability of the proposed action.

10.4 Washington Aqueduct Water Treatment Residuals Management EIS

Washington Aqueduct, a division of the Baltimore District U.S. Army Corps of Engineers, currently provides potable water to Washington D.C. and several surrounding communities, and has been since the capital had a central

Background on Case Studies

public water supply. On average, approximately 700,000 cubic meters of water a day is drawn from the Potomac River, the only water source capable of supplying that volume of water, and treated at Dalecarlia and McMillan Water Treatment Plants (U.S. Army Corps of Engineers 2005). The treatment consists of removing the solids, largely river silt, from the Potomac River water and disinfecting the finished water for distribution throughout the Washington Metropolitan area.

The treatment process removes an average of 125 tons of solids daily from the river water. For decades these residual solids were settled out as part of the water treatment train and then discharged back to the Potomac River, from whence they came. This process continued even after enactment of the Clean Water Act and the U.S. EPA establishment of the NPDES permitting system to limit discharge of pollutants to waters of the United States. However during the 1990s, there was pressure on the U.S. EPA to curtail the unconstrained discharge of solids from the two Washington Aqueduct water treatment plants to the Potomac River. EPA was reluctant to ignore this pressure because they felt compelled to regulate another government agency (i.e., U.S. Corps of Engineers) with the same vigor with which they regulated private industries and municipalities. Thus in 2003, they issued a new NPDES permit which severely limited discharge of water treatment residual solids to the point that Washington Aqueduct would be forced to intensely manage the solids and construct significant additional infrastructure.

The Washington Aqueduct had concerns over this new requirement for several reasons. The water treatment plants, although initially built in relatively undeveloped land, by 2000 were surrounded by single-family residential development, some of it high density and some of it occupied by influential Washington insiders. The construction and operation of the additional infrastructure could impact the aesthetics and land use in the neighborhoods and generate conflict and opposition. Also, under most scenarios after the residual solids were dewatered they would be trucked off-site, at a rate of up to 20 trucks a day. The truck traffic could not only generate unwanted noise but also significant traffic and possible safety concerns in the adjacent residential neighborhood streets. There was reluctance to impact the neighborhood and generate opposition if some benefit was not achieved by instituting an intense residual management operation.

The Washington Aqueduct and its parent organization, Baltimore Division of the Corps of Engineers, were not initially convinced that an intense residual solids management and disposal operation would achieve an environmental benefit that was commensurate with the adverse impact to the neighborhood generated by the action. In order to evaluate the question of adverse environmental effects from the discharge, they conducted extensive environmental tests on the residual solids, including chemical analysis and toxicity testing. The results of the testing demonstrated that the discharge would not result in severe impacts to the aquatic environment, which was not unexpected because the process just returned the solids back to the river. These findings combined with the anticipated neighborhood impacts and costs supported the Corps' concern over implementing an intense solids management operation. However the U.S. EPA prevailed and the Washington Aqueduct launched a program to manage the residual solids from their water treatment operations in compliance with the new NPDES permit restricting discharge to the Potomac River.

The Washington Aqueduct's action was subject to NEPA because they were a federal agency and there was potential impact to the human environment, primarily to traffic and aesthetics in the adjacent neighborhoods. Thus an EIS was initiated with the primary purpose and need to comply with the NPDES permit requirements. The purpose and need statement was expanded to support alternative development and selection to specify a solids management system which:

- Would not impact current or future production of safe drinking water reliably supplied to the Washington Aqueduct customers
- Would minimize, if possible, impacts on the neighborhoods and other stakeholders
- Would be cost-effective in design, implementation, and operation

This illustrates one of the interactions of environmental analysis and permitting as discussed in Section 9.2. The necessity for environmental analysis was driven by the need to meet an environmental permit condition and then permit requirements were incorporated into the purpose and need as well as the comparison of alternatives.

The EIS proceeded in line with the steps and procedures outlined in this book. Following the initial scoping, the engineering and environmental analysis team proceeded to develop, screen, and then analyze alternatives. Twenty six alternatives were identified and then screened. The screening resulted in a reasonable range of alternatives with the alternatives that managed the solids off-site being initially eliminated from detailed evaluation for a variety of reasons, including:

- Construction of a pipeline to transport the solids slurry would adversely affect properties on the National Register of Historic Places (like the Washington Monument).
- The low clearance of bridges across the Potomac River prevented barge transport of solids slurry offsite.
- Using the sanitary sewerage system to transport solids to the wastewater treatment plant for processing would exceed the system's capacity, resulting in sewage overflow to surface waters.

The alternatives screening resulted in a list of alternatives that met the purpose and need, and although they resulted in some impacts, the environmental analysis team felt they could be successfully mitigated.

A full scoping process, an objective environmental impact analysis of alternatives, and the NEPA process were proceeding smoothly. Then there was a hiccup when many stakeholders, particularly those in the adjacent neighborhoods, did not initially realize the magnitude and extent that some of the residual solids management alternatives could potentially affect their day-to-day life. Thus they did not pay close attention to the process until it was well along and approaching the draft EIS stage.

Once the stakeholders did weigh-in, it was with vigor. They forced reopening of the alternative screening and identified over 100 additional alternatives including off-site options that had to be preliminarily analyzed, including comparison to purpose and need. The input also forced reevaluation of the initial screening and several alternatives were added to the list for a detailed evaluation. As it turned out, none of the off-site alternatives fully met the purpose and need or proved to be superior to on-site alternatives.

However, the stakeholders' concerns and suggestions did prompt adjustments and reevaluation to some of the on-site alternatives. Working with the stakeholders, a layout was developed for on-site solids processing, including mechanical dewatering. The layout maximized the screening and aesthetic opportunities present on the site, by building the facilities into the side of a hill. There was also coordination with the stakeholders to modify the dewatered truck hauling operations to mitigate traffic impacts. The process produced an alternative that mitigated noise, aesthetic, and perceived odor impacts and was ultimately acceptable to the stakeholders. It is currently being implemented.

References

Maughan, J.T., R. Whitman, and J. Gendron. 1997. *Marine Community Risk from Submarine Mine Tailings Disposal Induced Reduction in Food Availability*. Society of Environmental Toxicology and Chemistry, San Francisco, CA.

U.S. Army Corps of Engineers. 2005. *Baltimore District Washington Aqueduct. Draft Environmental Impact Statement for a Proposed Water Treatment Residuals Management Process for the Washington Aqueduct*, Washington, D.C.

U.S. Coast Guard. 2008. *Final Environmental Impact Statement: U.S. Coast Guard Rulemaking for Dry Cargo Residue Discharges in the Great Lakes. DOT Document Number: USCG-2004-19621*. Washington, DC: Commandant U.S. Coast Guard Headquarters.

U.S. EPA. 1988. *Final Supplemental Environmental Impact Statement for Boston Harbor Wastewater Conveyance System*, Volume I and II. Boston, MA: U.S. EPA, Region I.

Index

A

AEC, *see* Atomic Energy Commission
Affected environment, 164; *see also* Environmental impact analysis
AJ Mine EIS-ecological risk assessment, 283–284
AJ Mine tailings disposal EIS, 365–366
Aquifer storage and recovery (ASR), 254, 255
Alternatives screening, 80, 84, 148–161
ASR, *see* Aquifer storage and recovery
Assessment endpoints, 269–271, 284, 285
Atomic Energy Commission (AEC), 25
Australian environmental program, 316–317

B

Basic Law for Environmental Pollution Control, 320
Benthic community structure evaluation, 205
Benthos, 206
Boston Harbor cleanup, 238–245
Boston Harbor cleanup alternatives, 213–216
Boston Harbor cleanup environmental impact statement, 359–362
Bureaucratic culture, 31–32

C

CAC, *see* Citizens advisory committee
California Environmental Quality Act (CEQA)
 process, 333–335
 unique features of, 335–337
Canadian Environmental Assessment Act (CEAA), 313
Canadian environmental program, 313–316
Case studies
 AJ Mine tailings disposal EIS, 283–294, 365–366
 Boston Harbor cleanup environmental impact statement, 238–252, 359–362
 DCR management mitigation, 221
 dry cargo discharge to the Great Lakes affected environment investigation, 182–184
 estuarine wastewater discharge-affected environment investigation, 171–178
 over-the-horizon (OTH) radar-affected environment investigation, 179–182
 RWA Lake Whitney mitigation, 221–223
 U.S. Coast Guard DCR management EIS, 81–85, 362–365
 Washington Aqueduct residuals management EIS, 77–81, 366–369
Categorical exclusion (CATEX), 29, 58, 86–96
 Massachusetts Environmental Policy Act (MEPA), 325
 multilevel environmental impact analysis, 231
 U.S. Coast Guard (USCG), 93–95
CATEX, *see* Categorical exclusion
CEAA, *see* Canadian Environmental Assessment Act
CEQ, *see* Council on Environmental Quality
CEQA, *see* California Environmental Quality Act
CERCLA, *see* Comprehensive Environmental Response, Compensation, and Liability Act

Citizens advisory committee (CAC), 127, 330, 365–366
City of Angoon v. Hodel, 76
City of New York v. U.S. Department of Transportation, 74
Civil and Voting Rights Acts, 12
Classic mitigation evaluation, 217
Clean Air Act, 2, 3, 35, 44–52
Clean Water Act, 2, 3, 21–22
   Australian protection, 317
   integrating environmental approvals, 343
   Section 201, 195, 198–202
   Section 404, 26, 194, 343, 346
   Washington Aqueduct water treatment residuals, 77
Coastal Zone Management Act (CZMA), 258
Coast Guard and Maritime Transportation Act, 83
Combined sewage overflows, 244
Combined sewer overflow (CSO), 237, 362
Comprehensive Environmental Response, Compensation, and Liability Act (CERCLA), 262
Conceptual site model, 267
Contaminant screening, 275
Contaminated sediment remediation, 302–304
Contamination of aquatic systems, 278–279
Council on Environmental Quality (CEQ), 236
   regulations, 15, 20
   website, 17
Cultural resources, 67, 129–131, 166, 179, 180–182, 198, 255
CSO, *see* Combined sewer overflow
CZMA, *see* Coastal Zone Management Act

D

Deposition area
   chemical characterization of, 205
   physical characterization of, 203, 205
Description of Proposed Actions and Alternatives (DOPAA), 133, 134
Description of the proposed action, 57

Dissolved oxygen (DO), 221
DO, *see* Dissolved oxygen
DOPAA, *see* Description of Proposed Actions and Alternatives
Dredging, 120, 302–304
Dry cargo discharge to Great Lakes, 81–85, 182–184

E

EA. *See* Environmental assessment
EC, impact level, *see* Environmental concerns (EC), impact level
Ecological endpoints, 272–273
Ecological risk assessment, 111; *see also* Environmental analysis tools
Ecological risk assessment conceptual model, 287
EISs, *see* Environmental impact statements
EMAP, *see* Environmental Monitoring and Assessment Program
Endangered Species Act, 2, 35, 269, 343
ENF, *see* Environmental Notification Form
Enhanced scoping approach, 132–136
Enrichment and nutrients, 205
Environmental analysis
   Canada *vs.* United States, 315
   categorical exclusion (CATEX), 86–96
   process, 114
Environmental analysis tools
   adapting tools, 309–310
   ecological risk assessment
      AJ Mine EIS, ecological risk assessment, 283–284
      effects characterization, 280–281
      and environmental impact analysis, 294–295
      vs. environmental impact analysis, 265
      exposure characterization, 277–280
      final problem formulation and detailed ecological investigation, 276–277
      history and development, 262–264
      vs. human health risk assessment, 264–265

Index

initial problem formulation, 267–271
process and approach, 264–266
risk characterization and
input, 281–283
screening level
assessment, 271–276
net environmental benefit analysis
(NEBA)
contaminated sediment
remediation, 302–304
environmental currency, 299–302
pollution abatement in Green
River, 304–308
process, 296–298
quantification of environmental
value, 298–299
Environmental approvals, 125–126
Environmental assessment (EA),
7, 68–71, 245–249, 319
Environmental Assessment Act, 314
Environmental concerns (EC), impact
level, 45
Environmental currency, 299–302
Environmental impact analysis
affected environment
dry cargo discharge, 182–184
vs. environmental
resources, 164–165
estuarine wastewater
discharge, 171–178
investigation summary, 184–185
mandatory first step, 165–169
original investigations, 169–171
over-the-horizon (OTH) radar-
affected environment
investigation, 178–182
alternatives
comparison and selection, 158–161
development of, 154–157
proposed action and, 150–154
screening of, 157–158
approach, 106–108
components, 163
critical success factors, 357–358
cumulative impacts, 226–227
hypotheses, 6
impact mitigation
classic approach to, 216–219
DCR management mitigation
case study, 221

integrated and proactive
approach to, 219–220
programmatic mitigation, 223–226
RWA Lake Whitney mitigation
case study, 221–223
impact prediction
of Boston Harbor cleanup
alternatives, 213–215
of DCR management in the Great
Lakes, 202–213
impact prediction conceptual
model, 186–189
impact prediction process, 189–191
impact significance
criteria, 192–202
integrating environmental approvals
and analysis, 342–350
international and individual state
programs, 313–338
managing
monitoring progress and
bad news, 355–356
project plan and delegation,
353–354
of work and truncated
analyses, 351–353
need for
environmental setting, 110
preliminary assessment, 110–113
proposed action, magnitude and
type of, 108–109
public outreach
benefits of, 143–145
commitment and extent, 139–143
tools, 145–148
scoping logistics and statement
basic scoping approach, 131–132
enhanced scoping
approach, 132–136
scoping statement, 136–138
scoping targets
social scoping, 126–128
technical scoping, 128–131
scoping topics
alternatives for
consideration, 122–123
environmental approvals, 125–126
environmental setting, 121–122
impacts and concern, 116–121
methods for, 123–124

Environmental Impact Assessment Directive, 317
Environmental impact prediction, 108
Environmental impact statements (EISs), 7, 15, 22, 25–26
    CEQ regulations, 60
    draft documents, 61–63
    vs. environmental assessment (EA), 70
    federal agencies filing, 48, 49, 50
    final EIS, 65–66
    of lack of objection, 51
    monthly filings of, 47
    planning and structuring, 58–61
    record of decision (ROD), 66–68
    review and comment, 63–65
Environmentally unsatisfactory (EU), impact level, 45–46
Environmental Monitoring and Assessment Program (EMAP), 168
Environmental Notification Form (ENF), 326
Environmental objections (EO), impact level, 45
Environmental policy, 18–19
Environmental Protection Agency (EPA), 7
Environmental Protection and Biodiversity Conservation Act (EPBC Act), 316
Environmental Quality Improvement Act, 35
Environment data collection, 171
EPA, *see* Environmental Protection Agency
EPBC Act, *see* Environmental Protection and Biodiversity Conservation Act
Equal Rights Amendment, 12
Estuarine wastewater discharge, affected environment investigation, 171–178
Existing conditions. *See also* Affected environment

European Union environmental program, 317–320
Executive orders, 36, 89, 138
    11514, 32, 35
    11991, 33
    12114, 28–29
    12316, 295

F

Federal agencies to facilitate environmental consideration, 23–24
Federal Clean Water Act, 269
Federal regulations, 36–37
Finding of no significant impact (FONSI), 71, 246
Fish and Wildlife Coordination Act, 35
Fish and Wildlife Service (FWS), 210
FONSI, *see* Finding of no significant impact
Food web, 280, 286
Fort Campbell programmatic environmental assessment, 245–249
FWS, *see* Fish and Wildlife Service

G

Goals, environmental analyses, 4
Graduate School of Oceanography (GSO), 199
Great Lakes, 81–85
GSO, *see* Graduate School of Oceanography

H

Habitat Equivalency Analysis (HEA), 299
Habitat Evaluation Procedure (HEP), 299
"Hard look," NEPA requirement, 15, 25–26, 97, 100, 206–210
HEA, *see* Habitat Equivalency Analysis
HEP, *see* Habitat Evaluation Procedure
Human health risk assessment, 263–265

Index

I

Impact mitigation
　classic approach to, 216–219
　DCR management mitigation
　　case study, 221
　integrated and proactive approach
　　to, 219–220
　programmatic mitigation, 223–226
　RWA Lake Whitney mitigation case
　　study, 221–223
Impact prediction approach, 203–206
Impact prediction conceptual model
　model development, 186–188
　USCG DCR environmental impact
　　analysis, 188–189
Impact prediction process, 189–191
Impact significance criteria
　Clean Water Act, 195–202
　environmental standards as, 193–194
　for specific environmental impact
　　analyses, 194–195
Implementation of environmental
　　impact assessment, 321
Integrating environmental approvals
　　and analysis
　environmental permits
　　and approvals, 325–345
　role of, 345–346
　support of, 346–350
Interim residuals management, 243
International implementation, 27–31
Isaak Walton League of America v. Marsh, 75

J

Japanese environmental program,
　　320–323
Japan's Environmental Assessment
　　Law, 322

K

King William Reservoir, 255–256

L

Lack of objections (LO), impact level, 45
Lake Whitney management plan, 119,
　　143, 160
Long-term residuals management, 244
Love Canal, 262–263

M

Major federal action, 56
Marine Ecosystem Research Laboratory
　　(MERL), 199, 200, 201
Marine Mammal Protection Act, 269
Marine mammal reproduction
　ecological endpoints, 287
Marine Protection, Research, and
　　Sanctuaries Act, 317
Massachusetts Environmental Policy
　　Act (MEPA), 237, 325–332, 361
Massachusetts Metropolitan District
　　Commission (MDC), 360
Massachusetts Water Resources
　　Authority (MWRA), 237, 361
MDC, *see* Massachusetts Metropolitan
　　District Commission
Measurement endpoints, 271
MEPA, *see* Massachusetts
　　Environmental Policy Act
MERL, *see* Marine Ecosystem
　　Research Laboratory
*Methow Valley Citizens Council
　v. Regional Forester*, 75–76
Migratory Bird Treaty
　　Act, 269, 343
Minimizing lake sedimentation, 120
Mitigated FONSI, 71
Modoc National Forest, 67–68, 179
Montana Legislative Environmental
　　Quality Council 2000, 324
Multilevel environmental
　　impact analysis
　approaches
　　inclusion by reference, 231–232
　　piggyback environmental
　　　analysis, 236–238
　　programmatic and tiered
　　　environmental impact
　　　analyses, 233–236
　　supplemental environmental
　　　impact analysis
　　　documents, 232–233
　Boston Harbor cleanup, 238–245

Multilevel environmental
 impact analysis (Continued)
 Fort Campbell programmatic
  environmental
  assessment, 245–249
 strategic environmental assessments
  (SEAs), 252–258
 U.S. Coast Guard dry cargo
  residue management
  tiered EIS, 249–252
Mussels
 endangered freshwater, 210–212
 invasive *Dreissena* spp., 129–130, 197,
  206–210, 211–213, 364
MWRA, *see* Massachusetts Water
 Resources Authority

N

National Contingency Plan, 262
National Environmental Policy
 Act (NEPA), 2
 CEQ regulations
  commenting and
   coordinating, 39–40
  public involvement, 34–35
  questions, 41–43
  record of decision, 40–41
  relationship to other
   statutes, 35–37
  user friendly, 38–39
 Clean Air Act, 44–52
 compliance process, 56
 contents of
  basic requirements, 19–22
  bureaucratic culture under, 32
  consultation and public
   disclosure, 26–27
  Council on Environmental
   Quality (CEQ), 31–32
  environmental impact
   statement (EIS), 22–26
  environmental policy, 18–19
  international implementation
   of, 27–30
  purpose and jurisdiction of, 18
 DCR discharge in Great Lakes, 81–85
 detailed statement, 15, 22, 25, 58–59
 enforcement, 96–99
 implementation, 52–53
 integrating environmental
  approvals, 342
 jurisdiction, 13, 18, 28, 56
 multilevel environmental impact
  analysis, 229
 process
  categorical exclusion (CATEX), 58
  defining the action, 56–58
  environmental impact statement
   (EIS), 58–66
  no significant impact, 68–71
  record of decision (ROD), 66–68
 purpose and need, 71–77
 Washington Aqueduct water
  treatment residuals, 77–81
National environmental policy,
 development of, 11–16
National Historic Preservation Act,
 35, 317
National Oceanic and Atmospheric
 Administration (NOAA),
 258, 296, 300
*National Parks & Conservation Association
 v. U.S. Bureau of Land
 Management*, 62, 77
National Pollution Discharge
 Elimination System (NPDES),
 21, 78–79, 151, 305, 344, 362
Natural Resources Conservation Service
 (NRCS), 210
*Natural Resources Defense Council
 v. Morton*, 74–75
Nature Conservation Law, 320
NEBA, *see* Net environmental
 benefit analysis
NEPA, *see* National Environmental
 Policy Act
Net environmental benefit analysis
 (NEBA)
 contaminated sediment
  remediation, 302–304
 environmental currency, 299–302
 pollution abatement in
  green river, 304–308
 process, 296–298
 quantification of environmental
  value, 298–299
NIMBY, *see* Not In My Back Yard

Index

NOAA, see National Oceanic and Atmospheric Administration
NOI, see Notice of intent
Notice of intent, 39–40
Not In My Back Yard (NIMBY), 145
NPDES, see National Pollution Discharge Elimination System
NRCS, see Natural Resources Conservation Service

O

Ocean and Coastal Resource Management (OCRM), 258
OCRM, see Ocean and Coastal Resource Management
Over-the-horizon radar (OTH)-affected environment investigation, 178–182

P

Pelagic organisms, 206
Piggyback environmental analysis, 236–238
Prediction of invasive mussel impacts, 206–210
Problem formulation, 267–271, 276–277
Professional Staff and Expert Opinions, and Scientific Analyses, 87
Programmatic and tiered environmental impact analyses, 233–236
Programmatic mitigation, 223–226
Public outreach, 138–148

R

Record of decision (ROD), 40–41, 208
Regional water authority (RWA), 118, 120, 221
Reversed Robertson v. Methow Valley Citizens Council, 75–76
Risk hypothesis, 267, 278
ROD, see Record of decision
RWA, see Regional water authority

S

Safe Drinking Water Act, 343, 345
Scientific method, 5, 105–106, 267–268
Scituate offshore affected environment, 177
Scoping logistics and statement, 131–138
Scoping process, 116
Scoping topics, 114–128
Screening level assessment, 271–276
SEA, see Strategic environmental assessment
Sediment alteration and deposition, 206
Sediment chemistry, 205–206
Sediment imaging camera, 176
Social scoping, 127–128
Sonar L-3Klein System 3000 Towfish, 182
South Weymouth Naval Air Station, 331
Strategic environmental assessment (SEA), 2
  multilevel environmental impact analysis, 252–258
Stressor, 280, 284–285
Superfund Amendments and Reauthorization Act, 262
Superfund law, 262
Supplemental environmental impact analysis documents, 232–233
Sweepings characterization, 203

T

TAC, see Technical advisory committee
Technical advisory committee (TAC), 167, 284, 365–366
Technical scoping, 128–131
Tennessee Valley Authority (TVA), 15
Title I, Section 101 of NEPA, 18
Title I; Section 102 of NEPA, 18, 20
TMDL evaluation, see Total Maximum Daily Load (TMDL) evaluation
Total Maximum Daily Load (TMDL) evaluation, 304
Toxicity tests, 205
TVA, see Tennessee Valley Authority

U

USAF, *see* U.S. Air Force
U.S. Agency for International Development (USAID), 138, 139
USAID, *see* U.S. Agency for International Development
U.S. Air Force (USAF), 232
U.S. Army Corps of Engineers, 75, 80–81, 210–213, 255–256
U.S. Coast Guard (USCG), 82–84
  of CATEXs, 93–95
  dry cargo residue management EIS, 362–365
  dry cargo residue management tiered EIS, 249–252
  live fire training, 112
U.S. EPA, 21, 44, 237–238, 263–264, 343–344
U.S. EPA Total Daily Maximum Load Program, 126
U.S. Fish and Wildlife Service (USFWS), 299
USFWS, *see* U.S. Fish and Wildlife Service
U.S. Navy, 88–92, 330–331
U.S. state environmental analysis programs
  California Environmental Quality Act (CEQA), 332–337
  Massachusetts Environmental Policy Act (MEPA), 325–332

W

Warranting protection, 269
Washington Aqueduct water treatment residuals, *see also* Case studies
  background, 77–78
  management EIS, 366–369
  purpose and need, 78–81
Waste cleanup, cost *vs.* ecological value, 296
Wastewater collection system, 238
Wastewater conveyance and treated effluent discharge, 244
Wastewater treatment plant (WWTP), 305, 307
Water chemistry, 205
Water fowl, 206
Water-quality impacts, 125
Watershed Initiative for National Natural Environmental Resources (WINNER) program, 224, 225
Water treatment plants (WTPs), 78
Weight-of-evidence approach, 123
WINNER program, *see* Watershed Initiative for National Natural Environmental Resources (WINNER) program
WTPs, *see* Water treatment plants
WWTP, *see* Wastewater treatment plant

CPSIA information can be obtained
at www.ICGtesting.com
Printed in the USA
BVHW04*0402020918
525864BV00006B/153/P